MARS
THE MEDIUM AND THE MESSENGER
McLUHAN

a biography

PHILIP MARCHAND

The MIT Press
Cambridge, Massachusetts

First MIT Press edition, 1998

© 1989 by Philip Marchand

All inquiries regarding the motion picture, television and related rights for this book should be addressed to the author's representative, Westwood Creative Artists, 94 Harbord Street, Toronto, Canada, M5S 1G6.

ISBN 0-262-63186-5

Library of Congress Catalog Card Number: 98-65303

The following material is used by permission:

Finkleman, Danny. "Marshall McLuhan and Tom Easterbrook,"
in *Speaking of Winnipeg*, edited by John Parr. Winnipeg: Queenston House, 1974.

Wolfe, Tom. "What If He Is Right?" in
The Pump House Gang. New York: Farrar, Straus and Giroux, 1968.

Gellner, Ernest. *The Psychoanalytic Movement*. London:
Grafton Books, a division of the Collins Publishing Group.

"Fashion Is the Medium." © 1968.
The Hearst Corporation. Courtesy of *Harper's Bazaar*.

Sloane, Leonard. "Advertising: A Different Look at Creativity."
© 1988 by The New York Times Company.

McLuhan, Marshall. *Understanding Media: The Extensions of Man*.
New York: McGraw-Hill, 1964.

Howarth, T.E.B. *Cambridge Between Two Wars*. London: William Collins.

Printed and bound in the United States of America

For Skaptidóttir, who has the courage
and faith of her ancestors

Contents

Foreword

I was honored to be asked to provide a foreword to Philip Marchand's splendid biography of Marshall McLuhan. It is, in fact, a double honor in that, first, it allows me to associate myself with such an honest and useful book and, second, it provides me with an opportunity to speak about McLuhan, to whom I owe so much.

The first time I encountered McLuhan was at Teachers College, Columbia University, in 1955. He had been brought from Toronto by Professor Louis Forsdale to address a class in which I was a student. After a brief introduction, he began to speak in what I took to be a stream of parables, beginning with an account of Edgar Allen Poe's story "Descent Into the Maelstrom." Each of McLuhan's parables led directly into a sort of proposition in which he connected one technological development with another or with some social trend. I recall his saying that the invention of eyeglasses in the 12th and 13th centuries led us eventually to the idea of genetic manipulation. I recall his saying that the invention of the photograph brought an end to conspicuous consumption, and that the invention of telegraphy created the idea of decontextualized information. Interspersed among these assertions were one-line jokes and puns, which he claimed were the natural form of humor in a culture where complex story lines have become irrelevant. "Though he may be more humble, there is no police like Holmes." That was one of his puns, and he offered it without attribution. I learned much later that the pun originated with E.W. Hornung, the author of *Raffles* and, incidentally, Arthur Conan Doyle's brother-in-law. McLuhan did not have much patience with the obligations of attribution, with the possible exception of the puns of James Joyce. But it didn't matter then (and it doesn't matter now). The point is that his performance was strange and wondrous and, as you'd expect, at various times during the performance students raised their hands to ask questions. Mostly, their questions wanted him to elaborate on the connections he had made, perhaps to say a few words in explanation of their meaning. I remember his reply as if it were yesterday. It was the same to every question. He said: "If you don't like that one, here's another one." And then he would proceed in rapid fire to give three or four more connections. I had never heard a professor respond that way to students' questions, and it did not surprise me that many of my fellow students were infuriated. But my friend Charles Weingartner, who sat next to me, and I were not infuriated. We were delighted. It was as if each of us had waited all our lives to experience such an event and such a person. It was clear to us that what McLuhan was doing was showing us a way of thinking about media, puns and jokes included, and that it was not to the point that the content of what he was saying needed to be justified. This was a lesson not so much in the aphorism "the medium is the message," but a more relevant one, at the time, "the method is the message." Later on, Charlie and I wrote five books together, and I believe our collaboration had its origin in our experience that day.

I must say that it never occurred to me to ask McLuhan for an explanation or an answer to anything I ever heard him say. The way in which he spoke, on that first day and afterwards, suggested that it was not his business to give answers. All of his assertions added up to a question. And he had supplied the question by the ancient literary device of constructing a grand metaphor, an invitation for us to play "let's pretend." His approach was not unlike the way in which Marx, Freud, and especially Nietzsche did their work. In one of his books, Nietzsche begins by asking the following question: "Suppose women are right. Then what?" It is a question that has many answers, depending on who asks it and when. In fact, not many people are now very interested in Nietzsche's answer. But they *are* interested in Betty Friedan's or Gloria Steinem's, and the answers of so many others who now address the question. Answers change. Good questions endure.

McLuhan's question (perhaps I should say, questions) may be put into somewhat similar form. Suppose that what we *are* is a product of how we represent the world, then what? Or, suppose that our sensorium, and therefore our psychic habits and ethical suppositions, are under the invisible control of our media, then what? Or, suppose the forms of expression are far more significant than any content, then what? Well, armed with such grand supposes, I was on my own and have been for the past forty years, during which I have tried to come up with many answers. And *that*, in a word, is what McLuhan has done for me — launched me on a lifetime project to find my own answers. For good or ill.

Incidentally, I don't think McLuhan would have entirely approved of the answers that are to be found in some of my books. *My* answers are too overtly moralistic to suit him. He repeatedly warned all investigators to stay clear of moral judgments since such judgments are apt to block one from understanding exactly what is happening. Nonetheless, on this point — McLuhan's claim of moral neutrality — I have always been more than a little suspicious, and I am not the only one. In an otherwise flawed little book about him, Jonathan Miller demonstrates persuasively that there is a strong moralistic strain in McLuhan's work, one might even say a definite religious component. I agree with this, but I would go even further than Miller, or at least in a parallel direction. In creating a grand metaphor, one must do more than invite us to pretend the world is as one sees it, or imagines it. You must also tell a story about what you have imagined, preferably a story with a beginning, middle, and end. And McLuhan did just that. It is a story that pays homage to his Catholicism, and reflects what must have been a profound optimism.

It is a story of loss, to be sure, but also a story of redemption. As I understand it, it goes like this: There was a time when human communication was holistic, when it made use of the fullness of our sensorium, when it caused no disconnect between brain and heart; a time of mostly, if not pure, orality, when myth mingled comfortably with fact, the past with the present, when speech was always infused with meaning, when piety, poetry, and history were intertwined, when the parable was not a mere ornament of speech but a conveyor of truth and wisdom.

And then that time ended and was supplanted by the time of the great extensions of man: picture writing, the alphabet, the mechanical clock, printing, telegraphy, and all the rest. But with each extension of our senses in one direction, there was an amputation of sorts in another. The clock, for example, dissociated time from human events. It made us time-keepers, then time-savers, and now time-servers. The authority of nature was superseded, and Eternity ceased to serve as the measure and focus of human experience. The Gutenberg Age, to take another example, gave us logic, sequence, and compartmentalism. But, as McLuhan put it, borrowing from James Joyce, it made us ABCED-minded — that is, absentminded. We became left-brain dominated. The power of the eye was enhanced, but the ear was amputated. We came to see the world through a single point of view and were blocked from seeing the whole picture, all at once. Our sensorium became fragmented, and it is possible that our penchant for violence is connected to the media-induced imbalance in our senses.

That is the loss part of the story. Here comes the redemption, and not a moment too soon. It comes in the form of the electronic revolution — the extension of our entire nervous system. Electric communication contains in its structure, that is, its speed, its volume, its multi-directionality, and its forms, the possibility of making us whole again, of retrieving the oral tradition of reclaiming the richness of multiple perspectives, and of reawakening the creativity of the right hemisphere of the brain. Once again, sight, sound, touch, and taste will blend graciously and healthfully.

Of course, this is nothing less than a story of Paradise, the Expulsion, and the Return. It is not for nothing that McLuhan was called the Apostle of the Electronic Age. McLuhan liked to refer to T.S. Eliot's remark that the content of a poem is like the piece of meat the thief throws to distract the guard dog. The poet uses content to distract the reader so that the magic of the poetic form can do its work subconsciously. McLuhan used this analogy to make the point that the content of media is a mere distraction. The medium itself is the main show. Well, here I should like to reverse all that and say: too much attention has been given to McLuhan's form. Yes, of course, he wrote in a zestful style, energetic and witty. There is not a ponderous sentence in the entire canon. Yes, he invited his readers to investigate these matters for themselves, and that is what first drew me to him and has occupied my attention and commanded my devotion to him all these years. And yes, he was the first writer I had ever encountered who could write a sentence in which the words Plato, Erasmus, Batman, and the Beatles could find a coherent place. Yes, McLuhan, for all his telegraphic obscurities was a delightful, provocative, even dazzling literary figure. But more than that, he had a message, and this time the message was to be found *not* in his manner, but in his matter. One might even say his manner was the distraction which obscured the matter. And his matter tells us that the days of the print-oriented bastards, of the tyrannical left brain, of the fragmented sensorium are numbered. The garden is within sight again. It is a global village filled with the joys of electronic gregariousness and communion. All we have to do is enter it.

You may notice that in all I have said of McLuhan, I did not include the word most frequently used by journalists to describe his work — originality. I have not used it because I do not believe it, and in saying that, I do not mean to diminish McLuhan's significance. For one thing, those who work in the field of social studies are never original in that they discover nothing new. They only rediscover what we once knew but have forgotten, and need to be told again. If the price of civilization is repressed sexuality, it was not Sigmund Freud who discovered it. If the consciousness of people is formed by their material circumstances, it was not Karl Marx who discovered it. If people are obedient to legitimate authority, it was not Stanley Milgram who discovered it. And if the medium is the message, it was not Marshall McLuhan who discovered it.

But beyond this, of all the literary virtues originality has the shortest shelf-life. Which is to say it is vastly overrated. Michael Jackson is original. So what? The Edsel was original, and it's gone, and most people have never even heard of it. Even, if I may be allowed to mention it, the Holocaust was original, and what are we to make of that? No, I don't think McLuhan was original. He was better than original, in part, because he stood on the shoulders of scholars like Harold Innis, Lewis Mumford and, to a lesser extent, Jacques Ellul. In the case of Innis, McLuhan was unfailing in expressing his intellectual debt to him. So much so that I have no hesitation in saying that were it not for McLuhan, Innis's work would still be largely unknown in America. Who knows, maybe even in

Canada. In Mumford's case, McLuhan was less generous, perhaps because Mumford was ungenerous to him. But the fact is that there are very few ideas in *Understanding Media* that you cannot find at least in embryonic form in Mumford's *Technics and Civilization*, published in 1934. As for Ellul, his magnum opus *The Technological Society* was published in English only a year before *Understanding Media*, but for many years before that, McLuhan knew of his work and was especially admiring of Ellul's concept of technique as a world view.

So, if McLuhan was not quite original, why then was it left to him to invent the modern field of communication studies? Innis's ideas were as profound; he wrote in the same telegraphic style as McLuhan, and he had the same intellectual distance from which to see things that, at least in America, we could not. Mumford was more learned than McLuhan, had a more certain grasp of the movement of history, and wrote a rich literary prose. As for Ellul, he certainly had a more realistic vision of how technology controls our ways of living, especially our politics.

There are probably many reasons why it was McLuhan and not the others to whom so many of us turned to learn about how to understand media. But here is the reason that makes the most sense to me: Innis, Mumford, and Ellul were all enemies of their own century. McLuhan was its friend. Innis could only see space-biased media leading to conquest. He despised American culture and because his own land was being colonized by America, he didn't think much of Canada either. Mumford saw, in the end, only the unstoppable megamachine, and Ellul was driven to such despair that he turned toward the only force that was not driven by technology. God. McLuhan saw something of what these men saw, to be sure. He spoke of the necessity of our having Distant Early Warning Systems to alert us to lurking media disasters. He spoke of institutions imploding, and of disruptions brought on by new media paradigms. And I suppose one can find in his work suggestions of an impending apocalypse. But there is no suggestion, as one finds especially in Mumford and Ellul, that the apocalypse *already happened* in this century. You can read McLuhan from beginning to end without getting more than a few hints that our century has been the age of mass murder on a scale unrivaled since the great Flood. I am not here accusing McLuhan of insensitivity, and certainly not ignorance; only of an awesome faith. As I have said, he was driven by the story he had to tell. In its structure, it is far from original. It is, in fact, quite familiar. But it is inspiring nonetheless and helped him to see, and helped others to see, the humor, the joyful confusion, the challenging contradictions, the creativity, and, above all, the happy ending of a century in the process of becoming what it had created. He did not despise the popular arts, he invited us to study them. He did not fear the megamachine, he prophesied its demise. He did not ignore centralized power but explained how it will become unglued. And, he was no technological determinist. To be sure, his story would have a happy ending, but only in the sense that there is a happy ending to life for a devoted and righteous Christian. For those who defy God, there is a different ending in store. And so, for McLuhan, the happy ending was not inevitable, not predetermined. Things could turn ugly if we did not understand what was happening. But if we did, we would learn how to control the media ecology, and teach the promised land. Like Moses, McLuhan was not permitted to see it but he left us with the assurance that if we built it, it would come. Perhaps I should change what I said a moment ago. McLuhan was not so much a friend of *this* century as of the next. He was an ameliorist, a futurist, a prophet of hope. Nothing is inevitable, he said, if we understand what is happening. No wonder young people were drawn to him.

Well, I was young once and was charmed, refreshed, inspired by McLuhan's story. I'm older now, but I think I never really believed in his story. Without meaning to put myself in the class of those men I mentioned before — Innis, Mumford, Ellul — I must say that in one respect I am like them. I, too, am an enemy of my own century, and if I make it there, expect to be an enemy of the next. I am also ABCED-minded, and I am an obedient servant of the left hemisphere of the brain — my own and those of others. And I might even say in passing and to show how old-fashioned I am that I regard some of the radical expressions of what is called post-modernism, especially those with a French accent, to be a form of mental illness. For all these reasons I was astonished, although deeply flattered, when McLuhan urged me many years ago to start a degree-granting program in media studies at NYU. He said that NYU would be a good place for it, and I a good man to lead it. Perhaps he thought I had bureaucratic skills, which is a depressing thought, or perhaps he trusted print-oriented bastards, since he was obviously one himself. Whatever the reason, he urged me to do it, and in 1971 I did. We gave the Department the name "Media Ecology," which is a term McLuhan later used in a letter he wrote to Claire Booth Luce. We like to think he got the term from us but I have been told that the phrase can be found as well in *Understanding Media*. I've read that book four times now, but I can't find it. Perhaps because, like McLuhan, I have become accustomed to reading only the left side of the page.

In any case, the department has flourished and has attracted young people from all over the world. They read most of the scholars McLuhan suggested they should — Walter Ong, Eric Havelock, Edmund Carpenter, Robert Logan, Goody and Watt, and, of course, Mumford, Ellul, Innis, Plato, the whole lot. Of course, they also read scores of scholars McLuhan couldn't have known about, but who write what they do because of McLuhan. I should add that many of these writers seem unaware that were it not for McLuhan, they would be mute.

As of 1996, over one hundred of our students have earned Ph.D.s and more than four hundred have earned MAs. I can assure you that all of them know they are the children of Marshall McLuhan. I, of course, consider myself to be one of his offspring as well, not quite an obedient child, but the kind of child who knows where he comes from and what his parent wanted him to do.

Neil Postman
New York
November 1997

Acknowledgments

Thanks are due, above all, to David and Stanley Colbert for their great efforts on my behalf.

Claude Guay at the Canada Council also made this book possible. It is a considerable pleasure to acknowledge the Council's financial assistance.

Ed Carson and Doug Pepper of Random House of Canada and Katrina Kenison of Ticknor & Fields deserve gratitude for their faith in this book and their work on its behalf. I thank Barbara G. Flanagan for her skillful manuscript editing. Judith Cumming and Dale Cameron at the National Archives of Canada were extremely helpful to me in my work there.

Special thanks to an old friend, Frederick Flahiff, and to Sheila and Wilfred Watson, who helped in many different ways; to William and Veronica Ellis, two stalwart comrades, and to Sandi, who lent me her burgundy turbo-charged Buick; to Claire Marchand and Samantha; to Tom Dilworth of the University of Windsor and Michael Power of the Canadian Catholic Historical Association; to Karin and Brian Long; to James and Betsy Imershein-Howe; and to David Staines of the University of Ottawa. It is unfortunately impossible to mention all those who talked to me about Marshall McLuhan and lent me invaluable materials for the preparation of this biography, but I am deeply appreciative of their assistance.

Thanks to the patron saint of Canada and of all those engaged in heavy labors, whose presence I encountered while examining the McLuhan papers in the National Archives. The late Barrie Nichol also extended important assistance to this project. Along with many others, I deeply miss the presence of this poet, teacher, and friend.

To Patricia, finally, my gratitude for her reading the manuscript and offering suggestions — and, above all, for her persevering with me in this long effort.

Introduction

I first heard Marshall McLuhan speak when I attended the opening class in his modern poetry course at St. Michael's College at the University of Toronto. It was the fall of 1968, and the room was overflowing with students waiting for the great man to appear. He was a celebrity at that time, perhaps the biggest celebrity most of us would ever see so close at hand.

He was called a "media guru," the oracle of the electric age, as *Life* magazine put it. Somehow he had attained this position by studying ancient Greek Sophists, obscure Renaissance humanists, and the baffling text of *Finnegans Wake*. Businessmen sought after the insights of this oracle, journalists like Tom Wolfe interpreted them to the public, and cultural commentators as cerebral and avant-garde as Susan Sontag testified to their stimulating properties.

It was possible he was a true intellectual giant, of the sort who forces subsequent generations to perceive the world in new ways. On the other hand, many serious academics and intellectuals dismissed him outright as a charlatan or a crank. To some, he belonged in that twentieth-century rogues' gallery of brilliant eccentrics who, in their search for truth, pursue false leads with ever-increasing obsessiveness — the gallery that contained Ezra Pound and his monetary theories, Wilhelm Reich and his orgone energy.

The great North American publicity machine, in spite of such critics, kept spreading his name across the continent. Henry Gibson kept asking, "Marshall McLuhan, what are you doin'?" on *Laugh-In*. Articles appeared on McLuhan in publications ranging from *Playboy* to *Mademoiselle* to the *Partisan Review*. John Lennon and Bob Dylan were intrigued.

The classroom where we waited for him was in a red brick and brownstone house built in the last century and once used as a residence for nuns. It exuded a sense of the musty and outmoded, an atmosphere that can sometimes be very pleasant in a university. But in 1968 there was less and less space on campuses for the outmoded. Change was gaining momentum in all areas of society, most visibly, perhaps, among the young on campuses. The man we were waiting to hear could tell us about the nature of this change. We were the first generation to grow up with television, and that potent device, it seemed, had effects on us that no one had dreamed of.

In the fifties it was feared the device would foster a generation of apathetic conformists, permanently lulled almost into narcolepsy by sitcoms and westerns. McLuhan had never been guilty of that fashionable doomsaying — doomsaying rendered ridiculous by the rise of the Woodstock Nation, as Abbie Hoffman termed it. McLuhan had not predicted the rise of the Woodstock Nation, but when it happened, it turned out to be his work from the fifties that explained the phenomenon more plausibly than anything else on the subject of television and its influence.

McLuhan made a striking entrance into this classroom filled with television-raised students. He was tall and pale — not with an academic pallor but with the complexion of a vital man fighting a serious illness — and he radiated nervous energy. We were more than a little intimidated. He was not exactly arrogant, but every crisp, assured, and effortlessly formulated sentence intimated somehow that he was a man who had not one minute to waste. It was almost as if he had run out of time. He had certainly run out of patience. Few, if any, in the class knew why this was so; but everyone recognized the feeling in the timbre of his voice.

He did not speak of the media in that first class but kept to his subject, modern poetry. He immediately launched into a discussion of Gerard Manley Hopkins's "The Windhover," reading the lines

> Brute beauty and valour and act, oh, air, pride, plume, here
> Buckle! AND the fire that breaks from thee then, a billion
> Times told lovelier, more dangerous, O my chevalier!

What, he asked, was the key word in those lines? (Correct answer: "Buckle.") The question did not seem strange to any of the students — we were all by this time familiar with the analytic techniques associated with the New Criticism. We did not know that McLuhan had lived through the revolution in literary studies that had enshrined Hopkins as a major poet, the same revolution that made it seem quite normal to ask what the key word in a line of poetry was and then give half a dozen reasons why the word was so crucial.

McLuhan had been there at the birth of the New Criticism, and it had made a deeper impact on his thought than anything else, except perhaps his conversion to Catholicism. The revolution had sharpened his perception that poetry was something profoundly vivifying, and dangerous. Later in the year when a colleague of his read to the class from Eliot's *Four Quartets*, McLuhan stood up at the lines "every moment is a new and shocking / Valuation of all we have been" and walked out of the room, as if fleeing from the sound of those words. McLuhan was not introspective in the conventional sense; still less did he ever dramatize his own emotions. But he was, in his own way, constantly reevaluating all that he was and had done, especially during that year, when he had experienced, during a grisly and prolonged operation, a foretaste of his own death. Compared with Eliot's poetry, such phenomena as television and the monstrous generation of children it had produced were almost trivial. Yet it seemed that Eliot and the other modern poets really had taught him something about all these things. As the year progressed and he became a little less intense, a little less formidable, McLuhan spoke more about the ideas that had made him famous. His comments were almost always in the form of asides, uttered with an unaffected casualness that by now had become second nature. He would ask, for example, why people persisted in regarding the computer as a *machine*, which it most definitely was not. His tone of voice suggested that he was actually irritated people did this.

On another occasion he mentioned the advice he had given to Richard Nixon's aides, who had consulted him on their candidate's use of the media in his current presidential campaign. Nixon, he had told them, should say as little as possible about anything during the campaign. (To us students, united in our disdain for Nixon, it seemed excellent advice, but hardly something that required a media guru to formulate.) McLuhan also pointed out Nixon's problem with his last name. The sound of "nix" had a highly unfortunate subliminal effect on the politician's audience. A person's name was a medium in itself, and it carried its own message.

McLuhan's classes always held the promise of permanently altering one's appreciation of some aspect of reality. No one could predict what aspect it would be. This unpredictability was part of what made his classes an adventure; oddly enough it was also part of the reason his colleagues dismissed him. They made the word "stimulating," as applied to McLuhan's teaching, sound as if it were something reprehensible.

The same reaction was noticeable in his critics. They regarded him as an irresponsible imp, attacking both left and right with totally unexpected *aperçus*. These critics, champions of ideological verities of all kinds, implied that McLuhan was not only mistaken but somehow morally deficient. His perfect mental equipoise as he sabotaged their most cherished truths was further evidence to them of charlatanism.

For many of his students and readers, however, McLuhan's comments had at least one virtue: they seemed to suggest that the world was more interesting than any of us had previously thought it to be. Without a trace of mysticism or obscurantism, McLuhan hinted that if the world really did seem to be dismal, it was perhaps because all of us were very far from perceiving it as it really existed. This is perhaps why, for those who knew him and those who studied him seriously, McLuhan has proved to be not only irresistible but irreplaceable.

1. Childhood on the Prairies (1911-1928)

> Elsie had more ambition than any other ten people I've ever met. Ambition.
> Motivation. Drive. Of course, that got her into problems. — Ernest Raymond Hall[1]

In the opening years of the twentieth century the last wave of pioneers in North America filled the prairies of the Canadian West with homesteaders. By 1900, the American West had been settled; and, as one Canadian historian remarks, "the great marching column of frontiersmen and settlers which had been travelling for generations over North America at last turned sharply north-west into Canada.... It was Canada's hour. Canada had captured the imagination of the Western World."[2] For ten dollars, any man over the age of eighteen could claim 160 acres of Canadian prairie. If he cleared part of it, planted a crop, built a house, and lived there for six months of every year, the land was his in three years. It was called betting the government ten dollars you couldn't be starved out.

In 1906 a man from Nova Scotia named Henry Seldon Hall, with his wife and two sons, claimed a section of land near the town of Mannville, about a hundred miles due east of the city of Edmonton, in what is now the province of Alberta. Hall thought his piece of land, set in rolling grassland and near a clear blue sheet of water, might support a ranch. As it proved, he was wrong; the land was too marshy to produce hay and the Halls eventually had to abandon it. Before they did, however, Henry's daughter, Elsie, an eighteen-year-old schoolteacher, joined her parents and brothers.

By the time Elsie arrived in 1907, her family had already made a good impression on the neighbors. The Halls were descendants of an Englishman from Bristol who settled in Nova Scotia before 1800; they were genteel folk devoted to books and education, staunch Baptists, abstainers from tobacco and alcohol, observers of the Sabbath — in every way respectable. As is the case with families, however, the view from inside the Hall domain was less happy than the view from outside. Hired hands, for example, never lasted long with Henry Seldon. They might have been edified by his habit of starting the day with a long prayer and a reading from the Bible, but they did not enjoy his other habit of kicking them afterward, when the mood seized him.

Henry Seldon Hall had the stubborn, domineering, and self-defeating qualities that mark the man of violence. He was the kind of farmer, for example, who couldn't resist the temptation to try to milk his range cows — the equivalent of riding a bucking bronco. There was something unpleasantly unpredictable about him.

His wife evidently put up with it. Explaining the decision to move from Nova Scotia, their son Ray mentions the romantic appeal of the West to Henry Seldon and adds, "Father was the dominating force. Mother, well, she just wouldn't obstruct any of Dad's plans."[3] Ray Hall, who had the kind of humorous, obliging personality sometimes seen in those who collapse under an overbearing parent, also put up with his father. His sister, Elsie, however, did not. On at least one occasion her father — whose nature had not been improved by a stint in the Canadian army, with its bare-knuckle methods of discipline — subjected her, for some trivial offense, to what her young son, Maurice, would delicately refer to many years later as "harsh treatment."[4]

Elsie was too spirited to submit in her soul to this treatment; but the resistance cost her. All her life she suffered from an inner tension and an emotional restlessness, as if the conflict with her father had created a subtle inflammation of her nervous system that could not be healed. It also rather complicated her feelings toward men. She could respect and even feel devotion for strong men, particularly those who had authority and intellectual competence — lawyers, ministers, and so on. At least she respected them as long as they kept their distance. Men of less worldly authority and strength tended simply to be objects of her contempt. If she could run roughshod over them, she would.

Before leaving Nova Scotia, Elsie had received a teacher's certificate from Acadia University, then a Baptist institution in the province. After joining her family in Alberta she taught school in districts neighboring Mannville for a few years, visiting her father's homestead from time to time. On one occasion, when the neighbors were gathered for a picnic by the lake she showed up with a tall, handsome young man named Herbert McLuhan. She had been boarding with his family while teaching school in the nearby town of Minburn.

As a single young woman in a pioneer community Elsie would have been in demand even if she hadn't been beautiful. As it was, with her slender form, delicate features, and extraordinarily expressive eyes and mouth, she was an object of universal admiration. She was also clever. "If there were a group of people, she was the center of attention," her brother Ray recalls. "If you were there in that group, you listened instead of talked."[5] Years later, when she returned to the area as a performer of dramatic monologues, she retained that conversational liveliness to the point where she impressed one local as being "reckless with the bullshit."[6] The young Elsie could afford to be choosy in her men; and Herbert McLuhan was the pick of the local bachelors. He was something of a ladies' man, possessed of considerable charm, and he was obviously intelligent and fun to be with. He could play the violin, carry on a lively conversation with almost anyone, tell stories, and do impersonations. He and Elsie amused themselves by mimicking the local people.

Herbert McLuhan and his parents had come to Alberta from Ontario. His grandfather, William McClughan, migrated to Canada from the family seat in Hillsborough, County Down, Northern Ireland, in 1846. According to family legend, William was dispatched to Canada by his parents in Ireland for the same reason so many other apparently worthless sons were shipped to the colonies — as a desperate measure to build character. Apparently William was a hard-drinking spendthrift; but the rigors of life in Ontario had, from all accounts, the desired effect. He prospered in the new land, along with his eldest son, James Hilliard McLuhan, who cleared a farm in the wilderness of southwestern Ontario near the present township of Conn and eventually became a local justice of the peace.

James McLuhan married a pious woman from Edinburgh who fit in well with the Ontario Scottish ethos, described by one of its products, John Kenneth Galbraith, as the God-fearing religiosity of a people who pray fervently but also "look after themselves, do not request the impossible and keep to an absolute minimum the number of purely ritualistic and ceremonial petitions."[7] The McLuhans were said to be good Methodists, but James rather lacked the piety of his wife. Like his father, he had a penchant for drink, but his greatest weakness seems to have been for debate. He had never been to school but he was a voracious reader of the "great books" and, like many autodidacts, was fond of mentally overpowering those who held different views, especially educated people like ministers and schoolteachers. To make his points clearer, he pounded the table with his fist.

In later years he mellowed, so that on his death in 1919 an Edmonton paper ran an obituary stating that

> Mr. McLuhan was a man of exceptionally high order of intellect, a
> genial personality and one who took a broad interest in the affairs of
> the community and the world at large. He was a man of wide reading,
> fond of good music, and keenly interested in astronomy.[8]

Forty-eight years later an English critic of his grandson remarked that Marshall McLuhan really belonged to the tradition of "the popular or populist sage, the cracker barrel Socrates, the lofty or ribald annunciator of values moral, national, cosmic, . . . the frontier publicist travelling the wide land with his grammars, recipe books, shreds of apocalypse and nostrums for spirit and bowels."[9] If Marshall McLuhan was indeed part of this tradition, he certainly came by it honestly.

In 1907 James McLuhan and his son Herbert arrived in Alberta. A younger son, Wallace, had already begun homesteading in the province, and Herbert staked a claim to his own 160 acres near the town of Minburn. Two years later, on December 31, 1909, he was married to Elsie Hall at the home of her father. The couple spent a year farming, but if Herbert McLuhan had any notions of making a career out of being a prairie homesteader, Elsie soon set him straight. It was not the life for her. A prairie winter meant withstanding blizzards and −40-degree cold; it meant surviving on leftover grain and rabbits caught in snare wires and hung up frozen in the shed, if winter work could not be found. After one year, Elsie insisted on a move to Edmonton, the capital of the province.

In Edmonton Herbert quickly entered the real estate business and formed a company with three other men (including his cousin Lou McLuhan) called McLuhan, Sullivan & McDonald. As a career move, it was all but inevitable. Edmonton was booming — in one year, 1912, its population increased 60 percent — and acres of empty prairie were being surveyed into streets and lots and sold as subdivisions at fabulous prices. The city was teeming with real estate salesmen and their "runners," who carried maps up and down the main streets trying to haul customers into the agency offices. At least one conductor had a real estate "office" in his streetcar, and a stranger getting off the train ran a high risk of being sold a piece of land by someone hanging around the railroad station.

McLuhan, Sullivan & McDonald prospered in these boom conditions, although most of its profits were strictly on paper. It seemed a happy hour in the fortunes of Herbert McLuhan when, on July 21, 1911, his first child was born and christened Herbert Marshall McLuhan. Two years later the McLuhans moved into the Highlands district of Edmonton, a prosperous neighborhood where Herbert had a comfortable two-story house built for his family and where Maurice Raymond McLuhan, Marshall's only sibling, was born in 1913.

Marshall later claimed that his earliest memory, dating from this time, was a view of the North Saskatchewan River in Edmonton, seen from a streetcar on a bank overlooking the river. According to this memory he saw horses in the distance and was profoundly impressed that they appeared small enough to fit into his nursery.[10] In view of McLuhan's later obsession with visual perspective as an invention of the print era and his almost visceral rejection of that perspective — in later years, the painter Harley Parker recalls, McLuhan seemed actually to believe that "things became smaller as they receded into the distance"[11] — the memory is almost too pat.

In any case, McLuhan always did feel that the prairie landscape was an antidote to conventional "perspective" and that it had a powerfully formative, if subliminal, effect on the psyche. As he told one interviewer,

> I think of western skies as one of the most beautiful things about the West, and the western horizons. The Westerner doesn't have a point of view. He has a vast panorama … he has at all times a total field of vision, and since he can take this total field at any time, he doesn't have to worry about goals.[12]

Perhaps more fundamentally, McLuhan felt that his upbringing on the prairies provided him with a kind of natural "counter-environment" to the great centers of civilization. He felt he had the advantage that any bright outsider brings with him from the boondocks when he comes to the big city: a freshness of outlook that often enables him to see overall patterns missed by the inhabitants who have been molded by those patterns. It was the advantage that he felt accrued to Canadians in general vis-à-vis the United States and Europe.

In 1914 the great Edmonton real estate boom ended and McLuhan, Sullivan & McDonald lost everything. The outbreak of World War I came almost as a relief. Although Herbert McLuhan was afraid of guns and would have been an extremely poor prospect for killing anyone, he enlisted in the Canadian army in 1915. Fortunately, he was put to work as a recruiting officer. He sent his family to live with the Marshalls in Nova Scotia, in the town of Middleton, near the Bay of Fundy. (McLuhan later claimed that the experience of the sea profoundly affected him for the rest of his life,[13] a curious claim given that he never resided anywhere close to the sea afterward and that his first major experience with the ocean, on board a cattle boat to England in 1932, was thoroughly unpleasant. Nor was he an avid swimmer in later life, describing himself on one occasion as "hydrophobic."[14]) Herbert McLuhan was not long in the army. Accounts differ as to whether he was discharged because he had flat feet or the flu; in any case he settled with his family in Winnipeg, capital of the province of Manitoba, after about a year.

Winnipeg, then the third largest city in Canada, was the financial and administrative capital of the West. After the railroad had reached Winnipeg in 1885 the city enjoyed a growth in size and population unparalleled by any other city in Canada before or since. Hard times lay ahead in the years following World War I, but when the McLuhans moved there, during the middle of the war, the city was at the height of its prominence and must have seemed a very likely place to rebuild the fortunes of the McLuhan family.

In one respect, the city was well chosen by Herbert and Elsie; it was, as Elsie discovered, the home of the Alice Leone Mitchell School of Expression. Alice Leone Mitchell was a product of the Emerson School of Oratory in Boston, one of the foremost schools in North America for training in the "principles of public reading." Sadly, these principles had seen their best days by 1920; audiences in North America were tiring of readers who specialized in "crystalline diction" and forceful expressions of emotion. Elocutionary performances were soon to fade from the scene, like torchlight parades and circuit chautauquas. Nonetheless, Elsie became one of Miss Mitchell's star pupils.

Her husband settled into being a salesman for the North American Life Assurance Company and, after a short period of moving around (Marshall McLuhan lived in at least seven houses in the first eight years of his life), found in 1921 a more or less permanent home on 507 Gertrude Avenue where Marshall spent the rest of his childhood and youth. The neighborhood was a pleasant residential district called Fort Rouge. At the time the McLuhans lived there, the area was an enclave of Scots and Irish, remote from the lively but less respectable ghettos of Jews and Ukrainians in the north of the city.

In some ways, Fort Rouge lodged in McLuhan's bloodstream like a pathogen that he could never quite isolate. As a boy on the verge of adolescence, he devoured *Tom Brown's Schooldays* and spent many happy hours fantasizing about children his age singing in the choir seats of Gothic chapels and roaming the English countryside. These daydreams, he later confessed, were a reaction to the "mean mechanical joyless rootless" life he grew up with.[15]

And yet Fort Rouge in the 1920s was not a bad place to be a child. Gertrude Avenue lay near the junction of the two rivers that intersect the city. As McLuhan later recalled,

> there was the Assiniboine River at one end of the street, a few hundred yards away; at the other end was the Red River. I had a boat on each river, a rowboat on the Assiniboine where I skied in winter and a sailboat on the Red. The opportunities for games and sports out of doors there [were] absolutely unlimited, and all within walking distance of your own house.[16]

The children played baseball on the street, using a sewer cover as home plate and boulevard trees as first and third bases. Marshall usually was the pitcher, which may have been more an indication of his status among the children than of his athletic ability. "Marshall as a boy always stood out as a leader.

Kids admired him and were partly in fear of him and encouraged his support," his brother, Maurice, recalls. "If an activity needed organizing, Marshall would organize it."[17]

As well as living in a setting advantageous to children, McLuhan lived in an age advantageous to children. It was an age in which, as one of his contemporaries remarks, "we weren't organized within an inch of our lives, and we were allowed to laze around and watch the clouds."[18] In such an atmosphere it was natural for Marshall and the other children to take pleasure in long walks. Sometimes Herbert led the way, entertaining the children with an endless fund of stories from his own childhood in Ontario. "All the children on the street were very fond of Herbert," one of those children now recalls. "He was a quiet, quiet, lovely man."[19]

Neighbors, noting that Herbert very often looked after his children, assumed he was henpecked. The assumption was unfair, since Herbert loved being with the children, but it was not completely unfounded. Elsie did not fuss over her children and for this, and no doubt for her independent spirit, she was generally disapproved of by the other mothers on Gertrude Avenue. But Elsie was unconcerned with what her neighbors thought. From her point of view, that was their problem. She herself was choosy about her social life, with a marked preference for professional and artistic types who could encourage and assist the dramatic talents she was developing at Miss Mitchell's school. Criticism from the likes of her neighbors was beneath contempt.

Herbert, certainly, did not fault her independence, even when in 1922 she began leaving home for extended periods to tour the province as an elocutionist and monologist. In fact, he defended her with an almost pathetic gallantry whenever anyone asked why Mrs. McLuhan spent so much time away from home. To all the people who bluntly asked him if he was experiencing marital trouble, Herbert concealed his outrage and replied politely that his marriage was just fine, thank you. In his letters to Elsie, meanwhile, he told her how lonely and upset he felt. She was everything to him, he insisted, and he begged her to be "a wee bit more patient" with life at 507 Gertrude Avenue.[20]

It was a futile request. Tolerance had been jarred loose from her a long time before she met Herbert. And in her worst moods, she could call many reasons to justify her impatience. She was, after all, proud of her descent from the English "bluenose" settlers of Nova Scotia and had a corresponding contempt for the Irish heritage of the McLuhans. Through Alice Leone Mitchell she tasted the high culture of Boston and, therefore, according to Marshall, she "looked down on Ontario from great heights"[21] — not to mention looking down on Winnipeg, which she thought represented the nadir of cultural life in Canada.

Looking down on the Irish or Ontario or Winnipeg might be a trivial form of vanity, but there was nothing trivial in the way Elsie finally grew to look down on her husband. At one point she told her sons, in exasperation, that their father was not a "man."[22] Herbert was almost universally well liked and well thought of by those who knew him, but nobody ever called him ambitious. He seemed maddeningly content to live on a level the rest of his family considered penurious, never owning a car, for example, or the house that he lived in, always scrimping and trying to make do with as little as possible. The coal bin would be down to the last chunk before he would order more. In a climate like Winnipeg's, that was serious business.

Left to himself, Herbert McLuhan would have happily forgone dreams of success and simply found satisfaction in various forms of socializing, for which he had considerable talent. He seemed never at a loss for an anecdote or some interesting bit of information. It was a talent suited to his profession and might have taken him a long way had he combined it with a single-minded determination to close a sale. Somehow, he never achieved that determination, although he tried to be conscientious about his work and even employed "psychology" in his sales calls. The twenties was a boom period, like our own, for pop psychology. The doctrines of Émile Coué ("Every day, and in every way, I am becoming better and better") and magazine advertisements for Dynamic Success Secrets found a receptive public. Herbert McLuhan was not slow to respond to the general themes. Working on the sound premise that people were most deeply interested in talking about themselves, he encouraged prospective customers to do just that. He thereby succeeded in establishing a "relationship" between himself and his customers, but he was not nearly as successful at inducing them to sign on the dotted line.

Herbert McLuhan's "psychology" tended to drift rather quickly into the mystical — he liked to discuss metaphysics with his son Marshall, and he once delivered a talk entitled "The Higher Plane" to the local psychology club.[23] When Marshall was twelve, both his parents discovered phrenology, the great pseudoscience of the nineteenth century, still alive and well in Winnipeg in the 1920s. Both Herbert and Elsie took courses in the art of studying an individual's character and mental capacity from the conformation of his or her skull. One evening a phrenologist they had invited over for dinner put his hand on Marshall's head and proclaimed that he was an outstanding boy, with great intelligence and a sizable organ of Adhesiveness, that is, capacity for loyalty and friendship.[24]

Regardless of the esoteric fields he wandered in, Herbert McLuhan always fell back on the bedrock of his spiritual and mental life, which was the Bible his mother had taught him. His children went to different churches — Baptist, Presbyterian, United — and when they were baptized as adolescents their friends were present for the occasion. Elsie McLuhan's attitude was not quite so orthodox. Her husband might find solace in biblical adages such as "Seek ye first the Kingdom of God and all these things will be added unto you," but she herself was not holding her breath. In her search for a spirituality of more immediate practical effect, Elsie dabbled in Christian Science and Rosicrucianism.

Meanwhile she sought to improve the family's position by offering elocution lessons. In the years immediately following the First World War, before the full advent of commercial radio, children were still encouraged to recite dramatic prose, oratory, and poetry for their elders, as they might play the piano or display some other accomplishment. Elsie, therefore, did not lack students, whom she required to memorize bits from Shakespeare and other poets and do breathing exercises to learn how to speak from their diaphragms.

Although she never formally taught them, Elsie's sons were also the beneficiaries of her elocutionary training: they developed habits of speech that remained with them for the rest of their days. The effects were more immediately noticeable in Maurice, who spoke often at church groups as he grew older and eventually became a minister, but they were more profound in Marshall. He memorized immense quantities of poetry and was familiar with the works of the greatest English poets before he entered university. As an eighteen-year-old McLuhan was uncertain of just how deeply he was able to comprehend these poets. "I often feel discouraged about the apparent lack of poetic taste or feeling within me," he confided to his diary. "I sincerely hope that the spring of that fount will shortly gush forth."[25] At one point, he even thought of memorizing *Paradise Lost* as a desperate measure to get those springs gushing.[26]

However he regarded his taste for poetry, he did thoroughly absorb in the home a sense of sophisticated verbal structures. In those days elocution comprised not only the arts of recitation but a rudimentary form of literary analysis. Textbooks of elocution discussed the "laws underlying literature," citing unity, principality (which specified that the literary work should have a "leading idea or figure"), movement (the work should "progress from beginning to end in an easy, rational order"), contrast, and climax. These "laws" are to contemporary literary criticism as the organs of amativeness and veneration are to contemporary psychology — they were one more thing the adolescent McLuhan had to outgrow. Still, exposure to them was not a bad way for a young intellect to begin focusing on the art of verbal composition, an art absorbed more by living with a woman like Elsie McLuhan — who performed no-holds-barred recitations of Milton and Browning while doing housework — than by any formal lessons in elocution. In a different way, the problems of verbal composition were made real and almost visceral when mother and son argued and debated — in a far from light-hearted manner — over the kitchen table.

Elsie arranged her professional tours through a network of churches that agreed to sponsor her and, despite the waning popularity of elocutionary performances as a result of competition from radio and the movies, she was successful. By the mid-1930s she had performed, mostly in church halls, in all of the major cities in Canada. Her performances consisted of "recitals" of Browning and Shakespeare and "plays (classic and modern), character sketches, musical monologues, humorous and dramatic stories" from more contemporary authors, now forgotten. A typical recital might include her one-woman performance of the trial scene from *The Merchant of Venice*, a monologue entitled "How the Larue Stakes Were Lost" (in which a young jockey sacrifices winning a horse race to save a child about to be trampled underfoot by the horses), and a playlet entitled "Are You Using Life or Is Life Using You?" — interspersed, perhaps, with a few numbers by the church choir or a violinist. She was very good at performing, and she was earning money, which she used to buy such items as a hardwood floor and Persian rugs for the house.

Satisfying as her professional endeavors were to her, her success did nothing to improve her domestic life. As Maurice remarks, "She was getting all the adulation and praise outside the home which her nature required. Dad was happy to sit back and socialize. His interests were philosophical rather than money-grubbing."[27] Herbert McLuhan took the admirable point of view that a man could be happy anywhere, in any situation, and proceeded to make himself as happy as he could by playing with his children, giving them his full attention, and generally acting in ways that today would be regarded as exemplary in a father. One of Herbert McLuhan's pastimes with the children was looking up obscure and interesting words in the dictionary — a find like *transubstantiationist* would be seized with delight. This educational habit was picked up in more intense forms by his elder son. Marshall, who later would remark that a single English word was more interesting than the entire NASA space program, tried to memorize three new words a day — words like *scaturient* and *sesquipedalian* — and use them in conversation. He collected lists of the words and in later life habitually pored over etymologies in the *Oxford English Dictionary* as if they were mystic runes.

Herbert McLuhan could find many consolations in his two sons. When they were in their early twenties Herbert reminisced, in a letter to his wife, about the elders in his church who had once prayed that he would marry well. He realized now why they had been so concerned for his future. They had been prophets, foreseeing that Herbert and his wife would produce sons who would be mighty servants of the Lord. Already, he reported, leading men in the city were amazed at the wisdom of Marshall and the eloquence of Maurice.[28]

Such consolation was powerful — and fortunate, given the other drawbacks of Herbert's domestic life. Through the 1920s Elsie spent longer and longer periods away from home. When she returned, there was usually a day or two of happiness and excitement before the quarreling resumed — if "quarreling" is the right word to describe a situation in which Elsie raged against Herbert and Herbert responded by trying to make light of it. *He* would not descend to the level of a shouting match over trivialities. It was a tactic guaranteed to rouse her to even greater rage. The voice that newspaper reviewers praised as an "excellent vocal instrument" lacerated nerve endings in the privacy of 507 Gertrude Avenue.

Elsie could be harsh with her sons as well as with her husband. A liberal user of the razor strop, she was one of those disciplinarians who sporadically blow up at children from some obscure and unrelieved frustration. A lapse in complete attention from one of her children when she was addressing them could trigger a blowup. Such domestic fury had differing effects on the two boys. Maurice, like his uncle Ray in the household of Henry Seldon Hall, collapsed under the tensions of living with a volatile parent. In his maturity he became one of the most genial and puckish of men, almost compulsively so. It was the legacy of a certain strategy of the class clown. He also became the favorite of his father, a fellow conspirator in the war against the harshness of the adult world. Herbert praised Marshall in his letters to Elsie, but it was Maurice, the young man gifted with eloquence and with a call to the ministry, whom he described to his wife in the proudest and most heartfelt terms.[29]

Marshall did not collapse under his mother's emotional violence. Like Elsie herself in her father's house, he never submitted in his soul to her outbursts. And, as it had in his mother, the stand exacted a price. In his diary at age nineteen, Marshall wrote, "Our domestic situation is such that I can rarely bear to reflect on it long enough to write a word about it."[30] The emotional turmoil he experienced in his home might account for the complaints of indigestion recorded throughout the diary, an indigestion he suffered for a great part of his life. But he resisted his mother beginning as a child, refusing to accede to even her modest demands. "If it was a picnic in the park, he didn't want to go," Maurice recalls.[31]

Aside from indigestion, this stubbornness seems to have stirred within him a permanent wellspring of nervous energy, a certain restless drive that he later harnessed in the service of the innumerable "projects" that were a source of awe and amusement to his friends. It also marked him early as an independent spirit, seemingly impervious to what others thought of him. He had a brittle kind of self-confidence; it could not be pierced by the worst barbs directed at it, and yet it was never so much a part of his self that he wore it lightly or easily. At times as well, his independence seemed to involve a spiritual distance from others that even friends could not overcome. "Marshall was very friendly and outgoing, but he was always holding something back," Una Johnston, a childhood friend, recalls.[32] As an adolescent, when he became a voracious reader, this distance sometimes amounted to rudeness. If he was reading when friends dropped by, he might make a perfunctory comment about the book and then keep on reading.

Elsie, paradoxically, respected him all the more for his stubborn opposition to her; and Marshall ultimately was far more emotionally attached to her than to his good-natured and even-tempered father. But this attachment did not prevent Marshall from passing severe judgment on his mother. Shortly before his nine-teenth birthday, he proclaimed in his diary that he had never really had a mother — someone "who would have softened, rather than developed my native and inherent harshness and weaknesses." With an eighteen-year-old's asperity, he noted that Elsie "has not many strong anchors... When the billows rage and hiss she knows no means to escape or palliate their wrath." Clearly, she was responsible, with her "boundless egotism," for the searing and profound rift between his parents.[33]

It is easy to forget, reading these comments, that Elsie probably displayed some indulgence toward her older son. This side of Elsie is not recorded, but it can be inferred from McLuhan's lifelong habit of dominating conversations — the sure mark of an indulged child. If Elsie McLuhan was not a model mother, there is no evidence of malice or purposeful neglect of her children, either. She was harassed in her spirit, but such love as she could give to Marshall she gave, mostly in the form of a fierce pride in his accomplishments and an encourage-ment of his intellectual faculties.

Partly because of this support from his mother, Marshall's attitude of supe-riority toward his father, who like James McLuhan had never graduated from high school, became painfully clear as he grew older. After Sunday school Herbert was in the habit of taking his sons for long walks in which he would dis-cuss moral and philosophical issues. By adolescence, Marshall felt that he had outgrown whatever perspective his father had to offer. If anyone was to be men-tor, it would be Marshall. He freely offered advice to Herbert on matters rang-ing from the bad effects of tea drinking on the digestive system to profitable reading matter.[34] His father seems to have taken it all seriously. On one occasion, as a university student, Marshall was startled by his father's espousal of social-ism; he had to administer a "tongue-lashing" and spend an hour and a half demonstrating to his father, from notes taken in his economics class, the fallac-ies of Marxism.[35]

Emerging from adolescence, with its unappeasable furies, Marshall shed much of his harshness. Pride always remained, but arrogance was diminished and a certain sweetness of temperament became far more evident. Before he matured, however, the young McLuhan could be very hard to live with. Maurice felt it most keenly. His clowning was unworthy of an intellectual being, according to Marshall — especially when people paid more attention to it than to what Marshall had to say. Of course, Marshall, as the responsible McLuhan male, also looked after his brother, which was not easy: Maurice the prankster was always getting into trouble. Whenever he did, the strict, paternal figure to whom he answered was Marshall, not Herbert.

In his diary, Marshall freely used an older brother's privilege of estimating, with a magisterial objectivity, the character of a younger. He described Maurice (or "Red") as "a great kid" who was acquiring the work habit and was waking up mentally; Marshall's assessment was that Maurice's sense of humor could have made him another Mark Twain if he had decided to become a writer.[36] On the other hand, Marshall felt Maurice was feckless and would have difficulty making his way past life's obstacles.[37] Red at one point squandered sixty-five cents, and Marshall was appalled: he shook his head over his younger brother's curious stubbornness, the kind of disguised obstinacy that is often found below the surface of amiable and ingratiating men.[38] Finally, he noted a tennis game played with Red and recorded with satisfaction that he once more triumphed over his younger brother.[39]

As children, according to Maurice, this physical superiority was also demonstrated on occasion by Marshall's manhandling him, particularly after Marshall had been disciplined by Elsie. The fact seems, in a way, incredible. McLuhan in later life quite lived up to the stereotype of the gentle scholar. And yet Marshall all his life had to struggle against a strong tendency to bully; it was not always easy for him to be gentle. As a child in a household that was at least verbally violent, it does not appear that he was gentle at all.

In an aggressive mood, however, his most devastating weapons of offense were certainly not physical. The arguments and debates with his mother left him more than prepared to handle any and all comers from outside the home. Even the most distinguished dinner guests often found themselves embarrassed by the mental and verbal agility of Herbert's elder son. "Here was this damned kid from high school who would challenge them in an argument and often, in his demand for accuracy, make mincemeat of them — unless they were people of real intellectual stature, in which case they would tend rather to encourage Marshall," Maurice recalls.[40] This habit of demolishing opponents in debate may explain a nightmare that the nineteen-year-old Marshall recorded in his diary; in the dream he was about to be lynched by a mob of churchgoers whom he had somehow antagonized.[41]

For all this agility he was not, initially, a good student. He certainly did not remember his teachers fondly; late in life he stated flatly that he had "never had a teacher who made me the slightest bit interested in anything I was studying."[42] He actually failed grade six; only the efforts of his mother secured his promotion that year. She went straight to the principal and, as a former schoolteacher herself and therefore well qualified to understand these matters, explained the school's mistake. Marshall was admitted to grade seven on condition that he could "handle" it. The seventh-grade teacher was at last competent in the classroom, particularly in English literature. "Things just seemed to turn for him then," Maurice recalls. "He seemed to meet his destiny in the experience of [his teacher's] interest in literature. The way she communicated it to her class stimulated Marshall to the point where he never left off."[43] Two years later, in grade nine, Marshall ended his school term as eighth in a class of thirty-seven, receiving in his final term A's in English, algebra, and music.

He was pointed on the path he was to follow, a path that Elsie McLuhan intended to lead to glory. Her own ambitions had resulted in her appearing before appreciative audiences in church halls across the country; it was not much, but it was something. Her sons, however, would be exalted. They would be presidents of universities. All the fine, poetic, cultivated things of the world would be theirs by right, and they would be honored above other men for their learning, their cultivation, and their moral, social, and intellectual authority.

2. University of Manitoba (1928–1934)

I remember when I was growing up. I couldn't feel the slightest respect for adults who were engaged in 1920 jazz forms of dancing. This seemed to me beneath all human levels of behavior. — Marshall McLuhan, 1967[1]

In 1928 when McLuhan entered the University of Manitoba, it was the third largest university in Canada, with about 3,500 students. The "campus," on the south end of the city, was largely a collection of Quonset huts. Instruction had something of the same rudimentary quality as the architecture. English, for example, was required of all students in the first and second years not so much to deepen an appreciation of letters as to impose a standard of literacy on the children of pioneers and immigrants. Of course, the university did possess here and there a distinguished scholar from England or the United States or eastern Canada; it was not entirely mediocre. Nonetheless, McLuhan later commented, "I never took university days in Manitoba seriously. All I knew was that I was not getting an education."[2]

McLuhan embarked on a five-year honors course that was intended to "lay a broad foundation" for the students and that included physical sciences and foreign languages as compulsory subjects. McLuhan resented the idea of required courses. He might follow the trail of an idea, if it captured his fancy, through arcane lore or unfamiliar subjects, but he did not like to be forced to study something alien to his basic interests. These interests, in his first two undergraduate years, were English and history.

His pertinacity in following the trail of his ideas sometimes caused his professors to view him as obstinate — the only real criticism they ever made of McLuhan as a student. But McLuhan, on his part, was not impressed by his English professors, and his perception of their mediocrity did not cure him of his stubbornness. In his first lecture in English at the start of his third year, for example, he noted that his professor, a W.T. Allison, spoke on Milton for an hour without telling him anything new. The first question for the class was the meaning of the word *imprimatur*. McLuhan, cashing in on his word lists, was the only one who could answer.[3]

In his diary, he claimed that it was impossible to imagine Allison returning again and again to a work like *Paradise Lost* or Gray's *Elegy* for aesthetic pleasure or for edification.[4] Clearly, McLuhan thought of literature as food for the soul, which hungered after truth and beauty; if Allison's soul had a limited appetite for it, McLuhan's was different. Only two members of the English department struck him as having any serious concern for literature; on the whole he judged, probably correctly, that his professors' reading, even in English literature, was limited and that he would not have difficulty surpassing them.[5]

His poor opinion of the University of Manitoba faculty was somewhat modified in his third year, however, by his acquaintance with two professors outside of the English department. R.C. Lodge, an Oxford man, introduced McLuhan to Plato and the heady vistas of philosophy. Although the mature McLuhan was allergic to philosophizing on a grand scale, as a young man he found it nearly as great a source of spiritual gratification as literature. In any case, Lodge's intellect was sturdy enough to push back when McLuhan shoved. He noted in his diary that Lodge did not try to slip away from him after lectures, an indication that other professors took evasive action when they saw McLuhan coming.[6] So grateful was McLuhan for Lodge's presence that in his last year he directed most of his energies to the study of philosophy, accomplishing what he considered his finest work as an undergraduate.[7]

The other professor who impressed him was a history scholar from Oxford named Noel Fieldhouse. McLuhan later called him the "most inspiring" teacher he had ever studied under.[8] Fieldhouse made no direct contribution to McLuhan's later intellectual development, but he did help to saturate his mind in history. If McLuhan later rejected conventional historical as well as philosophical approaches in his study of media, it was not for want of acquaintance with them.

Despite his overall dissatisfaction with the university's offerings, McLuhan displayed quite another attitude toward his alma mater when he once recommended to friends that they send their daughter to a small university; he had gained a great deal from attending a university as small as Manitoba's, he told them, where it was easy to have contacts with his professors.[9] This remark, made in the late 1970s, may have reflected McLuhan's dismay that the University of Toronto, a home to him for more than thirty years, had become swollen and bureaucratized, with the students functioning as trainee bureaucrats.[10] During that period he told some officials at the University of Toronto that they ought to hold the university in a pub where students and professors could loosen up and trade bullshit.[11] By the 1970s he believed that the university took itself too seriously as an institution for the instruction of the young. Other institutions were instructing the young at least as effectively as universities were. The real business of the university was loud, freewheeling debates and arguments among students and professors.[12] "One of his great theories," an associate recalls, "was that all the university taught you to do was to bullshit. And he thought that bullshit was a very high order of thought."[13]

It is doubtful, however, that McLuhan fully enjoyed the opportunity to "bullshit" at the University of Manitoba. Quite simply, he lacked partners who were very good at it, not finding in his first two years anyone as interested in literature as he was.[14] He was always hungry for people to talk to; he longed for someone as knowledgeable and contentious as he to chip away at the edges of his own assertiveness and thereby temper it; he thought that someday, although he was not irresistibly affable, he might attract a "very fine circle of intellects."[15] McLuhan recognized early that he was not a solitary thinker and that his thoughts took shape most easily when he could talk them out.

McLuhan spent his first two years at the university laying the broad foundations; he took English, geology, history, Latin, astronomy, economics, and psychology. In his first year he had a notion to major in engineering, a somewhat mystifying choice for a man who had the intellectual's typical uneasiness with mechanical devices. McLuhan was far from being a gadget lover at any time in his life. Once when his car overheated, he simply stood back and, helplessly awestruck, watched the clouds of steam coming from the radiator as if they had been conjured by a poltergeist. His father had to suggest that he cool down the radiator with buckets of water.[16] McLuhan's later remarks about human artifacts turning human beings into "servo-mechanisms" of those artifacts were heartfelt.

Nonetheless, he might have been interested in some *idea* connected with engineering. He was often intrigued by the implications of subjects that he would never have studied in a thorough or conventional manner. Astronomy and geology, for example, inspired thoughts in the undergraduate McLuhan of great cycles of human and divine history.[17] In the case of engineering, McLuhan later claimed that his initial fascination had to do with an "interest in structure and design."[18]

Whatever the cause of that initial fascination, however, it soon waned, and by his second year of university McLuhan was uncertain of what direction to take. He felt himself more talented than most of his peers and with more than the usual desire for honorable success in life, but success and talent in what field, precisely, was a mystery to him.[19] He wanted to be a Great Man; and he had a strong feeling that the road to greatness did not lie in being a professor.

He gave up the idea of engineering after one summer with a survey crew in rural Manitoba. He was bullied by some of his co-workers, who took a dislike to the lanky young man reading books in his tent while they went off on drinking sprees. By the start of his third year at the University of Manitoba he decided to focus on English and philosophy instead.[20]

The teaching of English at the University of Manitoba at that time was based firmly on the Victorian ideal of the cultivated amateur seeking aesthetic pleasure and edification from the acknowledged masters of literature: Spenser, Shakespeare, Milton, the great Romantic poets. The favored critics were Charles Lamb, William Hazlitt, Leigh Hunt, Walter Pater, Thomas De Quincey — nineteenth-century prose stylists who fashioned their ideas about Great Men and great literature into essays full of profundity, pathos, humor, and charm.

Those gentlemen did, in fact, write beautiful prose; their works deserve the title "belles lettres." But the belles lettres tradition in England was dead by 1930, killed by T.S. Eliot's *The Sacred Wood*, I.A. Richards's *Practical Criticism*, and Ezra Pound's scathing denunciations of English literary incompetence. No one in Manitoba seems to have been aware of this momentous shift in literary consciousness. In most North American universities, in fact, literary criticism had not advanced very far beyond the purely biographical and historical ("List the first five poems Milton wrote, including dates and the central theme of each") or the analytical techniques of the elocutionists with their "laws underlying literature" and their study of the "inherent quality of vowels and consonants that adapt them for vocal presentation of thought and emotion."

If McLuhan chafed at the limitations of his professors, he did not object to the tradition of belles lettres, the "adventures of the soul amidst masterpieces" that it embraced. In fact, he was enchanted by it. Such literature was *Tom Brown's Schooldays* on a higher level. He lived for the inspiration it provided.

Of all the great men and great writers McLuhan encountered as an undergraduate, the greatest in his estimation was the British historian and essayist Thomas Babington Macaulay, whom he discovered in his second year of university, the year in which he came into his own as a student. Macaulay's reputation, like that of many other Victorian literary giants, has steadily faded since the turn of the century. His prose remains admirable in its clarity and forcefulness, however, and something about his evocation of historical figures and events has a potent appeal to the intellects of bright adolescents.

McLuhan's adolescent intellect was besotted with Macaulay. He acknowledged in his diary that his love and admiration for Macaulay amounted to "hero worship,"[21] and he wearied his professors with continual references to the man. There was an urgency to his championing of Macaulay that went beyond any fascination for prose style. He confided to his diary, for example, that after reading Trevelyan's life of Macaulay, "I can say that my literary knowledge has been increased, my standards of real manhood raised, my standards of domestic affection set, my self conceit dispelled." In other words, he had entered a far different world than 507 Gertrude.[22] Macaulay's capacity for work — a specialty of the great Victorians — particularly daunted McLuhan, who was nagged throughout his life by the thought that he was continually wasting time. He felt a similar uplift reading accounts of the virtues and abilities of T.H. Huxley and Thomas Carlyle; and there was no stopping his visions of noble self-possession when he discovered Matthew Arnold's essay on Marcus Aurelius.

McLuhan's obsession with Great Men was at its most intense when he was most concerned with changes in his own life — the rapid, unsettling mental changes of a young man in the first years of university. He was disconcerted by the scope of those mental changes, fearful of what he would be left with when the period of rapid change was over and his adult character was more or less formed; he accused himself of terrible personal inadequacy in seeing his way through these changes. He could not do without the benign influence of his beloved writers. Conversely, he was harsh in judging the misdeeds of Dryden as recorded in a biography and scornful of the dissolute nature of Dryden and Pope's era.[23] The "sensual" nature of the poetry of Catullus and Spenser was also blamable.[24]

McLuhan's literary tastes were in keeping with his desire to be influenced for the good by great men and followed closely the canon of masterpieces as defined by scholars and critics in the late Victorian era. His favorite poem in the second year of university was Matthew Arnold's "Sohrab and Rustum," a ponderous mini-epic once a standard in high school literary anthologies.[25] At the start of his third year, stimulated by Macaulay's writings on Samuel Johnson and Joseph Addison, McLuhan resolved to specialize in English literature of the eighteenth century.[26]

The eighteenth century he was so fond of as an undergraduate was not the eighteenth century of Sterne, with his endless play of wit and satire, or Fielding, with his command of irony and tone — the tone that McLuhan in later years brilliantly summarized as that of an indulgent gentleman addressing an audience of dullards. The eighteenth century he laid claim to was that of the polite essay, in which Addison and Johnson educated their readers in a spirit of high decorum. Johnson's Latinate prose swelled McLuhan's vocabulary lists with new acquisitions and gave his intellect a pleasing workout.[27]

As for contemporary literature, he had heard of nothing worth mentioning. He felt the English-speaking world was experiencing a very low period and stood sorely in need of a second Macaulay to inspire other writers to greatness.[28] In later years, however, McLuhan would look back and see the literary achievements of the first decades of the twentieth century as eclipsing even the achievements of the Elizabethans. But in 1930 in Manitoba, it was as if Joyce, Eliot, and Pound did not exist. It is possible that their books were not even in the university library at the time or in any of the bookstores in Winnipeg. McLuhan could, in innocence, repeat with approval in his diary someone else's comment that a busy man could ignore anything written after 1842 with no hurt to his intellect.[29]

Following such advice would certainly have made his life easier. His reading lists were not short; and it was a bitter realization to him that life would never be long enough to read all that was worthwhile of English literature.[30] To help him absorb what he did read, he devised various expedients. For English literature in general he conceived the idea of listing important writers chronologically, accompanied by thumbnail sketches of their lives and works. For novels he wrote lists of characters with their roles in the narrative. For his readings in the *Reader's Digest* he wrote a single brief comment at the end of interesting articles, developing his skill at concise summary. For his other reading he indexed, at the back of each book, all the references he thought important, rather than underlining sentences, a temptation he thought should be manfully resisted.[31] This indexing habit he retained all his life.

With all his conscientious labor he felt desperately rushed for time. If a bout of flu laid him up for a few days he was frantic; the distraction of parental quarrels (he lived at home for his five years at the university) made him think wistfully of lifelong bachelorhood.[32] (The worst of the quarrels ended the year he graduated, 1933, when his mother left Winnipeg and her husband for good.) He hoped, after all, to leave some intellectual mark on the world, given some peace and quiet.

An inkling came to McLuhan of what he might do. While sitting on the toilet one evening in April 1930, he hit on an idea that he thought might be the germ of a great work: the realization that all of life — mental, spiritual, and physical — was governed by laws that are still largely unknown to human beings. If a person unwittingly violates these laws, he is thwarted in his doings. If he obeys them, he prospers. "Death, sickness and sin" might well disappear, in fact, in a world where these laws — laws based on a psychological and metaphysical understanding of Christ's precepts — were finally elucidated.[33]

The idea, which probably owed a great deal to his mother's Christian Science, excited McLuhan. He thought it might be a good subject for a Ph.D. thesis in philosophy.[34] It would supplement rather than displace any of the tenets of orthodox Christianity, which he was anxious to preserve. He was never scornful of fundamentalist Christianity and was particularly impressed by predictions of the end of the world based on biblical prophecies. Still, his unpleasant experiences with his two Bible-reading grandmothers and their arid spirituality made him restless with this form of Christianity. The wider metaphysical view was just the thing to make Christian belief palatable.

With such notions on his mind, it is no wonder that Marshall McLuhan sometimes struck other students as a trifle serious. He was, according to one female student, one of the "morons." If you wanted to find McLuhan, the place to look was the library. "I certainly went to a lot of parties and he wasn't at any one of them," his former classmate recalls.[35] In fact, McLuhan did go to parties, although likely not the kind favored by the University of Manitoba's flashiest undergraduates. He confessed in his diary that he was an onlooker rather than one of the boys. By the time he entered university, his old circle of friends seemed less than stimulating, marked as they were by such signs of intellectual degradation as a tendency to play cards.[36]

But McLuhan was not entirely cerebral. He agonized, for example, over his skinny physique — at eighteen, he stood six feet one and a quarter inches tall and weighed a little under 140 pounds — and was desperately concerned about putting on weight.[37] He swam frequently at the YMCA in an attempt to build up his frame. Notwithstanding his reputation as a "moron," he also loved to dance at the occasional university social, an inheritance perhaps from his grandfather James McLuhan, who was famous even as an old man for his agility on the dance floor.

Socially, McLuhan considered himself to be rather shy with casual acquaintances, particularly girls, even though he sometimes struck other people as self-confident to the point of arrogance. He yearned for female company and entertained the usual quota of adolescent daydreams, of a chaste and idealistic nature, in which he appeared as a young hero worthy of girls' affections.[38] Infatuated with a famous American teenage aviatrix, Elinor Smith, McLuhan cherished a photograph of her and wrote poetry in her honor.[39]

In no respect does McLuhan differ more from a late-twentieth-century university student than in his disgust at the carnal worm in the core of these romantic yearnings. "It makes me sick to think that one is attracted to a girl entirely as a result of sex stimulus," he wrote in his diary. "Yet such is clearly the case, on analysis. For my own part I have never truly admired a girl that ever inspired such emotions in me." A further entry in the diary mentions his strong attraction to a girl named "Babe." "There is no perceptible sex appeal about her for me," he notes approvingly. His high-school sweetheart Gwen, however, arouses him to an alarming degree. She "is the worst offender in that particular of all my acquaintances. I can't explain it and certainly it is not loose behaviour or speech or thought on her part."[40]

Strong religious training, a good dose of northern Ireland in his blood, and the fact that the Victorian era took a long time to fade from Winnipeg, Manitoba, all contributed to this rigorous attitude toward sex. He really did hate sensuality and coarseness, as Macaulay and Arnold might have phrased it. In fact, he was

so repulsed and so disgusted by the lubricious talk of some of his male acquaintances that he wished he had no sex drive at all.[41] He was infuriated by wet
dreams, which he thought were brought on by his father's occasionally sleeping
in the same bed with him and thereby causing an excess of bodily heat.[42]

At the advanced age of nineteen, McLuhan noted in his diary that his stu-
dious habits and his elevated perspective on sexual matters had safely carried
him through the worst period of youthful lusts; he knew himself too well, he
thought, and had too much wisdom to form any passionate attachment to a
girl.[43] Not that he intended to remain celibate forever. He thought he might
marry at the age of thirty, unless he was overwhelmed by true love before then;
his wife would have to be intellectual, although he had not yet had the good fortune to meet any woman of real intellectual stature.[44]

McLuhan's views of women were outdated even for 1930. In his view, the
real differences between the sexes became evident when boys were taught by
women in the unfortunate institution of coeducational schools. In an article he
wrote ten years after he graduated, he wondered how women teachers — who
did things by rote, had low intellectual horizons, and tended to take everything
personally — could ever inspire their pupils.[45] Yet he was aware that his perspective was skewed by his family situation: he was profoundly attached to
and also profoundly repulsed by the woman he knew best, his mother. He loved
her and he fought her off with the intensity of a man whose life was in the balance. Under the circumstances he frankly lamented that he had no sisters.
Lacking them, he lacked any understanding of girls, any appreciation of their
outlook, and he experienced difficulties getting on with them. He also realized,
with some prescience, that the woman he would marry would not have a very
easy time of it.[46]

McLuhan's extracurricular activities at the university were limited to a few
sports — rugby and hockey — and to a far more congenial pastime, debating.
The elocutionist's son, seasoned by kitchen table disputes with his mother, soon
discovered a dangerous talent for constructing arguments on any side of a question — and the more the arguments defied the weight of evidence, the more
exhilarating McLuhan found them.

What he termed his "haranguing" was not limited to official university
debates.[47] One mild-mannered professor, Lloyd Wheeler, found his classes
taken over by McLuhan. Any casual encounter with other students could also be
the occasion for McLuhan to take the floor — and hold it. Shortly after he discovered his metaphysical approach to Christianity, for example, he found himself eating lunch with several anti-evolutionists who were freely expressing their
contempt of Darwin. Though sympathetic to their fundamentalist Christianity,
McLuhan had read too much of T.H. Huxley and was too excited by his own

new vision of Christianity to let their remarks pass. He leaned forward over his homemade cheese and butter sandwich, straining for a chance to pounce.

He soon found the opening. In the smallest pause between their sentences, he announced with a nineteen-year-old's infuriating certainty that they were wrong and then proceeded to explain that God did not work by fiat, incomprehensible to humans, but through intelligible laws, including the laws of evolution. One of the students gave McLuhan a pitying smile, took out his Bible, and counterattacked. Three hours later McLuhan was still talking. As soon as he made a point, one of his five or six antagonists would fire back an objection. McLuhan returned them one for one, yet he could not make his opponents understand that he was merely building on the basic tenets of their faith, not replacing them. One of them kept shaking his head sadly and saying, "You are lost, my boy." The group finally split up with smiles and handshakes all around, but McLuhan noticed that his stomach, always delicate, was upset from such strenuous argument.[48]

Before he left the University of Manitoba, McLuhan discovered a sparring partner, a fellow student named Tom Easterbrook, who could hold his own against him. Arguments between the two went on until the early hours of the morning. They would walk the streets of the city all night, fighting over the existence of God and other interesting questions. "We had an absolute agreement between ourselves to disagree on everything," McLuhan recalled.[49] Easterbrook took the view of an empirical scientist, always challenging McLuhan to produce the evidence for his assertions — a tack that never failed to irritate his opponent.

Reminiscing years afterward, McLuhan spoke of his good luck in attending the university during the Depression since, as he put it, "we weren't tempted to go job-hunting in the summer. We used our summers to study and chat, and we made huge progress intellectually."[50] The recollection may be slightly disingenuous; in reality, McLuhan, along with most other students, was desperate for a summer job and would have taken virtually anything, short of another stint with a survey crew.

During the summer of 1930, he worked on the mosquito campaign, pouring oil on insect breeding grounds for forty cents an hour. The job was physically exhausting and dirty in the extreme. Another summer he worked on a farm with Easterbrook, where he confused a farmer by talking about a "sine qua non." (He was still trying to learn three new words a day.) He even took temporary jobs during the school year for a few extra dollars, including selling coats at a one-day sale in a Winnipeg department store. The night after the sale he was tormented with dreams of rapacious customers demanding overcoats; the next morning he decided that a career as university professor was not such a bad idea.[51]

Summer jobs were either unavailable or unnecessary in 1932, however, and McLuhan and Easterbrook took off for England. The trip did not start well. They sailed as crewmen on a cattle boat, looking after cows from 4:30 in the morning until 2:00 in the afternoon. Their meals were monotonous affairs of meat and potatoes, and McLuhan soon found himself craving fruit and vegetables and drinkable water. When the weather turned foul, he was agonizingly seasick for three and a half days and later reckoned he vomited at least 150 times.[52] His bunk room became a little purgatory, and nausea rose in his gorge every time he retired there: the air was full of tobacco smoke and the smell of vomit. The nausea intensified when he heard the men whooping and shouting as they divvied up his share of the meat and potatoes, which he himself could not eat. Some kind soul gave him castor oil and, best of all, apples near the end of the three days, which helped his recovery. Before long he was well enough to argue (victoriously, of course) about politics with his working-class mates.

Once he set foot on the soil of England, McLuhan was in heaven. He had equipped himself with a bicycle and a copy of Palgrave's *Golden Treasury*, the poetry anthology that was a monument to Victorian taste in literature and the despair of poets like Pound and Eliot. Its verse added a mellow haze to every ruined abbey and meadow he encountered.

Easterbrook and McLuhan survived happily on a hundred dollars each for three months. McLuhan was in his element, visiting the homeland of all the great men he had read at university and finally hearing for himself the larks and nightingales of Shakespeare and Keats. The streets of London held a faint aura, still, of Johnson and Hazlitt and Dickens.

He was twenty-one years old, and he saw in poetry and literature a noble protest of the human soul against the mechanical and the commercial, against the vulgarities of modern life. Modern life wholly dominated Winnipeg, as it did all of North America; but in England one could still find echoes of something cultivated and literary that had once set the tone of civilization. "It was like going home," McLuhan later said. "The place I had grown up imagining was my headquarters."[53]

The other personally noteworthy event of 1932 also occurred in the company of Tom Easterbrook. The two had a habit of browsing in secondhand bookstores; one day McLuhan — always curious about the general outlines of disciplines quite removed from literature — picked up a book on economics that promised more enlightenment than his university course on the subject. Easterbrook bought a twenty-five-cent copy of *What's Wrong with the World* by G.K. Chesterton. The two quickly became bored with their purchases and swapped books. It was a transaction, McLuhan maintained, that changed his life.[54]

What's Wrong with the World seems an unlikely book to change anyone's life. The mannered prose repels as many readers as it attracts. T.S. Eliot could hardly read Chesterton without feeling intense exasperation, although he was in entire sympathy with Chesterton's social, political, and religious views. Toward the end of his life, McLuhan himself confessed that he was slightly "disillu-sioned" when he went back to Chesterton and found that he couldn't read him.[55] (Nevertheless, he always defended *What's Wrong with the World*, recommend-ing the book to a friend in 1974 as more relevant and instructive than ever.)[56]

In any case, unlike the two other enthusiasms McLuhan contracted while at the University of Manitoba, Thomas Macaulay and George Meredith, Chesterton did have a lasting impact on his thought — in part because Chesterton, amid his rich Edwardian sentences, frequently flashed out brilliant one-liners. Such one-liners became, in turn, McLuhan's own specialty. More important, Chesterton's thinking was, according to McLuhan, "analogical" rather than "dialectical" or "logical" — a distinction that later became of great importance to him.

But none of this struck McLuhan immediately on reading *What's Wrong with the World*. The book was a magnificent defense of certain notions that McLuhan was already predisposed to champion: personal liberty, the sanctities of the family, the traditions of Christian Europe that were opposed to both socialism and rampant capitalism. Chesterton did not build his case through any methodical presentation of ideas, but rather by means of what McLuhan later called "percepts," as opposed to concepts. In Chesterton's case, "percepts" meant simply the free play of an insightful mind over clichés and the supplying of missing and unexpected contexts to those clichés. Citing the cliché "You can't put the clock back," for example, Chesterton remarks,

> The simple and obvious answer is "You can." A clock, being a piece of
> human construction, can be restored by the human finger to any figure
> or hour. In the same way, society, being a piece of human construc-
> tion, can be reconstructed upon any plan that has ever existed.[57]

Statements like this may have prompted McLuhan, some sixteen years after reading the book, to describe Chesterton as a supremely reasonable man.[58]

Chesterton's shower of percepts prompted his opponents to accuse him of "a certain intellectual recklessness that made him indifferent to truth and real-ity";[59] this was, of course, exactly the charge that would be leveled against McLuhan himself in the years of his notoriety. Chesterton's method also involved a certain disdain for "experts," as reflected in his statement "What ruins mankind is the ignorance of the expert."[60] Few remarks would have had more wholehearted agreement from McLuhan. The ruinous authority of experts, which he compared to a flashlight aimed at one's face, was his lifelong theme. As far as McLuhan was concerned, what the expert did *not* know, the questions he had *not* asked, were the most important aspects of his expertise; once one knew those, it was possible to arrive at some real insights.

Chesterton also pointed McLuhan in the direction of Roman Catholicism. Chesterton's writings constantly reaffirmed the Catholic belief that the world, although very complex and not easily explained, was real and ultimately reasonable: it was not a tangle of deceptive appearances and it was not to be reduced to any intellectual or mathematical formulas. Also, the world was good, since it was created by God.

In its goodness can be seen all sorts of reflections of God. The medievals, for example, saw the faithfulness of the dog guarding the sheepfold as a reflection of the faithfulness of the priest guiding his flock, which in turn was a reflection of the faithfulness of Christ to his people. But the dog, claimed Chesterton, was not just a "symbol" of faithfulness. The dog *was* faithful — faithful according to its being, as the priest was faithful according to his being, that is, his human being; and both were related by analogy to the faithfulness of Christ, who was Being itself. The dog and the priest were real, and they were distinct; and yet they participated in the one Being of God.

Such was the analogical reasoning of St. Thomas Aquinas, which was replaced in later Western thought by the logic of Descartes and the dialectics of Hegel. There never was any doubt about which kind of reasoning Marshall McLuhan preferred. Analogical thinking shaped his view of life in the way that laws of probability color existence for mathematicians. It was a mode of thinking that became as ingrained and unremarked for McLuhan as his accent. Inevitably, however, this mode of thinking caused confusion for McLuhan's readers, who had no way of knowing that he was always speaking of analogies rather than equations. For example, McLuhan might say, "When man is 'on the phone' or 'on the air,' moving electrically at the speed of light, he has no physical body. He is translated into information, or an image" — a statement that makes sense only when it is understood as a series of analogies.

However bewildering or fanciful his analogies became, McLuhan never lost sight of what Chesterton taught him — that all things were real and lovable and ultimately coherent because God had created them. So profound was the impression made on him by this simple truth that McLuhan in his later years maintained that the real heresy of the twentieth century was not materialism but gnosticism — the denial of the reality, reasonableness, and goodness of God's creation in favor of some intellectual construct or of the One, an esoteric and hidden reality denied to human senses.

Chesterton's influence permeated McLuhan's experiences in his last year at the University of Manitoba, as did an event that he had been half longing for and half avoiding. That year, for the first time in his life, he fell in love. The young woman was a medical student at the University of Manitoba named Marjorie Norris.

At the beginning of 1933, finding himself helplessly obsessed with her, McLuhan poured out his feelings on paper in an attempt to give them some coherence. His account, written in the privacy of his diary, began with an assessment of his own character. He emphasized his aloofness from the society around him, his intense views and emotions, and his stringent morals, especially regarding relations between men and women, which were gradually souring his outlook.[61] He also had, he confessed, a very strong awareness of beautiful women; he had yearned for the gentle and wholesome touch of the female to balance and complete his own astringent and hyperintellectual nature. Marjorie was the first woman he'd met who possessed not only incomparable beauty but a stainless character sufficient to arouse all of his latent chivalry, all of his nobler aspirations.[62]

McLuhan had been powerfully attracted to Marjorie for some time but had steered clear because she had a boyfriend. By the spring of 1933, she was free of her suitor, although not before McLuhan had suffered agonies of jealousy. After a few dates he submitted himself to the hallowed rituals of a young man in love, writing poems about her and daydreaming endlessly. On a mild April night the two strolled along the banks of the Assiniboine River, sat on a tree stump, and gazed at the moon. McLuhan was possessed by the desire to kiss her. He had never kissed a girl before. He suggested that she look at the moon and close her eyes to see if there was an afterimage. As soon as her lids closed, he held his breath and suddenly moved his head toward her, but she opened her eyes at that moment and he drew back.

She laughed. It was not a laugh meant to make him feel silly; in fact, it was an encouraging laugh. Later in the evening, he finally did snatch a kiss. A little shaken after the attempt, he joked, "Fair Christian, turn the other cheek." But she did not seem in the least perturbed. As they walked home, he asked her whether, if he threw her a kiss from the sidewalk just ahead, she could catch it. She said no. He asked if she could catch one from where he was standing, a foot or so away from her. No. "Are you going to be a Christian, after all?" he asked. She said yes. McLuhan was terrified, but he took her in his arms and kissed her properly. Afterward he babbled like a happy idiot about streetcar routes or something of the sort, and when he finally left her that night he could hardly breathe from his fear, excitement, and pleasure.[63]

As the relationship deepened, McLuhan was tormented by the fact that

Marjorie was two years older than he. Since he would have to undertake at least four years of graduate work before he could think of getting a teaching position and marrying, she would, in the meantime, waste the best years of her life as a medical student. It was horrifying to him that the woman he loved would have to suffer years of such an indelicate occupation, and he raged against a world in which women were compelled, or felt compelled, to pursue a career. His feelings on this point did not bode well for their future, but they made a commitment, and thereafter McLuhan spent almost all of his waking hours either in Marjorie's company or in front of a book. As time went on, Marjorie Norris became not only a beloved object but something of a captive audience as well: Easterbrook recalled a date in which McLuhan simply read poetry to her.[64]

In 1933 McLuhan obtained his bachelor's degree from the university and won a University Gold Medal in Arts and Science. That fall he began work on a thesis for his master's degree entitled "George Meredith as a Poet and Dramatic Novelist." Like Macaulay, Meredith is a nineteenth-century literary figure whose reputation, immense at his death, has shrunk drastically. When he died in 1903 he was thought to combine the genius of a poet and novelist with the insights of a sage, expressed in a suitably "difficult" style.

For a student who had majored in both English and philosophy, Meredith seemed a marvelous choice for a thesis. McLuhan was convinced that he was a figure of great significance and that most of the existing critical material on him was worthless. He took up the challenge of literary rescue with enthusiasm. However, after finishing the thesis, McLuhan rarely referred to Meredith again.

The thesis itself is remarkably polished, especially in light of the fact that the author's later prose was almost universally deplored as unreadable. In fact, the writing is sometimes tinted with a fin de siècle mauve, as when McLuhan states that Meredith "has all the joyous freshness that belongs to the morning of the world, besides the inspired common sense that is sadly associated with that twilight in which the owl of Minerva commences its flight."[65] He observes that Meredith possessed a hearty, unaffected love of human greatness in both life and literature — a love so similar to the young McLuhan's.[66] He provides a lucid summary of the ideas behind the work of this very intellectual writer, including Meredith's Hegelian belief in "creative evolution," which McLuhan was beginning to reject even as he wrote about it.

In 1934, McLuhan wrote a series of articles on general topics for the student newspaper *The Manitoban*. His first published article had appeared in that publication in the fall of 1930, under the title "Macaulay — What a Man!" (Macaulay's genius, McLuhan urged, was sufficient to inspire "anyone that has a spark of verve in him.")[67] Since then, under the influence of Chesterton, McLuhan had traveled a far distance from Macaulay's assumptions that the political and economic adventures of the English bourgeoisie constituted a magnificent success story.

In one of his 1934 articles, entitled "Tomorrow and Tomorrow?" McLuhan characterized the modern world as hopelessly sunk in corruption and cited with approval Victorian thinkers such as Carlyle, Ruskin, and William Morris, all of whom derived inspiration in some measure from the Middle Ages. (He later told a colleague that as a youth he had rejected the twentieth century as "totally unfit for human habitation.")[68] His article also anticipated with uncanny precision the theme of his first book, *The Mechanical Bride* — not published until seventeen years later — when he complained of capitalist industrialism distorting human life and sexuality. He did not look to Marxism for solutions to the problems wrought by capitalism; then and later, he believed Marxism to be beneath contempt. The contemporary political movement he mentioned with some guarded approval was fascism; aware of their numerous errors, he nonetheless approved of the Fascists' diagnosis of the ills of the modern world. The Fascists, in urging a return to heroic enterprises, in rejecting the dull, "emasculating" utopias of socialism as well as the rapacious appetites of capitalism, seemed to him to be on the right track.[69]

This theme of the impasse reached by modern civilization was reinforced in other articles. In an imagined interview with Dr. Johnson, for example, McLuhan's Johnson ridicules notions of progress based on science and technology. He praises the Irish statesman Eamon De Valera for his policy of national self-sufficiency based on agriculture and small-scale industry. And in an article on Meredith, McLuhan criticizes his explicit feminism while approving of his creation of female characters who are "symbols of the sweetness and health and sanity of Earth."[70]

These articles — which borrow freely from the ideas and sometimes the very phrases of Chesterton — provide excellent ammunition for critics of McLuhan who maintain that his later work is a screen for basically conservative and ultra-Catholic tendencies. But to take a vigorous stand against the modern world and to resurrect visions of an older and a better time, when men and women lived more intensely and in deeper communion with each other and with the world, is not mere romantic nostalgia. Such visions — in the work of Yeats, Pound, Eliot, and Lawrence — animate some of the best writing of the twentieth century.

McLuhan, in the meantime, realized that if he was to develop further, he would have to study somewhere other than the University of Manitoba, which accepted his master's thesis in the spring of 1934. Elsie, still charmed by the thought of Boston, urged him toward Harvard. McLuhan, however, leaned toward Oxford and Cambridge. Two events decided McLuhan against Oxford. The first was his failure to win a Rhodes scholarship for studies at Oxford.

Evidently McLuhan ruined his chances when he so far forgot himself as to enter into a vigorous debate with the professors on the scholarship committee. The professors, who no doubt saw their role as judging the merits of candidates rather than defending their own opinions, were not pleased. The second event was a conversation with Professor Fieldhouse in which the Oxford alumnus convinced McLuhan of the superior merits of Cambridge by accurately describing the enfeebled state of English studies at his alma mater.

McLuhan's gold medal, along with recommendations from professors such as R.C. Lodge — who called McLuhan his "most outstanding" student — ensured that he would have no problem being accepted at Cambridge. Money was the only obstacle. Help came from an aunt, Ethel McLuhan, a chiropractor living in Richmond, Virginia, who was reputed to have extraordinary healing power in her hands and who practiced her art with great success. On the occasions when she visited her nephews in Winnipeg, she made a particular impression on McLuhan — he wished that she could have been his mother instead of Elsie.[71] Ethel must have been equally charmed by him, for she contributed to his Cambridge education with a series of loans.

McLuhan was also assisted by a scholarship worth $1,600 from the Imperial Order of Daughters of the Empire, although he almost ruined his chances for this one, as he had ruined his chances for a Rhodes. One of the IODE's adjudicating professors, A.J. Perry, head of the English department at the University of Manitoba, had suffered numerous humiliations at the hands of the intellectually more agile McLuhan. Fortunately, Herbert McLuhan saved the day by calling on Perry and telling some charming fibs about his son's immense regard for him.[72]

That scholarship secured, the only other complication McLuhan faced in going to Cambridge was his relationship with Marjorie Norris. He was unhappy to be away from her, of course, but the two promised to correspond faithfully. The promise was kept, and Norris sent McLuhan such tokens of her affection as a photograph and a sweater she had knit for him.[73]

Throughout his first year at Cambridge, McLuhan still insisted he would marry her.[74] But the effects of time and distance on the romance were evident by his second year there. At the end of that year, McLuhan, not sure why he was doing it, issued Norris an "ultimatum": either she visited him at Cambridge or the relationship was off. Norris dutifully took the next boat to Great Britain, and McLuhan realized he was not happy she was actually coming. The two had a pleasant time touring together, but McLuhan was relieved when she left for Winnipeg, and a few months later he wrote her terminating their "engagement."[75] Forty years later, when he heard of Norris's death, McLuhan noted the occurrence in his diary and called her a "perfect woman."[76]

3. Cambridge (1934–1936)

Tight-lipped Calvins of Art, teaching the young to love literature by first loathing
nine-tenths of it, and carrying their white and lofty foreheads with the self-important
anguish of waiters staggering under towers of exquisitely brittle crockery.
— F.L. Lucas, on his colleagues F.R. Leavis and Q.D. Leavis[1]

During the two years he spent at Trinity Hall at Cambridge University in the
thirties, McLuhan virtually unlearned everything he had absorbed about English
literature at the University of Manitoba. He became, literally, an undergraduate
again. (His exact status at Cambridge was that of an "affiliated student," that is,
a student who was allowed one year's credit toward a three-year Cambridge
B.A., in lieu of whatever degrees he had obtained elsewhere — in McLuhan's
case, his B.A. and M.A. from the University of Manitoba.) In 1962, recalling his
decision to go to Cambridge, McLuhan told a reporter, "One advantage we
Westerners have is that we're under no illusion we've had an education. That's
why I started at the bottom again."[2]

That statement may give the twenty-three-year-old McLuhan a touch more
credit for realism and humility than he possessed at the time. Certainly,
Cambridge University was a sobering experience for a young man who with
good reason thought of himself as the brightest person in any given classroom.
At Cambridge, he encountered a far-reaching intellectual authority, not only
from books but from teachers who didn't necessarily leap to the conclusion that
he was their best student.

It was also a new social universe. Years later McLuhan remembered the
occasion when the famous novelist and poet Charles Williams appeared before
some Cambridge students: Williams nearly fell at the feet of his audience in
shame over his lack of a public school accent. McLuhan was astounded that such
a man should be so embarrassed in front of students who were not fit to change
his typewriter ribbon.[3] His own Canadian accent was a lesser liability than, say,
that of a grocer's son from Leeds. Nevertheless, he felt soon enough that he was
a hick. Other undergraduates treated this outsider with gracious and affable

condescension, but the more McLuhan attempted to impress his opinions on them, the more their affability was tried. Nevertheless, they were always polite to him.

In short, McLuhan's fellow students were not altogether different in spirit from those University of Manitoba undergraduates who had classified him as a "moron." The difference was that the Cambridge students had a more serious attitude toward their subjects. They were not trying to "lay broad foundations" for learning; if they chose to study English literature, they did not waste time in a chemistry lab. They possessed a self-assurance and a coolly professional commitment to their area of study that very much impressed him.[4]

Not that this dedication necessarily translated into brilliance. Perhaps in self-defense, McLuhan soon satisfied himself that most of the students were basically mediocre; what he termed his "firm conviction of my superiority" remained unshaken.[5] That conviction did not entirely cancel out the unpleasant sensation of being considered an outsider, however. It was a sensation he always remembered. In later life, "yokel" was a term of abuse he employed with the relish of a man who knows how badly it can hurt.

McLuhan later maintained that life in the prairies, or in Canada for that matter, provided the advantage of a "counter-environment" for those inspecting life in the cultural centers of the world; such an advantage, however, was not always lightly or easily assumed. As his friend the English writer John Wain comments, "Marshall was a country boy from the prairies, and in a way his obsession at being up-to-date, his delight at being able to manipulate the modern world, always contained a bit of the rustic."[6]

The heart of McLuhan's Cambridge experience had nothing to do with any question of social inferiority, however. The real challenge to McLuhan was the teaching of English at Cambridge. By February 1935, he had heard enough from his lectures to realize that the weighty judgments he had formed about English literature in Manitoba were completely worthless.[7] Sobered by this realization, he applied himself to study in a manner more docile and less outspoken than at any other time in his life. Compelled to devalue drastically his heroes Macaulay and Meredith, McLuhan himself felt humbled.

His tutor, Lionel Elvin, comments, "When McLuhan came to me I wouldn't say he coruscated at all. It would not be quite fair to say he was plodding, but he was *earnest*."[8] Of course, he still had his smile and, according to Elvin, a "light in his eye"; he was, as ever, "full of his own ideas," but he was not trying to overpower anyone in debate, and he was not disposed to take over his conversations with Elvin the way he had taken over Professor Wheeler's class in Manitoba. In turn, Elvin regarded McLuhan as a willing and receptive but not necessarily brilliant student.

McLuhan had much to absorb. He had made the right decision in choosing Cambridge: no other institution in the world could have offered what
Cambridge did to the student of English literature at that time. Cambridge virtually pioneered modern literary criticism. At Oxford, for example, English
scholars spent more wrestling with Old Norse roots and dialect forms than with
actual poems, particularly if the poems were written after the lifetime of
Geoffrey Chaucer. Although there were brilliant Oxford dons, such as C.S.
Lewis, the experience of McLuhan's Canadian contemporary Northrop Frye,
who was studying at Oxford in the thirties, is typical. According to Frye, the
teaching of English at Oxford was "still under the spell of the old nineteenth-
century philology, which was really an imperialistic conception. The primary
thing was the language — the literature was derived from that."[9]

Cambridge, on the other hand, had established the King Edward VII
Professorship of English Literature in 1910, for the explicit purpose of subject-
ing English literature to critical analysis rather than philology. Following the
establishment of the chair, Cambridge attracted a host of remarkable teachers
and scholars. McLuhan's exposure to such men provided intellectual stimulation
beyond his wildest expectations.

The most notable of his mentors was I.A. Richards, the university's
uncrowned king of English studies. McLuhan attended his lectures on the philosophy of rhetoric during his first year at Cambridge. Many things about this
professor made an impression on McLuhan, even the way he read poetry aloud.
There was a great emphasis in the Cambridge English school on this activity,
but prior to hearing Richards McLuhan assured his mother that her way with
verse was still the best he'd had the pleasure of listening to.[10] He was not so sure
after hearing Richards, who attempted strenuously to recover the voice of the
poet rather than to leave listeners full of admiration for the beautiful tones and
expressive delivery of his reading — a rather novel elocutionary approach.

Not everything about Richards impressed McLuhan favorably, however.
He was disgusted by his atheism and his clinical psychologist mentality, which
tended to base all of human sensibility on such things as "stimuli" and
"impulses." He also had no use for Richards's attempt, in the Matthew Arnold
vein, to turn poetry into a sort of substitute religion. McLuhan eventually overlooked all of these faults, however, because of the boldness of Richards's
approach to criticism, as expressed chiefly in his books *Principles of Literary
Criticism* and *Practical Criticism*.

The latter, first published in 1929, relates the results of an experiment in
which Richards placed some extremely good poems and some extremely bad
poems before students and asked for comments. None of the poems were
signed. Even today the results make painful reading. Students heaped scorn on

poems by writers whose names they worshipped; they were captivated by some of the worst poems. The experiment was so revealing that Richards continued the practice for years after.

McLuhan himself participated in a similar exercise devoted to unsigned prose passages. Following the exercise, Richards, as was his wont, read a few of the students' comments in class. This could be a humiliating experience, even under conditions of anonymity — it was no fun hearing a class laugh uproariously at one's opinions. McLuhan was spared the humiliation, for when Richards read his comments in class, he did so with approval.[11]

To remedy the massive critical incompetence revealed by these classroom experiments, Richards proposed "exercise in analysis and cultivation of the habit of regarding poetry as capable of explanation."[12] To analyze a poem, Richards instructed readers, they must pay much closer attention to the actual *words* of the poem and how they worked; no more chasing after the phantoms of truth and beauty, no more adventures of the soul amidst masterpieces, no more looking at poetry as the expression of an age or of the personal life of the poet or as a statement of truth embellished by all sorts of poetic devices. In Richards's view, a poem is simply a supreme form of human communication. A reader analyzes a poem to see how it is able to achieve its effects — that is, to communicate an experience. English studies are themselves nothing but a study of the process of communication.

What particularly interested McLuhan was Richards's view of words. Richards attacked what he called the "proper meaning superstition" — the belief that "a word has a meaning of its own (ideally only one) independent of and controlling its use and the purpose for which it should be uttered."[13] Rather, Richards insisted, words have multiple meanings, dependent largely on their contexts; if the context is very complicated, the reader more or less subliminally juggles meanings in his head even as he reads the words. As McLuhan wrote in his notes from a Richards lecture in March 1935, "Words won't stay put."[14] Almost all verbal constructions, therefore, are highly ambiguous.

One of Richards's pupils at Cambridge, William Empson, pursued this line of reasoning in his 1930 work *Seven Types of Ambiguity*, which McLuhan read with great enthusiasm in his second year at Cambridge.[15] The book taught him, finally, how to read poetry — and incidentally showed him that his earlier anxieties about his ability to appreciate poetry were partly due to the Shelleyan notions of poetry as rhapsodic uplift and food for choice spirits.

There was very little rhapsodic uplift in Empson's book. Rather, it was a relentless analysis of poems in terms of the ambiguities swarming in even the simplest lines. "I am treating the act of communication as something very extraordinary, so that the next step would be to lose faith in it altogether," Empson declared.[16] McLuhan, who was later fond of observing that communication of any kind between human beings was almost a miracle and that "all people tend to misunderstand each other almost totally all the time,"[17] took that statement to heart.

Richards and Empson together were the godfathers of what would become known as the New Criticism; in his later years McLuhan marveled that he was the only student of their work to perceive its usefulness in understanding electronic media.[18] In retrospect, it is not difficult to see how McLuhan used the approaches of Richards and Empson as an entrée into the study of media, though it took many years of reflection and reading before he was able to carry over their approaches successfully. If words were ambiguous and best studied not in terms of their "content" (i.e., dictionary meaning) but in terms of their effects in a given context, and if those effects were often subliminal, the same might well be true of other human artifacts — the wheel, the printing press, and so on.

All these other artifacts were also ambiguous. A simple statement of what they did barely began to reveal the significance of their effects on human beings, particularly as some of the most important effects were also subliminal. Empson's comment that "the process of getting to understand a poet is precisely that of constructing his poems in one's own mind"[19] was also the germ of a later notion of McLuhan's that the "content" of any poem is the reader of that poem. McLuhan extended the insight to mean that the content of any medium or technology is its user. He would counter the demands of Canadian nationalists for more "Canadian content" on Canadian television, for example, by pointing out that a show such as "Bonanza" became Canadian content the moment it was viewed by Canadians.

The man who gave McLuhan the first hints that the New Criticism might be a fruitful approach to studies of the entire human environment was the other great figure in Cambridge English studies — and a former student of Richards — F.R. Leavis. Leavis at the time was nearly as influential in English studies as Richards; he exercised great sway in the university through his hold on undergraduates. One of the most loyal was McLuhan.

The two first met in May 1935 at a gathering hosted by Leavis and his wife, Queenie (Q.D. Leavis). Afterward McLuhan described Leavis as "an uncompromising idealist, tactless, impatient, vain and affected."[20] It was not a bad description. On this occasion, Leavis quarreled with the venerable scholar Arthur Quiller-Couch and walked out in a huff. At that time, Quiller-Couch was practically a national cultural monument, editor of the 1900 edition of *The Oxford Book of English Verse* and an accomplished essayist in the old belles lettres tradition. He stood for everything Leavis despised. Leavis, who pronounced the phrase "belles lettres" as if it referred to a sexually transmitted disease, wrote in a flat, almost banal, style that was a defiant rebuke to Quiller-Couchian purple prose. He also attacked, in his criticism, the lush Victorian and Edwardian verse Quiller-Couch doted on.

McLuhan soon joined the Leavis camp. For at least a decade after he received his Cambridge B.A. in 1936, McLuhan echoed and amplified the critical judgments of Leavis with the fervor of a true believer. Leavis's enemies were McLuhan's enemies — and Leavis, who had a touch of paranoia, did not lack for enemies at Cambridge. After leaving Cambridge, McLuhan corresponded with Leavis, who remembered McLuhan fondly as the student "from the wilds of Manitoba."[21] Mrs. Leavis was less indulgent. "McLuhan impressed me," she once commented, "as a rather loud, aggressive person, always running around arguing with everyone."[22]

F.R. Leavis's book *Culture and Environment*, written with Denys Thompson and published in 1933, showed Leavis's Canadian disciple how the analytic powers of the critic could be exercised not only on literature but on the social environment. Leavis adopted a tone of moral urgency as he lamented the passing of what he termed the "organic community," in which people were educated in folk traditions, crafts, and ways of life based on the soil and on cottage industries. This was entirely congenial to McLuhan's outlook, though in Leavis's case the attitude had been inspired not by Chesterton and his medievalism but by D.H. Lawrence.

McLuhan eventually shed both the moral tone and the explicit assumption that the old rural ways were superior. The more lasting effect of *Culture and Environment* on McLuhan was its suggestion that practical literary criticism could be associated with training in awareness of the environment: "Practical Criticism — the analysis of prose and verse — may be extended to the analysis of advertisements (the kind of appeal they make and their stylistic characteristics) followed up by comparison with representative passages of journalese and popular fiction."[23]

This suggestion bore stunning fruit in McLuhan's case. He had been intrigued by advertising as early as 1930, when he noted in his diary that current ads might ultimately prove more interesting to future readers than other forms of contemporary literature.[24] His first book, *The Mechanical Bride*, published in 1951, consisted almost entirely of an analysis of advertisements. (His last book, *City as Classroom*, published in 1977 and, like *Culture and Environment*, intended for the schoolteacher, also contained a section on advertising.) *Culture and Environment* helped nudge McLuhan away from being a purely literary critic in the manner of Richards and Empson to becoming a student of society and eventually of the media.

A book published by Mrs. Leavis in 1932 was also seminal in McLuhan's thinking. *Fiction and the Reading Public* took a novel approach to the examination of fiction: rather than viewing it as an entity generated from the brains of writers in a vacuum, Q.D. Leavis studied it as a kind of response to the various reading publics or audiences that demanded fiction. The cultivated and highly literate public of the eighteenth century, for example, called forth from its writers a very different fiction than did the readers of the cheap mass periodicals in which Dickens's novels were serialized. Here was a suggestion that McLuhan also developed in his later years when he began to insist on the audience as a *cause* of any work of art, a cause that should be studied almost as carefully as the work of art itself.

Of course, McLuhan was also impressed by what most struck other followers of F.R. Leavis — his overhauling of the canon of English literature. In his book *Revaluation* (1936) Leavis pushed forward the process, begun by poets like Pound and Eliot, of changing people's minds about which poets were worth studying and which were not. The latter were most of the poets idolized at places like the University of Manitoba English department — Milton, Shelley, Tennyson. (One professor at Manitoba had suggested to McLuhan that only the best scholars, after a lifetime of effort, could truly appreciate Milton, as if no finer purpose could be imagined for the intellect than the enjoyment of this poet.)[25] Leavis rudely informed such academics that they were mistaken; the finest values of English poetry were to be found in poets such as Donne and the other seventeenth-century Metaphysicals.

As for Pound and Eliot, Leavis wrote *New Bearings on English Poetry* (1932) to haul them into the twentieth-century canon. McLuhan quickly saw the magnitude of his delusion, fostered at Manitoba, that only inferior literature had been produced so far in the twentieth century. He was properly grateful to be set straight. McLuhan always insisted, echoing that supreme pedagogue Ezra Pound, that the job of a teacher was to save the time of his students by putting them in touch with what they needed to know. Leavis, by emphasizing the importance of poets like Hopkins, Eliot, and Pound, did this superlatively well.

By the end of his first year at Cambridge McLuhan had decided that Eliot far surpassed other modern poets.[26] It helped, of course, that Eliot was also a Christian and a Radical Tory, but McLuhan's lifelong love affair with Eliot and other Leavis favorites had more behind it than their religious and political sympathies. He credited these poets with being the real inspiration of his media studies. Certainly they reinforced the Richards view of literature as the supreme form of communication. Pound and Eliot, for example, were variously influenced by the French Symbolist poets of the nineteenth century, who in turn had been influenced by Edgar Allan Poe and his essay on poetry "The

Philosophy of Composition." In that essay, Poe described, not very convincingly, the way he set about composing his poem "The Raven." His account was never taken seriously in England and America, but the essay had a profound effect on the French Symbolists, who were intrigued with Poe's notion of the poet consciously striving for effects. This striving radically de-emphasized the content of poetry, to the extent that the content became, in fact, the act of writing poetry.

If poetry was the sum of its effects on the reader, would not the same be true, McLuhan later reasoned, of any human artifact? No longer did one have to examine a poem in terms of what it had to *say*, or to examine a machine in terms of what it *did*. Better to examine a poem, as Richards taught, as a far-reaching arrangement and clarification of impulses in the reader's head. Similarly, one could examine a machine as a far-reaching arrangement produced in the lives of its users. One did not understand a photocopier by grasping that it reproduced documents. One began to understand it when one grasped the sum of its effects, which included the destruction of government secrecy (by making it easy to leak documents) and the conversion of writers into publishers.

The old way of understanding proceeded by means of concepts, or static mental pictures divorced from a rich sensory life — what Eliot termed the "dissociation of sensibility." The new way proceeded by "percepts," or mind fed by perception, and was as agile as the human nervous system. Percepts did not dwell on the content of either poetry or technology. As Eliot remarked,

> The chief use of the "meaning" of a poem…may be…to satisfy one
> habit of the reader, to keep his mind diverted and quiet, while the
> poem does its work upon him; much as the imaginary burglar is
> always provided with a bit of nice meat for the house-dog.[27]

In *Understanding Media*, that remark became "the 'content' of a medium is like the juicy piece of meat carried by the burglar to distract the watchdog of the mind."[28]

These writers and teachers also gave McLuhan the first hints of what later became a key element of his ideas: the notion that human perception varied greatly according to which senses were predominant in the perceiver. The poets of interest to McLuhan, for example, emphasized the role of sound in their poetry, as exemplified by Yeats's declaration "I have spent my life in clearing out of poetry every phrase written for the eye" and Hopkins's admonition "Take breath and read it with the ears, as I always wish to be read, and my verse becomes all right." More significant was Eliot's well-known advocacy of the "auditory imagination," which he defined as

the feeling for syllable and rhythm, penetrating far below the conscious levels of thought and feeling, invigorating every word; sinking
to the most primitive and forgotten, returning to the origin and bringing something back, seeking the beginning and the end.[29]

All this suggested that there is a visual mode of perception entirely different in character from an auditory mode and that the auditory mode does not
accord with a strictly logical, linear, sequential way of thinking. McLuhan
always remembered the day when I.A. Richards described the structure of
Eliot's *The Waste Land* as a "music of ideas." Suddenly, McLuhan recalled, students realized that the poem simply dispensed with logical and narrative links.
Readers had to understand it in an entirely different way.[30]

The third Cambridge professor to exercise permanent influence on
McLuhan was the founder of the Cambridge English school, Mansfield Forbes.
McLuhan thought the first Forbes lecture he attended, in October 1934, was his
most exhilarating intellectual encounter yet. Forbes announced to the students
that he would not "cover" the course. "I shall cover no ground — I shall teach
you to dig in the most fertile parts." Thereupon, according to McLuhan, Forbes
"skipped over a 1001 things."[31] His characteristic style in so doing, according
to one account, was to offer "a sweep of paradoxical generalizations, cross-
fertilizing references from widely separated fields, tangential wit, and the explo-
sive compression of meaning in puns."[32] Such a style became McLuhan's own
signature, both in writing and in lecturing — to the delight, and sometimes the
annoyance, of his students.

At the same time he was absorbing the impact of these important teachers,
McLuhan continued to study at his own college at Cambridge, Trinity Hall, a
medium-sized college known more for legal studies than literary scholarship. At
the time McLuhan lived there, it was somewhat overshadowed by the larger
King's College and Trinity College next door. A decade earlier, Trinity Hall
(not to be confused with Trinity College) had been notorious as the home of
wealthy students with more interest in partying and pursuing the traditional
sports of the rich, rowing and hunting, than in studying. Since then the college
had undergone a reform that had improved its academic standing, although in
1934 there still was a great divide between rowing and nonrowing students.
Athletic prowess was greatly esteemed at Cambridge in the thirties, and Trinity
Hall in particular was proud of its rowing teams. Robin Maugham, nephew of
Somerset Maugham and one of McLuhan's Trinity Hall contemporaries, recalled
"a tough group of rowing hearties" at the college.[33]

The routine of the undergraduates consisted largely of academic work in
the mornings (attending university lectures or working on papers), some form

of outdoor exercise in the afternoons, and work or socializing with other students in the evenings. McLuhan's forms of exercise were rugby and rowing. In
both sports he performed adequately. A rugby teammate recalls him as not particularly vigorous unless goaded by physical abuse. "The theory was that if you
pulled the hair on his legs in the scrum, it would get him cross and then he'd
play well."[34]

He never made the fastest two or three boats in the college, but in the Lent
term of 1936 he rowed on the fifth boat, which did sufficiently well in the races
that each member was entitled to keep his oar, according to Cambridge tradition. The rowing was one way that McLuhan compensated for his feelings of
being an outsider. "We very much enjoyed his company," one of the boatmen
recalls. "Rowing every day, eight men in a cox, you get to know each other a bit,
and I have memories of a very comradely sort of person."[35]

As at the University of Manitoba, so at Cambridge, he led the straitlaced
life of his Baptist forebears. It was not until his second year at Cambridge, for
example, that he became, for the first time in his life, tipsy. (A few months
earlier he had recorded in his diary the experience of smoking his first ciga-
rette.) His chief recreational indulgence at Cambridge seems to have been the
local cinema.[36]

Most of the time, of course, he worked. The focus of an undergraduate's
work at Cambridge was the weekly session with his tutor, at which he read
and discussed a paper on a subject assigned by the tutor. McLuhan's tutor,
Lionel Elvin, was described by Maugham, who also studied under him, as "an
ardent humanist and Socialist. He treated us as if we were his equals in age; he
never condescended."[37] Elvin therefore was courteous but unimpressed when
McLuhan expressed sentiments such as "the Reformation was the greatest cultural disaster in the history of civilization" during their tutorials. This was
Chesterton talking. For McLuhan, Catholic culture had produced Cervantes
and Chaucer and peasants dancing around the Maypole. Protestant culture had
produced Milton, Tennyson, and Winnipeg, Manitoba.

Elvin remembers McLuhan as "very friendly, but a shade intense." Of his
work he remembers little, except for one bit of Forbes-inspired word play and
punning. McLuhan wrote about Ben Jonson and Jonson's former collaborator
and personal enemy Inigo Jones. In describing the process whereby Jones
ousted Jonson from the favor of King James I's court, he described the pair as
Inigo Jones and Outigo Jonson.[38]

McLuhan maintained his intellectual interests outside of English literature.
He actually met his hero Chesterton at a dinner in 1935. And he also began reading another Catholic writer, the Thomist philosopher Jacques Maritain. As far
as McLuhan was concerned, Maritain's best work was *Art and Scholasticism*,

which anticipated the assumption of the New Criticism and of poets such as Pound and Eliot that art and literature were not the products of some mystical inspiration or transport of emotion but rather of perception and intellect working at a high pitch of intensity, "a flashing of intelligence on a matter intelligibly arranged," in Maritain's words.[39] The book confirmed McLuhan's growing sense that art was valuable not for the concepts that could be milked from it (as one could be forever studying the concepts of a writer like Meredith) but for the world that art created, a world that could be contemplated and explored but not understood scientifically.

Maritain's chief significance for McLuhan, however, lay less in specific insights and more in his strengthening of McLuhan's interest in Catholicism. McLuhan was later fond of observing that at Cambridge a student was either a Roman Catholic or a Communist. This was a gross exaggeration, but he was keenly sensitive to the fact that Communist influence at Cambridge was at its height in his years there, even though only a handful of students actually joined the party.

According to McLuhan, the Cambridge of his day was also full of homosexuals.[40] This was another exaggeration, although Cambridge homosexuals, like the Communists (and the two were often the same), wielded an influence out of proportion to their numbers. Homosexuality at Cambridge tended to express itself in a sort of hyperaestheticism, which McLuhan found almost as distasteful as Marxism. If social life often boiled down to a war between the "hearties" and the aesthetes, McLuhan, with his rowing oar, ultimately preferred the hearties. (When he had tea with fellow Trinity Hall student Robin Maugham, for example, he was appalled at Maugham's arty friends.)[41] Certainly from Cambridge onward, he had an absolute disregard for aesthetes and removed himself from any temptation to play the role of lover of beauty. He was never interested, for example, in the "music of words" in poetry, and his tastes in painting and music were doubtful at best.

More than thirty years after McLuhan left Cambridge, he questioned whether students appreciated their days on campus. Almost certainly they would never again enjoy such freedom to explore new mental worlds, to play with ideas.[42] If McLuhan himself ever experienced any such freedom and playfulness, it was at Cambridge. That he did not experience the problems with digestion that had plagued him in Winnipeg is perhaps the most telling indication that he was happy there. At one point, he confessed in his diary that his love of life had never been greater.[43]

Moreover, his two years at Cambridge, although initially unsettling, left him with even more self-confidence than when he arrived. He had encountered the foremost scholars and some of the foremost artists of his time. As he later noted, such figures appeared intimidating from a distance. Seen close up, however, they rather made one feel it was not so difficult to be a Great Man after all.[44]

McLuhan was not slow to assert himself in the company of the great. He enjoyed the distinction of being threatened by Gertrude Stein when that formidable lady spoke at Cambridge on the topic "I am I because my little dog knows me." McLuhan, from the back of the room, interrupted her talk with a rude comment to the effect that her prose style was rather childlike, if not infantile. Stein, who realized this student had been reading her mortal enemy Wyndham Lewis, expressed her annoyance in a direct fashion. She grabbed her umbrella and made her way through the audience to McLuhan. Everyone stared at the robust Stein approaching the tall McLuhan as if two land animals from the Mesozoic era were about to engage in combat. Stein, however, did not use her umbrella. She merely asked McLuhan, "What are people like you doing here at Cambridge?"[45]

McLuhan's self-assurance received its most severe test when he took the "tripos," the final exam in English literature at the end of his two years as an undergraduate. He was determined to obtain a first on this final.[46] But, as most students do when faced with such a crucial exam, he began to feel that he was less and less prepared the closer it approached. In the last weeks before the exam, he felt uncharacteristically depressed and even blamed the lingering effects of the University of Manitoba for his lack of preparation.[47]

As it proved, McLuhan received an upper second — just about what his tutor had expected. "He would not have struck me as a first but as a good second," Elvin comments.

> I rather pictured him then as a respected professor of English — not at
> Harvard or some university of that prestige, and not as somebody on
> the level of, say, Northrop Frye, but quite respected and conscientious
> and producing two or three useful books. Someone second rate, or
> rather second order.[48]

McLuhan professed himself both relieved and dissatisfied at the results of the tripos.[49] An upper second was good, but it was most unhappily not a first. A first from Oxford or Cambridge is a credential that traditionally has meant a great deal in English life, a sort of promise that the holder will enjoy a great and glorious career. In that sense, of course, it is somewhat overrated; and McLuhan could console himself, among many similar instances, with the example of John Ruskin, who received a fourth at Oxford.

Whether or not he ended up in the top rank of students, McLuhan was influenced by Cambridge to such an extent that it is hard to imagine any of his classmates absorbing more thoroughly what was taught there. McLuhan's years at Cambridge permanently set the foundations for almost all of his subsequent intellectual work. For that, he remained grateful to the ancient university. When he left, he took his oar from Trinity Hall and kept it hanging in his office for the rest of his life. Morton Bloomfield, a professor of English at Harvard who met McLuhan when they were both studying in England, recalls, "McLuhan always spoke of his Cambridge years as the great years of his life. This was partly because he had been a boy from the boondocks and he had been overwhelmed by the elegance which he found there."[50]

4. Apprentice Professor (1936–1940)

In no Western society is the intellectual prestige of Catholicism lower than in the country where, in such respects as wealth, numbers and strength of organization, it is so powerful. — D.W. Brogan, quoted in *Time*, February 9, 1962

Where are the Catholic intellectuals? — *Time* cover, February 9, 1962

When McLuhan left Cambridge University in 1936, he was faced with the problem of finding a teaching job at a university in the depths of the Depression. He had realized for some time that the university was his natural home, and in December 1935 he had gone so far as to apply for a job at the University of Manitoba, in the English department he had once so heartily despised. There was a new head, Professor E.K. Brown, who McLuhan felt was a respectable figure. He told Brown that eight years of university, during which he had depended financially on his family, was quite enough — it was time for him to start teaching and earning money.[1]

McLuhan ended up not at his alma mater, however, but at the University of Wisconsin, as a teaching assistant in the English department. At that time, the University of Wisconsin was one of the better state universities in the United States, with an English department rivaled among state universities only by those of California and Michigan.

The student body at Wisconsin was animated by a tradition of social consciousness and progressive politics; among the graduate students there was even a nucleus of Communists, mostly composed of Jewish New York intellectuals who found Wisconsin one of the few universities welcoming them in the thirties. "There were no Republicans among the graduate students in the department, unless McLuhan was one," a fellow teaching assistant, Kenneth Cameron, recalls.[2] The atmosphere among both students and faculty was lively, leftish, and intellectual. "Most Canadians regard Americans as an underprivileged and even inferior group socially and politically," McLuhan once commented. "I grew up that way. When I arrived at ... Wisconsin I found that there was probably more culture in that town of Madison than in the whole of Canada. I had to jettison my views of the United States and do it in a hurry."[3]

As a graduate teaching assistant, McLuhan earned $895 a year. In return for this salary, each assistant was given a section of freshman English to oversee. That meant meeting with about twenty-five students three times a week, discussing and grading their assignments, which were partly exercises in appreciation of literature — mostly American — and partly exercises in composition. Since the course was compulsory, the level of work and enthusiasm was, to say the least, uneven. Reading some "impromptu themes" he had assigned to his classes on the first day, he noted in his diary, "It is heartbreaking to read some of them but the level of expression is not contemptible."[4] There was no mistaking that he was engaged in apprentice labor — the academic equivalent of carrying the master's toolbox.

McLuhan had a congenital unwillingness to carry anybody's toolbox. Instead of performing the usual chores expected of a freshman English teacher, he following the suggestions laid down in Leavis's *Culture and Environment* and transformed his section into a survey of contemporary culture based on advertisements, newspapers, popular fiction, and so on. In later years McLuhan regarded his teaching experiment at Wisconsin as an effort to reach his students, the first Americans he had ever confronted in a classroom, by way of their culture. "There was a language barrier," he once recalled. "Either the students had to learn mine or I had to learn theirs. I decided it would be better if I used their idiom — though not necessarily for their ends."[5] More likely, he undertook the experiment so he wouldn't be bored talking about symbolism in Hawthorne or wrangling with students over subordinate clauses.

The spirit of his survey of popular culture was, of course, in the spirit of Leavis — that is to say, censorious. "I find most pop culture monstrous and sickening," McLuhan declared in the sixties. "I study it for my own survival."[6] Except for the aggrieved tone, his sentiment would have been the same in 1936. From the experience of his original survey McLuhan probably came up with his simple and ingenious answer to the perpetual moaning of educators about the influence of pop culture on their students. If teachers really wanted to kill student interest in comic strips (or television or rock music), he suggested, all they had to do was put these things on the curriculum and test students on the content — in short, give popular culture the same treatment as Shakespeare or Dickens. The results, he felt, would be similar: lifelong alienation from the subject.[7]

McLuhan continued to delight in argument, and soon after his arrival on campus he set about initiating an "informal talk club" among his fellow teaching assistants — the first of such clubs McLuhan would form throughout his academic life.[8] In discussions with his colleagues, he frequently stirred up the

left-wingers among them by championing right-of-center positions. Fellow
teaching assistant Kenneth Cameron remembers, for example, that McLuhan
dismissed evolution by claiming that it was easier to believe in a bearded
Jehovah than in a homo sapiens who was the product of randomly muted
genes.[9] Another teaching assistant, Morton Bloomfield, recalls McLuhan promoting G.K. Chesterton's Distributist vision of a society composed of sturdy,
independent, and God-fearing peasants.[10]

What struck Cameron and others was not so much McLuhan's views as his
manner of delivering them, honed and perfected by the elocutionist's art. "If
you had tape recorded his remarks and transcribed them, you would virtually
have a perfect essay," Cameron recalls. "He was extraordinarily articulate."[11]
Until he was robbed of speech by a stroke near the end of his life, McLuhan
always retained his ability to speak extemporaneously in perfectly fluid, coherent, and grammatical sentences.

Another striking aspect of McLuhan's arguments was his use of Chester-
tonian paradoxes and pithy summaries of great intellectual questions. "He was
always interested in general theory," Cameron remarks.

> I don't think details interested him. He was interested in ideas, in toss-
> ing them around like a juggler tossing balls. You never knew whether
> he took them seriously or not. His attitude seemed to be that nobody
> knew anything anyway, so one might as well play with ideas. Toss
> them around. Maybe some of them were true, who knows?[12]

Sometimes his playing with ideas became frenetic. If Cardinal Newman
was right that "representations of any kind are in their own nature pleasurable,
whether they be true or not," McLuhan found such pleasure almost a compul-
sion. On October 17, he confessed in his diary that he was "overstimulated mentally," a condition he would experience frequently in the years to come. He
wondered, too, if there was not a bit of the dilettante in his rapid juggling of
ideas and subjects.[13]

At the same time, McLuhan was moving toward one kind of quiet certainty.
In 1936 he wrote an article on Chesterton for a quarterly published by Dalhousie
University in Nova Scotia. Father Gerald Phelan, the president of the Pontifical
Institute of Medieval Studies at St. Michael's College at the University of
Toronto and a friend, oddly enough, of Elsie McLuhan, admired the article and
wrote to its author. There followed a correspondence between Phelan and
McLuhan that finally nudged McLuhan into becoming a Catholic. On his
Christmas visit to his mother in Toronto in 1936, McLuhan met Phelan, and the
priest examined McLuhan about the state of his beliefs. It was a satisfactory

examination for both parties. On Holy Thursday, March 25, 1937, McLuhan was received into the Church. Thereafter he never failed to note the anniversary of this epochal event. For McLuhan, the life of a believer, fortified by the teachings of the Church, was the supreme human adventure. As he remarked to an interviewer in the 1970s, such belief altered existence, "making it mystical and converting a leaden uninspired human into something lyrically super-human."[14]

McLuhan informed his parents of his conversion in a long letter, carefully explaining that he had no intention of hurting them by his decision, which he had made only after two years of prayer. His father took the letter to a minister, who was impressed by McLuhan's concern for his parents' feelings and by the two years of prayer and advised the perplexed father to give this remarkable child his blessing. Herbert McLuhan thereupon felt much better.

Elsie, however, was inconsolable. She had long been concerned over her elder son's leaning toward Catholicism and bitterly attributed it to bad McLuhan genes — adducing as an example a brother of Herbert's who had become a Jehovah's Witness. Her feelings were based not so much on philosophical differences with Catholicism as on a supposition, not far-fetched at the time, that the religion was a serious social handicap in North America. "Mother was broken up," Maurice McLuhan recalls. "She felt Marshall, a bright, potentially outstanding scholar, was going to lose every opportunity he might otherwise have had in the academic world. She saw Harvard going down the drain."[15] McLuhan did not need his mother to tell him that being Catholic would never win him votes in academia. Throughout his career, he remained circumspect about his religion in public; in private, however, he was as Catholic as only a convert from Protestantism can be. He said the rosary, went to Mass almost every day, prayed to Saint Jude (patron saint of lost causes) in exceptionally trying circumstances, and was particularly devoted to Mary, the Mother of God.

Mary was, he sometimes claimed, an unfailing help in his work — not merely as some vague or indirect source of inspiration, but as an actual intellectual guide. "He had a direct connection with the Blessed Virgin Mary," an associate recalls.

> He alluded to it very briefly once, almost fearfully, in a please-don't-laugh-at-me tone. He didn't say, "I know this because the Blessed Virgin Mary told me," but it was clear from what he said that he was interrogating her about his ideas and that one of the reasons he was so sure about certain things was that the Virgin had certified his understanding of them.[16]

In the same way that McLuhan avoided conventional philosophical approaches to the study of earthly phenomena, he also avoided theology in his religion, at least after his decision to convert. Making the decision itself settled all theological questions for him; they no longer had to be reasoned out or defended in his mind. After his conversion, in fact, he seems to have adopted the time-honored Catholic habit of leaving theology to the professionals, as if investigation into matters of divinity was dangerous to the rank and file.

For McLuhan, prayer and the seven sacraments were all that a good Catholic need cling to — the supreme sacrament being the Eucharist, the Body of Christ consecrated in the Mass and partaken by the faithful. McLuhan maintained that daily reception of this sacrament was as necessary to him as his daily bread. It was the most efficacious channel of divine grace for a Catholic; and intellectually the sacrament was a keystone in McLuhan's thought. The fact that real bread and real wine were transformed, through the actions of the priest, into the real Body and the real Blood of Christ was the ultimate refutation of both materialism and gnosticism, the denial of the supernatural world and the denial of the natural world. It also meant that Christ blessed the very senses of the human body, giving humans an advantage even over the angels.

McLuhan described prayer as ideally consisting of "a constant, nonstop dialogue with the Creator." He was emphatic that life without this dialogue was unthinkable.[17] Part of the urgency of prayer for McLuhan was his recognition of the consequences of the loss of God's friendship. He believed that hell existed, and he was rather puzzled that the Church did not play this supernatural trump card, as it were, more often. He frequently noted that, just as one needed bad news ("real" news) to sell the good news (advertising) in newspapers and journals, so one needed bad news (hell) to sell the good news (the Gospel) in religion. It was the job of the Church, he said, "to shake up our present population. To do that you'd have to preach nothing but hellfire. In my lifetime, I have never heard one such sermon from a Catholic pulpit."[18]

Heaven and hell were as real to McLuhan as the planet earth, and he did not exempt himself from the possibility of eternal misery. In the 1970s he told a friend that he was worried about his own fate after death.[19] But if his religion increased anxiety about the hereafter, it decreased anxiety for the world. Catholics could, McLuhan believed, see the world more clearly than those who lived in and for the world — just as strangers can often see a place more clearly and visibly than those who have lived in it all their lives. One could simply enjoy and explore this created universe, confident of its ultimate coherence and intelligibility, unconcerned that it fell short of perfection. Whether most Catholics saw the world in this way was another question. Early in his life as a Catholic, McLuhan recognized that the Church in North America got along basically by trying to blend into the landscape as much as possible instead of being the radical alternative to North American society he thought it should be. In fact, sometimes he doubted whether any American Catholics really existed. Something in American life leached the vital substance of their religion.[20] Individual Catholics around him, he was always disappointed to note, were no different from other people. He complained that as far as his own work was concerned his fellow Catholics were, if anything, more hostile and uncomprehending than non-Catholics.[21]

Immediately after his reception into the Church, however, McLuhan was eager to join a Catholic environment in which he imagined he would find the truly congenial "circle of intellects" he had been seeking since his undergraduate days. He was tired of Wisconsin. The constant political wrangling with other graduate students was bothering him, despite his love of arguing. As well, even a good state university like Wisconsin seemed to him to be a mediocre establishment, where students submitted to impersonal instruction far removed from the tutorials of Cambridge. McLuhan felt he had somehow put a halt to his progress and mismanaged his life by finding himself there.[22]

The institution he turned to as a way out was St. Louis University, a Jesuit institution then reputed to be the finest Catholic university in America. Its English department was something of a Cambridge stronghold: its chairman, William McCabe, S.J., was a Cambridge Ph.D. and had a continuing interest in Cambridge graduates. McCabe hired McLuhan as an instructor in the English department in 1937, after McLuhan had inquired about positions there on the advice of Father Phelan.

McCabe was everything McLuhan could have hoped for in a department head — vigorous, bright, well informed, cheerful, open-minded, and loved by virtually everyone he met. He and McLuhan got on especially well. Unfortunately, like all Catholic universities in the United States at that time, the university itself was decidedly inferior to the better secular universities. Its Catholic status meant, for one thing, that it was starved for money. Salaries were minimal. McLuhan never earned more than $2,500 there, even after he obtained his Cambridge Ph.D. in 1943. The appalling condition of the library at the university shocked McLuhan, especially compared with the Cambridge library, which had struck him as a marvel, and he was forced to borrow books from nearby Washington University.

The Catholic affiliation of St. Louis University also meant that the university was ruled by the dead hand of clericalism. In 1937 Catholic universities had not completely escaped the idea that they were basically seminaries for the training of priests. It was some time after 1937, for example, before the Jesuits realized that a good layperson made a better department head than a mediocre Jesuit. There was a musty Jansenist air about American Catholic thinking as well. A prevailing attitude in the philosophy department at St. Louis University was that female students, those daughters of Eve, and the teaching of ethics could not and should not be mixed.[23]

Despite all this, St. Louis University had one strong point, that same philosophy department. (Half its students at the time McLuhan was there were philosophy majors.) The philosophy began and ended with St. Thomas Aquinas, but at least the scholars in the department were looking at the writings of the angelic doctor in a fresh light. One of those scholars, Bernard J. Muller-Thym, became a close friend of McLuhan's and an important influence on his thinking. Muller-Thym had received an M.A. in philosophy from St. Louis (where, just for fun, he had written his thesis in Latin) and had gone on to the Pontifical Institute of Medieval Studies in Toronto to complete a Ph.D. on the work of the medieval mystic Meister Eckhart. His teacher at the institute, the great French medieval scholar Étienne Gilson, spoke of Muller-Thym as the cleverest student he had ever had.[24]

In many ways, Muller-Thym acted the role of the eccentric genius. At spaghetti dinners in his apartment in St. Louis he handed out bath towels — sometimes still wet — instead of napkins. Legend has it that, in a fit of absent-mindedness, he walked out of his apartment one morning without his pants on. He could be stubborn and garrulous as well as generous and likable. McLuhan always remembered life with Muller-Thym and his wife, Mary, in St. Louis as his real university. At this university he was initiated into a rich world of metaphysics, fine cooking, music, and liturgical renewal.[25]

McLuhan and Muller-Thym spent hours walking together during the evenings, discussing the affairs of God and man. "No generalizations were less than cosmic," Muller-Thym later said of these discussions.[26] Muller-Thym gave McLuhan his first real introduction to the life and philosophy of the Middle Ages, his first clue that the period was far richer and more complicated than he might have suspected from reading Chesterton and the pre-Raphaelites. Muller-Thym also conveyed to McLuhan, from his studies in medieval thought, a fairly sophisticated theory of the human senses at work. He published an article in the university literary magazine, for example, arguing that the act of listening to music was the act of contemplating the movements in one's own senses, not the movement in the music itself. This act of contemplation he compared to a discourse. Such notions reinforced the hint McLuhan had already picked up from poets like Hopkins, Eliot, and Yeats that a person experienced the world according to the mix of senses he used to apprehend that world.

In general St. Louis University completed McLuhan's education in the history and development of Western civilization by immersing him in the thought of the Middle Ages and the Renaissance. The Jesuits might have been intellectually narrow in many respects, but one thing they did know was Latin

— and a vast number of classical, medieval, and Renaissance works written in that language. What McLuhan, in turn, brought St. Louis was a faint British accent and the Gospel according to Leavis.

In his first year he taught not only freshman English but also courses in Milton, Shakespeare, the English Renaissance, and something called "Reading and Discrimination." McLuhan introduced the poetry of Gerard Manley Hopkins — still little known but being strenuously promoted by Leavis — to his students. He also introduced them to the latest word in literary criticism as propounded in Leavis's magazine *Scrutiny*. "What he said came as a revelation to me, and it seemed as if it had been a revelation to him, as well," one former student, Maurice McNamee, recalls.

> He said it as if he had just thought it up yesterday, as if it had come
> out of the blue and he couldn't wait to communicate it. When we
> came across *Scrutiny*, however, we found that a lot of what he had said
> came word for word from that periodical.[27]

St. Louis University in the late thirties and forties was an interesting place for a Catholic intellectual. Comparatively recent figures on the American scene, Catholic intellectuals were still trying to fight the American stereotype of a Catholic as an immigrant with a scandalous number of children and the cultivation of a saloonkeeper. The scholars of St. Louis University were educated Catholics a generation or two removed from those immigrants, and they felt ready to make their mark on North American culture. Around them they saw a godless world devastated by war and economic depression. Surely, they reasoned, that world was ready once again for the truths of Aquinas and the other thinkers of the Middle Ages, revitalized by their own scholarship.

In the university literary magazine, *Fleur de Lis*, and the university philosophical review, *The Modern Schoolman*, they issued their challenge to modern society, offering the "perennial philosophy" of Aquinas "as a remedy for the ills of our war-torn world."[28] The editors of *Fleur de Lis* confidently predicted in their first issue, in 1937, that "with the decline of Protestantism and the rise of innumerable anti-christianisms, drowning western civilization will soon be forced to clutch at Catholicism."[29]

McLuhan joined in the challenge. In a 1938 *Fleur de Lis* article entitled "Peter or Peter Pan," he condemned advertising, industrialism, big business, Marxism — already familiar McLuhan targets — for their irreligion and their undermining of the economic and social basis of the family (a favorite theme of the Distributists). These forces would ultimately present Western man with a choice between Peter (the Catholic Church headed by the successor of

St. Peter) and Peter Pan (the condition of perpetual fantasy and emotional immaturity fostered by modern civilization). Already, McLuhan noted with some satisfaction, Peter Pan was taking his lumps in Franco's Spain and in Nazi Germany. (McLuhan's attitude toward the latter was ambivalent. He seems to have regarded Nazism in this period as, at worst, a characteristic bit of excess on the part of what he called "the sentimental, forest-minded Teuton.")[30] This basic line of thought, a kind of wedding of themes from Chesterton and Leavis, would characterize his social criticism for at least another decade.

Apocalyptic struggles between the Church and the modern world aside, McLuhan seems to have enjoyed himself during his first few years at St. Louis. A senior colleague at Wisconsin had described McLuhan as "a very likeable young man, friendly, warm-hearted, out-going, easy to work with."[31] He was the same at St. Louis and enjoyed a full social life. There, as at Wisconsin, he was usually seen at parties surrounded by a knot of people arguing with him. "He was the type of person who would take you off to the side and more or less start lecturing you," Vernon Bourke, a philosophy professor, recalls.[32]

Occasionally these intellectual amusements were varied by other innocent pastimes, such as charades or dramatic readings of Shakespeare, long a favorite party activity of McLuhan, son of Elsie. Karl and Addie Strohbach, a couple close to McLuhan in those years, recall one New Year's Eve when he arrived at their party bearing copies of Shakespeare for all the guests. "He arranged the parts," according to Karl Strohbach, whose memory of McLuhan as a "born antagonizer" is still tinged with a sort of good-humored outrage. "I had a part consisting of three words. He had the leading part, of course."[33]

McLuhan felt at home with the Strohbachs because Addie, in particular, enjoyed arguing with him. She never tried to prove how unsound his ideas were — McLuhan would have more than enough of that in his long career as an academic — but playfully twitted his earnest assertions. "If you had a sarcastic sense of humor you could twist something he said around, and he loved that," she recalls. "Then he would twist what *you* said around and show where *you* were wrong. He liked that kind of thing because it gave him a second platform for his argument."[34]

The summer after his first year at St. Louis, McLuhan went to the Huntington Library near Los Angeles to do research on Thomas Nashe, the Elizabethan writer he had decided to make the subject of his Ph.D. thesis. His mother was studying at the nearby Pasadena Playhouse and had made the acquaintance of another student across there, a young woman named Corinne Keller Lewis. Elsie McLuhan made up her mind that this was the woman her son would marry.

Corinne Lewis was born in Fort Worth, Texas, on April 11, 1912, and graduated from Texas Christian University. Since 1935 she had been working as a drama and speech teacher in the Fort Worth school system. Graceful, intelligent, full of fun, she was beautiful with a sort of stateliness that ages very well. The Lewises were proud of their heritage as an old and distinguished Southern family; by 1938, however, the family had seen better days (not surprising since the family business was based on the manufacture of carriages), and there was an air of faded gentility about them. Unlike McLuhan, Corinne Lewis always remained proud of her bloodlines and retained throughout her life both her Texas accent and a gracious manner redolent of the Old Plantation.

McLuhan was immensely impressed by her. The two enjoyed an idyllic trip to Catalina Island, watching youths dive for quarters as the boat approached the island and dancing in a ballroom overlooking the bay, with Mexican musicians and dancers. That evening they climbed a hillside and looked out over the waters of the Pacific under a full moon, until the dawn mists started to rise and a pale light appeared in the eastern sky.[35]

To Corinne, McLuhan was an exotic creature, undeniably dashing. (One of McLuhan's female students at Wisconsin had compared him to Fredric March.)[36] McLuhan was wretched when Corinne had to leave California, and when he returned to the university he began an ardent correspondence with her. Corinne possessed, in a way, all that McLuhan considered desirable in a woman — a fine intellect, beauty, a liveliness combined with an air of delicacy and reserve and high-mindedness. It was also clear that she would never compete with him in his chosen domain of ideas or try to manage his life the way Elsie had tried to run Herbert's.

For his part, McLuhan was not a man who was in love with being in love. He was not about to relive the adolescent rapture of his first kiss with Marjorie Norris. "It's a pity you didn't know me before I went to Cambridge, before I was 'spoiled' as you say," he wrote Corinne in one letter. "*Then* you would have had at *least* one poem a day, celebrating your perfections, real, dubious, and imaginary. You would have been showered with gorgeous epithets, and it would have served you right. Fatal fare, my fairly faerry fair!"[37] In September 1938 he assured himself, in his diary, that he was not yet in love with Corinne. The reassurance was in vain. He realized shortly afterward that he was obsessed with Corinne, and in early November he wrote to her, "practically proposing." Corinne wrote back practically accepting, and McLuhan was in heaven.[38]

The exchange of letters continued; a lovers' quarrel was quickly resolved and repented of, in writing. Corinne, perhaps sensing McLuhan's ambivalence, was annoyed that he was not more romantic. But McLuhan continued to press his proposals of marriage. At the Christmas break he visited her in Fort Worth, staying at a hotel, and made his proposal for the eighth time, according to his own estimate. Corinne, however, was now experiencing her own ambivalence.[39]

For one thing, her mother, whom McLuhan described as having "the will and torso of Napoleon," had never allowed Corinne to have men friends unless their families were personally known to her, and she was unrelenting in her hostility to McLuhan. At one point, she had demanded that her daughter show her these letters that kept coming from St. Louis. Fortunately, the combination of McLuhan's chicken-scratch penmanship and his tendancy to discourse at length on intellectual matters repelled her curiosity. Still, she was not happy. As far as she was concerned, there were three stikes against him: he was Catholic (they were Episcopalians); he was a college professor and therefore doomed to poverty; and he was Canadian, that is, a species of Northerner. She did consent to have a social gathering at her home for McLuhan, but she also invited several of Corinne's gentlemen friends, including one young man with serious hopes of marriage.

The ploy backfired. At a New Year's Eve ball at the Fort Worth Club, McLuhan showed the Lewis family and Corinne's gentlemen friends that this Northerner had more dash and style than they had bargained for. Dressed in tails, top hat and white gloves — "Fort Worth had never seen tails," Corinne McLuhan remarks — he was the center of attention on the ballroom floor. "We cut a swath, Marshall with his long legs, striding across the dance floor," Corinne McLuhan remembers. His rivals were reduced to gawking at the couple.[40] No wonder that McLuhan, on his return to St. Louis, shook his head, so to speak, over the feeble state of American manhood.

The romance by mail continued, as did Corinne's hesitations. She was enough her mother's daughter to worry about his comparative lack of worldliness: a university professor in that era, especially at a Catholic college, had opted virtually for life in the cloisters. Even McLuhan had severe — and, as it turned out, justified — doubts that he could adequately support her and a possible family.[41]

She was also sensible of her mother's objections to McLuhan, and she had no desire to leave Fort Worth. Neither was she a woman who would easily abandon her reservations under the spell of a man, no matter how prepossessing he might be. Despite her seemingly endless patience and deference to others, she was fully as stubborn as her mother and her friend Elsie McLuhan. Once her determination was fixed, nothing in the world could destroy it.

At first McLuhan was resolved not to return to Texas until Corinne had definitely agreed to marry him. But that resolution faltered as his longing for her continued. Corinne herself discouraged another visit; neither was eager for another social evening with Mrs. Lewis. Nonetheless, the two met again, in

Austin, where Corinne was working, in June 1939. McLuhan pressed the point, as he had done on previous occasions, that he was going to Cambridge in the fall to begin work on his doctorate. Unless Corinne married him before he left for England, he would end the relationship. He did not want to repeat the Marjorie Norris experience. On the last night of his visit, after a week of tension, indecision, and misery, she agreed to marry him before the end of August.

On July 17, Corinne arrived at the train station in St. Louis. A colleague who was present at their meeting was surprised at the coldness of the greeting between the two. Corinne, at least, was having second thoughts. She was fighting not only her family's disapproval but her whole genteel Southern Episcopalian upbringing. Nonetheless the couple spent three weeks in St. Louis attending the opera with the Muller-Thyms, dining with other colleagues, and agonizing over the wedding date. By the end of that stressful three weeks, Corinne noted that she had lost ten pounds. On August 3, Corinne, unable to handle her anxieties over the marriage, decided to go home. The thought of packing her bags, however, triggered a powerful reaction — sufficiently powerful for her finally to decide that she did want this man. The next day, after a frantic search for a wedding license, rings, and a passport for Corinne, the two were married in the Catholic cathedral. The Muller-Thyms and a cousin of Corinne's were the only witnesses.

After spending their wedding night at the New Hotel Jefferson in St. Louis, the couple parted for a few days, Corinne returning to Fort Worth to pack for their long stay abroad, Marshall staying in St. Louis to make final arrangements for their trip. When they rejoined in Fort Worth, the atmosphere was not festive.

Mrs. Lewis, who knew all the head buyers at the principal Fort Worth department stores, arranged for those stores to send their best clothes to the house, where Corinne could pick and choose for her trousseau. But she never reconciled herself to the marriage. In later years, she only visited the McLuhan household once, after Corinne told her that her husband was away and she needed help with the children. It was a fib — Marshall was present when Mrs. Lewis arrived. The visit proved amicable, but it was not repeated. Neither of Corinne's parents ever really understood their unusual son-in-law — in truth they made no great effort to understand him. He was, after all, a damn Yankee.

After their stay in Fort Worth, the couple returned to St. Louis and then made their way to New York, to board an ocean liner bound for Italy. So impressive were these newlyweds — Mrs. McLuhan arrayed in the best Fort Worth department stores had to offer — that the captain of the ship gallantly arranged for their stay in a first-class suite, though they had booked tourist class.

On the voyage, Mrs. McLuhan could relax for the first time in weeks, and enjoyed herself greatly in the ship's ballroom, artfully camouflaging her wedding ring with a chiffon dance cloth so that one handsome passenger in particular persisted in cutting in while Corinne and Marshall were on the floor. Marshall finally had to cut back in and inform the surprised gentleman that he wished to have an uninterrupted dance with his wife. It was an episode that the bride may have felt was salutary for a new husband who had so recently triumphed against long odds.[42]

In Rome the couple visited the graves of Keats and Shelley and saw the Colosseum by moonlight. In Venice they spent an evening drifting over the moonlit canals, while their gondolier sang operatic arias. Such experiences went a long way toward easing the tensions of the preceding months. In Paris they discovered that the Louvre was closed because of the threat of war, a threat McLuhan dismissed as if it were an exercise purposely mounted to deceive the public. "There will be no war of course," he wrote his mother.[43]

War, in fact, broke out just as the McLuhans settled in Cambridge in September 1939. They took lodgings near the university and settled down for a two-year stay, financed in part by a Canadian government bursary. McLuhan was now a research student at Cambridge, and the supervisor of his thesis on Thomas Nashe was a young Elizabethan scholar named Muriel Bradbrook, a very bright and agreeable junior member of the faculty who was responsible for guiding his academic work and reporting on its progress to the faculty board. She herself was working on a book on Shakespeare and was immersed in the world of the Elizabethans. Even at the height of his career as a media guru years later, McLuhan would frequently send Bradbrook urgent queries about arcane matters of Elizabethan literature, as if all the fuss about media was a distraction from the great work they had started together in 1939.

Bradbrook's role, however, did not entail any close collaboration with or supervision of her student. She more or less stepped aside and allowed him to do his research on his own. It was a wise move. McLuhan was past the point where he could usefully be supervised by anyone. In that regard he was fortunate to obtain the help and critical judgment of F.P. Wilson when Bradbrook left Cambridge for war service in 1941.

Wilson had been an Elizabethan scholar at the University of London when he was evacuated to Cambridge during the war. He was "the best scholar of his generation," Bradbrook recalls, the kind of professor, trained in the meticulous Oxford style of literary scholarship, who enjoyed arguing about the doubtful spelling of a word in one of Shakespeare's less important sonnets. McLuhan appreciated his good nature and his scholarly expertise.[44] While not precisely sympathetic with McLuhan's intellectual flights, Wilson was generous and open-minded and was able to properly value the originality of McLuhan's thesis, for which he served as an examiner.

McLuhan began work on his dissertation by spending long hours in the Rare Books Room of the Cambridge library, poring over the works of Nashe and his contemporaries. Corinne was usually by his side, copying out long quotations. Despite the inconveniences of wartime — and McLuhan's initial dismay, standard for those beginning work on their doctoral theses, at the amount of work he saw ahead of him — it was a pleasant year for McLuhan. "We lived pretty near the bone," Bradbrook recalls. "But the sense of a really vivid, magnetizing intellectual life was enough to compensate for this. We were very, very unworldly — but also very sure of ourselves. My impression about McLuhan, in particular, was that he was very happy here and very much at home here."[45]

It is possible, in fact, that McLuhan was never so happy again in his life. He did not have to deal with faculty animosities, dull students, or any of the usual academic routines. He could renew acquaintance with his old mentor Leavis (one of the first things he did, in fact, on arrival at Cambridge was to drop in on the Leavises for tea, with his bride). And most important, everything he studied was fascinating to him and could be absorbed in an unhurried and unpressured manner. In the evenings, he and Corinne relaxed by reading aloud to each other, Corinne knitting and McLuhan smoking his pipe while the other read. In the daytime, McLuhan frequently rode on his bicycle through the southern counties of England, memorizing important material for his thesis.

McLuhan's initial idea was to write a thesis entitled "The Arrest of Tudor Prose." Its premise — first articulated by R.W. Chambers, whose lectures at Cambridge McLuhan had greatly enjoyed as an undergraduate — was that English prose had suffered a severe setback with the execution of the great English humanist and Catholic martyr Thomas More. It was an idea consistent with the notion McLuhan had presented to his first tutor at Cambridge, Lionel Elvin, that the Reformation was the greatest cultural disaster in the history of civilization.

By the time he actually arrived at Cambridge, however, McLuhan had read enough Tudor prose to realize that it had continued to flourish despite Henry VIII's headsman. In fact, there was an incredibly rich prose literature throughout the entire period. One of the most interesting of the writers McLuhan found was Nashe, a satirist, journalist, and polemicist.

Nashe had enjoyed a mild vogue at Cambridge when McLuhan was an undergraduate. His high-spirited, colloquial style, his indulgence in seemingly endless and pointless literary horseplay, and his appetite for words and puns made him very attractive to people like Empson and other New Critics who were more interested in the way writers handled language than in their "ideas."

No less a writer than James Joyce took Nashe as a model. And Nashe was con-genial to McLuhan. As Bradbrook recalls, "McLuhan had an instinctive sympa-thy with the vivid, inconsequent, but very powerful style of Nashe — and with Nashe's impulse to be new, to be rhetorically arresting, to be, in fact, a sort of sociological journalist, or journalistic sociologist. I couldn't help feeling that there was a Thomas Nashe inside McLuhan, dying to get out."[46]

McLuhan decided, then, to write his thesis on Nashe. The more he read Nashe, however, the more he kept running into the profound influence of the ancient theory and practice of rhetoric on Nashe's supposedly off-the-cuff, purely journalistic and satirical writing. McLuhan started to read up on this theory and practice. As he did, an astonishing world opened up for him. With mounting excitement he began to realize that nearly the entire cultural history of the West could be read in terms of an ongoing and often hidden war between the proponents of each of the three branches of the trivium — grammar, logic, and rhetoric — the curriculum (together with the sciences of the quadrivium) of ancient Greece and Rome and the Middle Ages. Nashe remained his ostens-ible subject, but he was entirely overshadowed in the thesis by McLuhan's dis-covery of the significance of the trivium.

McLuhan found, for example, why people like Nashe hated the philoso-phers of the Middle Ages. Nashe was simply involved in a quarrel between the Renaissance followers of a tradition of rhetoric going back to Cicero — the tra-dition of wisdom as eloquent speaking — and the "schoolmen" of the Middle Ages, who were followers of dialectics — wisdom as pure logic. The quarrel, in fact, had originated in the scorn of Socrates (the dialectician) for the Sophists (the rhetoricians). Occasionally joining the rhetoricians in the war against the dialecticians were the grammarians, the students of literature and of the science of words.

Each of the three branches represented a different way of looking at the world. Dialectics, or logic, was the employment of critical reason, a highly abstract form of argument. Rhetoric was the fivefold art of persuasion, of mov-ing the human imagination to accept truth. Like the Sophists, the rhetorician had his certainties, his truths, already in hand before attempting to convey them to others. McLuhan later termed this attempt "putting on the audience" and con-sidered himself an expert at it. In December 1944, for example, he was thinking out complicated strategies for reading two papers he was scheduled to give at the annual meeting of the Modern Language Association, which he was attend-ing as a job hunter. He told a friend that he had his mask prepared: the mask of a humble and guileless junior academic, unconsciously crying out for a mentor to guide him. Technically, he was using the fifth part of rhetoric, pronuntiatio, or delivery.[47]

It was the third part of the trivium, however, that interested McLuhan most, both when he wrote his thesis and for the rest of his life. Grammatica, or the study of words, was based on the belief that all human knowledge inhered in language. To the ancient Stoics, who developed this science, the universe itself was the Logos, or divine word; the order of human language and the order of reality were closely related. In biblical terms, Adam, who named the creatures of the earth before his expulsion from Paradise, used names that captured the creatures' very essence. To crack the secrets of language would be to penetrate deep into the heart of the universe.

The ancient grammarians studied poets such as Homer and Virgil with a technical passion for categorizing figures of speech, for analyzing prosody, and for explaining the derivations of the meanings of words. McLuhan displayed the same passion, at least for etymology. In his later years a day hardly passed in which this scholar, who had tried to learn three new words a day as an adolescent, did not consult the *Oxford English Dictionary*. So great was his passion for the roots of words that he often irritated his colleagues by presenting them with completely bogus etymologies that he was utterly convinced were true and that he used to bolster whatever point he was making. He would discuss the nature of violence, for example, by claiming that the word *violence* was derived from the Latin word for crossroads and hence had to do with people getting in each other's way — a comment that had more to do with McLuhan's own hatred of being jostled by crowds than with any known derivation of the word.

McLuhan took this process to absurd lengths by applying it to people's names. It was as if he shared the primitive belief (and the belief of the ancient grammarians) in the potency of names: if you knew the name of a thing or a person, you had direct power over that thing or person. Such a belief has a distinguished pedigree in Western literature — it is discussed seriously in Plato's *Cratylus* and comically in Sterne's *Tristram Shandy*, for example. McLuhan's own restatement of the thesis occurs in *Understanding Media*, where he notes, "the name of a man is a numbing blow from which he never recovers."[48] (It was no accident that McLuhan dropped his own first name, which was, of course, the same as his father's.)

At the very least, a person's name gave clues to his real identity. On one occasion McLuhan was discussing the question of the honesty of Albert Speer, the convicted Nazi war criminal. He maintained that Speer was not honest about his knowledge of the Holocaust. To prove his point he cited a book about Nazi involvement with the occult, which had nothing to do with Speer, called *The Spear of Destiny*. That the name Speer, in English, was a homonym of *Spear* was no coincidence in McLuhan's view.[49]

Such eccentricities apart, McLuhan saw that it was impossible to understand

Western culture without understanding the shifting relationships between grammar, rhetoric, and dialectics. In the High Middle Ages, for example, the schoolmen had emphasized dialectics by developing a logic so strong that it threatened to overwhelm every other intellectual discipline. With the Renaissance, however, rhetoric and grammar made a comeback. Then, in the seventeenth century, Descartes once more upset the balance between the disciplines by emphasizing logic — an emphasis so pronounced that the world, in his philosophy, was turned virtually into a set of mathematical equations.

In the twentieth century, grammatica, though still under attack from mathematicians and logicians, had returned to favor, McLuhan believed, with the advent of the New Criticism and its emphasis on verbal analysis of literary texts. The age's best writer, James Joyce, studied Skeat's *Etymological Dictionary* and learned that almost every word he put down on paper had layers of meaning linking the present to unimaginably distant eras of human history. In later years, McLuhan associated grammatica with the acoustic world — the world of Eliot's "auditory imagination" — fostered by the electronic media, while dialectica belonged to the visual world fostered by the phonetic alphabet and the invention of printing.

Once again, McLuhan's work at Cambridge permanently and profoundly altered his mental world.

5. In Search of a Home (1940-1946)

I know not whether...the happiness of a candidate for literary fame be not subject
to the same uncertainty with that of him who governs provinces, commands armies,
presides in the senate, or dictates in the cabinet. — Samuel Johnson[1]

The return of Marshall and Corinne to St. Louis University in 1940 marked the
beginning of the dreariest period, in some respects, of McLuhan's academic life.
The problems he would face in the coming years were not apparent at first,
however. He carried on with the enthusiasm of his work at Cambridge, immersing himself in the world of ancient rhetoric and grammar, the world of Cicero,
Quintilian, Seneca, and the Greek and Roman Stoics. The more he worked the
broader his paper seemed to become and the more all-embracing its implications. Like many a Ph.D. student before him, he seemed to sink into a bottomless pool of scholarship. Even then McLuhan had no instincts for specialization
to help limit his descent. He was incapable of doing what most academics
do: staking out a manageable section of intellectual turf and then working it
to death.

Instead, he initiated a practice, which he would continue throughout his
life, of enlisting his students as collaborators. One student was a Jesuit named
Maurice McNamee, then working on a Ph.D. thesis on Francis Bacon. Bacon
interested McLuhan as one of the giants of the Renaissance who was steeped in
the art of rhetoric, in particular the art of the aphorism. Bacon maintained that
the aphorism — the pithy, arresting statement — was useful precisely because
it did not explain itself. In its incompleteness and suggestiveness, it invited "men
to enquire further" into a subject. The McLuhan who later became famous for
his aphorisms — notably "The medium is the message" — was intrigued by this
use of language.

McLuhan directed McNamee's thesis, though the direction was minimal.
He cared very much about the fate of those students whose theses he supervised, but students who wanted a great deal of explicit guidance never sought
him out. To be sure, no one was better at pointing students toward important
books to read, and nobody tossed out more valuable ideas and suggestions —

it was, in fact, almost impossible for *any* colleague or student to have a casual conversation with McLuhan and not come away with several ideas for books, theses, or articles.[2]

But McLuhan would not closely supervise students. In a curious way, he treated students as his equals, which is to say as potential collaborators. As McNamee recalls, McLuhan "directed" his thesis by "coming into my room and throwing himself on my bed and talking about his own dissertation."[3]

As far as McLuhan was concerned, the best way to explore any subject was to talk. "I have to engage in endless dialogue before I write," he once told a reporter. "I want to *talk* a subject over and over."[4] He was never so happy as when he was talking. For McLuhan, conversation had "more vitality, more fun, more drama" than writing and was the chief, almost the only, way he had of arriving at insights and conclusions. "I do a lot of my serious work while I'm talking out loud to people," he proclaimed. "I'm feeling around, not making pronouncements. Most people use speech as a result of thought, but I use it as the process."[5]

The one-sidedness of these verbal exhibitions bothered many of his friends and colleagues. McLuhan's ability to talk was well developed; the same was not true of his powers of listening, a fact of life that would drive away some of his collaborators. Hugh Kenner, who was close to McLuhan in the late forties, commented, "What Marshall always really needed was a stooge.... I think he liked to have someone else in the room while he thought aloud."[6]

This is a harsh assessment; in fact, if someone pricked his ears with an interesting fact or anecdote, he was immediately and completely attentive. One had a better chance of accomplishing this feat by citing some new discovery, say, about the technique of used car salesmen, than by trying to make some important intellectual point, especially if it bore a suspicious resemblance to McLuhan's own ideas. Nor was he very tolerant if there was a hint that the speaker was about to bring up some intricate personal problem. Usually, in fact, a speaker had about thirty seconds before McLuhan either cut him off or heard him out, depending largely on the speaker's level of energy.

If McLuhan's talks with McNamee were more monologues than conversations, McNamee did not find them any less valuable for that. "I got a great many insights from him," McNamee recalls. "Sometimes he would come in with some of his grand generalizations which he had gotten from secondary sources about certain authors or texts, because he couldn't read the original sources in Latin or Greek. Afterwards you would go to those original sources and find that he was absolutely right."[7] (McLuhan did eventually become reasonably proficient in Latin and Greek through his studies of these languages at St. Louis University and later at the University of Toronto.)

He was also helpful to a second outstanding Jesuit student, Walter Ong. McLuhan directed Ong's M.A. thesis on Gerard Manley Hopkins in the same manner he had directed McNamee's, by talking to Ong about what was on his mind. As McLuhan had no interest in the matter of Ong's thesis, which was a technical study of Hopkins's prosody, he did virtually nothing more than sign the papers approving it. McLuhan did, however, point out to Ong the historical figure who would provide the material for the start of Ong's long and distinguished scholarly career.

That figure was an obscure Renaissance theologian named Peter Ramus. Ramus was a key proponent of dialectics, of utilitarian logic, as opposed to the rhetorical tradition of Nashe, Bacon, and Cicero. Tipped off by McLuhan, Ong later devoted years of research to Ramus and eventually produced his classic study, much quoted in McLuhan's *The Gutenberg Galaxy*, entitled *Ramus, Method, and the Decay of Dialogue*. It was while working on this book that Ong hit on the basic notion underlying *The Gutenberg Galaxy*, namely that Western culture in the Renaissance had shifted from a primarily auditory mode of apprehending reality to a primarily visual mode and that the vehicle for this shift was the invention of printing.

However profound and lasting McLuhan's influence on graduate students such as Ong and McNamee, he did not always inspire undergraduates at St. Louis. "He was especially good," McNamee recalls, "with the better students who could really appreciate him." He had avid followers among undergraduates, graduate students, and faculty, but there were always some who were totally bewildered by him and therefore at least mildly hostile. Of such non-admirers, he would have plentiful experience throughout his career.

In addition to courses in the Renaissance, he taught Leavis-style Culture and Environment studies and Richards-style Practical Criticism. In his Culture and Environment course McLuhan contrasted contemporary civilization and its advertising, newspapers, and so on with what he considered to be the more rationally ordered sixteenth-century society. He also passed on to his students his discovery that some of the techniques of the Renaissance rhetoricians had been appropriated by twentieth-century advertisers.

The Practical Criticism course was an attempt to get the student of literature actually to look at a given text and see how the writer achieved his effects, rather than boning up on the author's biography or figuring out what he was trying to "say." Like Richards, McLuhan gave his students unsigned poems and asked them to evaluate them. McLuhan believed that he, virtually single-handedly, was pioneering the techniques of Richards, Empson, and Leavis in the United States. His classroom analysis of texts made the Great Books program of Mortimer Adler and Robert Maynard Hutchins seem puerile by comparison.[8]

Adler and Hutchins's Great Books program at the University of Chicago had long been a bone sticking in McLuhan's throat. McLuhan despised the program not because he didn't think students should be reading Aristotle or Spinoza — quite the contrary — but because of the mechanical way in which the program ground up and processed these venerable authors. As well, he felt the Great Books proponents were turning their backs on any attempt to study and analyze the contemporary world.

McLuhan could not, however, fault Adler and Hutchins for one thing: they attracted publicity, always a feat that commanded his respect. In so doing they focused some much-needed public attention on the academic world. After the novelty of his first few years of teaching had worn off, McLuhan had begun to resent the comparatively low status of his occupation. It was, after all, an occupation he was now fully committed to. Despite his early misgivings as an undergraduate about the professoriate, the mature McLuhan never considered anything but teaching as a vocation. If ever there was anyone born for the academy, it was McLuhan.

He was naturally unhappy, however, about its low pay and lack of prestige. In 1944 he wrote a brief statement urging the recruitment of more professors into the College Teachers Association (CTA), a professional organization that he hoped would raise the prestige of teachers in the same way the American Medical Association had raised the prestige of doctors. He recognized that Americans traditionally had a mild contempt for the figure of the college professor, and he argued that this attitude lowered the quality of teaching. In reality, professors were men of altruism, lifelong dedication, and intellectual attainments equal to or surpassing those of any other profession. He proposed that each member of the CTA should enjoy a minimum salary of $5,000, which would at least certify the professor as a valued member of society.[9] That figure was twice his salary as an instructor at St. Louis.

Although underpaid, McLuhan was fortunate to obtain a comfortable, spacious three-bedroom apartment on Maryland Street, in a pleasant neighborhood not far from the university and downtown St. Louis. He and Corinne even employed a maid once a week. McLuhan, who had no taste for luxuries except for a choice brand of tobacco for his pipe, did not otherwise chafe at his material circumstances. Their accommodations remained comfortable even after the birth of their first son, Thomas Eric McLuhan, in January 1942.

McLuhan always enjoyed the idea of being a father and a family man, but his wife was more realistic when she told friends he should have remained a celibate. (McLuhan himself, on the occasions of his entrance into the Church, had maintained that if the entrance had occurred five years earlier he would have become a priest. To Corinne, however, he insisted that he never had felt the claims of any such spiritual vocation.)[10] In any case, McLuhan simply did not possess what is perhaps the prime virtue of a father — patience. He did have affection, but that alone was not sufficient; and he regarded Eric, for much of the boy's growing up, as a puzzle and a source of vexation. Moreover, McLuhan — who was a firm believer in corporal punishment — was ill-served by the puritanical philosophies of child raising that he had inherited, with their emphasis on opposing the "obstinacy" and "willfulness" of the young. That McLuhan himself, one of the most willful of men — indeed, a man who had needed every ounce of his native willfulness in order to get where he was — should adopt this approach is a familiar irony of parenthood.

Especially with his first child, then, McLuhan adopted a stern approach. Some of his expediences for dealing with the child were startling. One time he took a screen off a window and screwed it on top of the crib so Eric couldn't climb out. When Eric was four years old McLuhan noted, with some relief, that the boy had finally become less demanding. The perpetually crying, squirming irritant to his father had evidently settled down a little.[11]

McLuhan's friend John Wain accurately observed in a memoir of McLuhan, "Marshall was not very much attuned to children. He was a kindly man, but his thoughts and perceptions were elsewhere. Like all people of his kind, he tended, in the presence of children, either to ignore them altogether or to give them fitful bursts of attention that were slightly overdone."[12] He certainly did not take after his own father, who loved nothing better than to take walks with children and tell them stories.

McLuhan's concern for children showed itself chiefly in his attempt to shelter them from the influence of media. When he discovered that the Muller-Thym children were staying up on Friday nights to listen to "The Inner Sanctum," a radio show featuring eerie dramas, he solemnly assured them that they were ruining their minds and offered them a nickel a week for every week they didn't listen to the program. McLuhan maintained this concern throughout his life. As a grandparent, he advised his son Eric to limit the time his young daughter spent watching TV: television, he wrote Eric, in language he permitted himself only in private, was a "vile drug which permeates the nervous system, especially in the young."[13] Ideally, he felt, the young should be limited to one hour a week viewing time; when his own children were young they were not allowed much more than this.[14]

To counteract media influence as his family expanded and his children grew older, McLuhan tried to stimulate discussions of a more or less edifying nature at the dinner table, asking his children to select a quotation from *Bartlett's* as a topic for conversation, for example. Or he would try to involve them in his work through the curious method of waking them up at three or four in the morning to write down a "breakthrough" he had just experienced in his thinking. This was also the time of night that McLuhan tended to read his Bible, and he shared that too with his children. "He used to get us up at 4 o'clock in the morning and read us the Bible in four languages," his daughter Teri once told a reporter. "That was our only emotional contact with him. He wouldn't bother with us again for the rest of the day."[15]

It is almost certain that if McLuhan had not been Catholic and if the pill had existed in 1942, Eric would have been an only child. McLuhan was obsessed with his work; only his relationship with God took precedence over that obsession. His family had to be content with third place in his priorities — a state of affairs McLuhan never tried to deny or conceal.

If his children suffered, his wife also found life with McLuhan hard to take, especially in the early years of their marriage. Corinne had had more than her share of beaus in Fort Worth and had become accustomed to a certain amount of gallantry and consideration from men. But McLuhan, whose unromantic approach had bothered Corinne before their marriage, was hardly about to live up to this Fort Worth standard after the marriage. He wanted Corinne to be happy, of course, but seemed unable to comprehend that this might involve paying serious attention to her, showing affection in small but graceful ways, and in general observing the courtly standards of behavior expected of fond husbands. Instead, he displayed a lamentable tendency to take her for granted. Corinne could not refrain from expressing a sense of disappointment to her friends in St. Louis.

It did not help that McLuhan, around this time, had embraced another of those intellectual themes that tended to stay with him for months and years, becoming a sort of hobby horse he rode to exhaustion. This particular theme was based on the cartoon character Dagwood Bumstead. In McLuhan's view, Dagwood was the model of a whole generation of American males who were helpless sad sacks dominated by their more resourceful and tough-minded wives. McLuhan had perused the comic strip as part of his culture and environment studies, guided by his central conviction that modern commercial civilization had destroyed traditional family life and traditional masculine pride in work. McLuhan's analysis of Dagwood, given in full in *The Mechanical Bride*, shows a great deal of brilliance. It also owes something to McLuhan's experience of watching his mother cut his father down to size. In any case, having Dagwood on the brain did not make McLuhan more sensitive to the needs of his wife.

In some form or other, the theme stayed with McLuhan for the rest of his life. In 1974, for example, he told an interviewer, "A woman can have all the influence she wants without ever leaving her armchair."[16] The comment suggests that McLuhan never really stopped living with Elsie and that Corinne took over her role in one sense at least — in maintaining the standards of cultivated behavior that McLuhan males, Herbert and his son Marshall, were always threatening to abandon. Corinne's husband, after all, was a man who, in her absence, would sometimes not wear socks, would eat off other people's plates, or would keep an old piece of cheese in his briefcase (it got better with age, he would claim). It was as if McLuhan was bound to confirm the ancient suspicion of women that men, if freed from feminine vigilance, degenerate into hogs.

In another sense, though, Corinne was very different from Elsie. She never shirked her duty in running the house and raising the children. If she hadn't handled these responsibilites, her husband certainly wouldn't have stepped in. Fortunately, her domestic skills were considerable. In St. Louis she became famous in their circle for her cooking. It was not unusual for Corinne to spend two or three days preparing various delicacies for an important dinner to be presented, with some pride, on the best of her family linen and china. She also served as McLuhan's faithful typist until at least the mid-sixties, typing the draft versions of both *The Gutenberg Galaxy* and *Understanding Media*. McLuhan was not unappreciative of her talents and her support; until his death he remained proud of the woman he married — of her wifely skills, her loyalty, and her beauty. For all his obsession with thinking, McLuhan was not entirely a cerebral creature. His wife once startled a seminarian at the University of Toronto by pointing to a double bed the McLuhans were selling and remarking that springs did not last long on any bed Marshall shared with a woman.[17]

Career problems began to make McLuhan's life in St. Louis considerably less pleasant than it had been before his departure for Cambridge with his bride. One concern was the replacement of the sympathetic William McCabe as chairman of the English department by a Jesuit of entirely different disposition, Norman J. Dreyfus. McLuhan found Dreyfus to be capricious, arbitrary, unimaginative, and full of nervous agitation.[18] According to McLuhan, Dreyfus hated anything to do with McCabe and therefore had it in for McLuhan, who was one of McCabe's protégés. The result was a "campaign of attrition" against McLuhan, in which Dreyfus assigned him to "the K.P. duties of the dept." (freshman English) and did his best to alienate students and faculty from McLuhan's work. McLuhan claimed that the only reason he stayed on under this harassment was to irritate Dreyfus. (Also, he had not yet obtained his Ph.D. and was therefore in no position to apply to other universities.)[19]

McLuhan almost certainly exaggerated the extent of Dreyfus's malice toward him. Throughout his academic career he had an unfortunate tendency to attribute the most Machiavellian characteristics to other people. Few human beings are capable of the calculated, persistent, and malevolent maneuvers

McLuhan sometimes saw in those who thwarted him. Dreyfus was in reality prone to feuds with people in his department, but the true cause of his antipathy to McLuhan was not so much personal dislike as simple lack of understanding. Dreyfus was a meticulous scholar trained in the Oxford tradition of philology. He was obsessed with getting facts right and had little in common intellectually with McLuhan, who sometimes wrote and talked as if he agreed with Faulkner's dictum that facts and truth don't really have much to do with each other. When Dreyfus refused to allow McLuhan to use in his classes a literature textbook he had compiled based on a combination of Leavis, Richards, and his trivium studies, the enmity between the two men hardened.

Also deeply affecting McLuhan's life in St. Louis in the 1940s was the entrance of the United States into World War II. After the Japanese attack on Pearl Harbor, members of the English department found themselves in Quonset huts teaching nursing students how to write reports. It was torture for McLuhan. His teaching schedule went from ten or twelve hours a week to sixteen. Moreover, he felt the situation placed him even more under the thumb of Dreyfus, since the alternative to teaching at St. Louis was now joining the army — a prospect that filled him with horror.

His feelings were not relieved by any patriotic fervor. Never for a moment did he experience any enthusiasm for the war aims of the Allies. Not only were those aims compromised in his view by the Soviet Union's involvement with the Allies, but the prolonged slaughter inevitably accentuated the mechanical organization of society, the unholy collaboration of industry and science in devising new ways to enslave humanity. The war had nothing to do with the four freedoms of Roosevelt and Churchill; it was simply, as one of McLuhan's friends put it, "a joy ride to inferno."[20]

As the war progressed, McLuhan became increasingly fearful he would be drafted, despite having a wife and child. That fear came to a head in December 1943, when he was classified I-A by his St. Louis draft board. With some urgency he began to look around for an escape. His task was made easier when he finally received his Cambridge Ph.D. in that same month and was in a better position to look for a job in Canada. (It was his fifth degree, counting a titular Cambridge M.A. awarded him in January 1940.) Earlier that year he had hurriedly completed his thesis and mailed it off to F.P. Wilson, without even typing the last seventy pages. Entitled "The Place of Thomas Nashe in the Learning of His Time" and running 456 pages, it was basically a review of the trivium and its function in literature from ancient Greece and Rome to the era of Nashe. Wilson was impressed and wrote McLuhan that he had learned more from his thesis than from any other he had had the pleasure of reading. It dealt, he said, with matters never before raised, in a highly original fashion.[21]

With his thesis completed and his promotion from instructor to assistant professor assured, McLuhan turned his attention not only to thoughts of Canada but also to hopes of getting published in the journals most likely to advance his job prospects. He prepared an article entitled "Francis Bacon's Patristic Inheritance," based largely on his thesis, and submitted it to the *Journal of the History of Ideas.* McLuhan had high hopes for the *Journal* and thought publishing his article in it would be a real coup. The editors sent him a letter praising the article for "materials and ideas of great value," but they were clearly not as impressed as McLuhan with the significance of the trivium and suggested that he rewrite the article, concentrating on Bacon. This was the kind of "sound" academic advice that McLuhan was incapable of listening to. He was on the track of much bigger game than a few original insights about Francis Bacon; he was trying to articulate a comprehensive vision of Western culture.

The editors also suggested that McLuhan "clarify the present confused structure" of his piece.[22] They were too late. McLuhan's prose was already veering off into the regions of compression and elliptical reference that would characterize his later work. Leavis had purged the last influences of Chesterton from McLuhan's prose style. McLuhan sometimes worried about his new style as "uncontrollable and unaccountable," but he did nothing to restrain it.[23]

As the forties progressed, McLuhan would increasingly look for inspiration to the prose criticism of Ezra Pound, which made McLuhan look a model of lucidity. But he felt that the vigor and intensity of prose like Pound's more than compensated for the lack of what he called "dialectic and persuasive charm" (i.e., the clear, connected structure and the smooth prose of the traditional polite essay).[24] Though the absence of such charm no doubt hurt him badly with potential publishers such as the editors of the *Journal of the History of Ideas,* McLuhan absolutely refused to employ it. Carefully linking points in clear sequence, making explicit the significance of all of the allusions and references in one's prose — this McLuhan referred to contemptuously as "tatting," as if it were a species of finicky lacework unworthy of a robust intelligence.

This was not affectation on McLuhan's part. His prose became increasingly dense at this time partly because he was trying to cram numerous ideas into a short space; each article he wrote was almost a condensed book. For the typical scholarly monographs built around one or two proud little *aperçus* he had little use.[25]

Instead of the *Journal of the History of Ideas,* McLuhan found himself published by *Columbia,* a magazine of the Knights of Columbus, a Catholic fraternal organization. In January 1944 the magazine ran his article "Dagwood's America," an articulation of his Dagwood theme and his first major published exercise in social criticism via analysis of a popular art form. In the article, McLuhan pointed to Dagwood's childlike and ineffectual nullity and Blondie's resourceful supremacy as evidence of how far America had been feminized in recent years. "Blondie and her children own America, control American business and entertainment, run hog-wild in spreading materialism into education and politics," he insisted. The traditional order in which man upholds a rational authority and is sustained by the emotional support of his wife had been overturned. The only cure was for Dagwood to reclaim "the detached use of autonomous reason for the critical appraisal of life" — in other words, to invoke the combined spirits of St. Thomas Aquinas and F.R. Leavis — and to engage in "healthy and fructifying work," that is, to give Mr. Dithers his notice and sign up in the wheelwright's shop that Leavis so admired in *Culture and Environment*.

McLuhan was happy to see the article exposed to a relatively large readership (half a million circulation), but he was worried that it appeared in a Roman Catholic magazine. He wanted to prove himself in the larger world of scholarship before making a public appearance under such auspices.[26] To become identified as a Catholic intellectual at the start of his career seemed unwise. Far better to act, like his opposite number, the dedicated Communist, as a sort of mole in the academic world, swaying minds that might be repulsed if one's affiliation were made obvious.

McLuhan definitely wanted to sway minds with this theme. He had visions of expanding "Dagwood's America" into a book, which he tentatively called *Sixty Million Mama's Boys*. He thought such a book might prove a hit as a piece of "impudent" pop sociology along the lines of *Generation of Vipers*, Philip Wylie's 1942 best-selling attack on "momism" — though of course McLuhan's book would be far more profound than Wylie's.[27]

The implicit antifeminist message of this book (which was never published) and of the Dagwood theme in general consorted, in those years, with McLuhan's persistent suspicion of homosexuals, overt and closeted. In 1949, Felix Giovanelli wrote him about Gershon Legman, a mutual acquaintance who was an authority on Freud. Legman, Giovanelli reported, was as wary of homosexuals as McLuhan was and held the same belief that they lurked everywhere in places of cultural influence.[28] Giovanelli implied that Legman was another ally in the struggle to alert the public to this unhappy situation, which was first hinted at in a 1931 book McLuhan had studied, *The Diabolical Principle*, by the painter and writer Wyndham Lewis. In it, Lewis suggested that the "homosexual cult" was a direct result of the feminist revolt.

In McLuhan's eyes, Dagwood, the mama's boy, and the homosexual were closely related figures. Grand sociological theories aside, he had been horrified by the "homosexual cult" since his Cambridge years and retained throughout his life the repugnance of a strait-laced Methodist child for homosexual activity. During his later years he was very careful to keep his opinion to himself, but throughout the 1940s his distaste was transmuted into a significant, if discreet, motif of his social criticism. He feared that male homosexuality was rampant in the age, the result of Blondie's emasculating Dagwood in front of Cookie and Alexander, their children. Such homosexuality was probably the chief threat to contemporary morality.[29]

McLuhan tried to publish his social criticism in general-interest magazines. During the war, he submitted an article entitled "Is Postwar Polygamy Inevitable?" to *Esquire* magazine. His argument was that the increasing prevalence of divorce, sexual promiscuity, adoption of illegitimate children, and artificial insemination had already removed any aura of sanctity from monogamy and the bearing of children by monogamous parents. In addition, the dynamics of industrial and commercial life, reducing both men and women to the status of wage-slaves and consumers, were making the traditional monogamous family an economic luxury. Men could no longer support families in proud independence. McLuhan's prediction was that women, who were "constitutionally docile, uncritical and routine-loving," would take over the running of commerce, industry, and government, collectively supporting a class of male loafers and warriors, in the manner of certain primitive tribes.

The argument is not entirely clear — it was certainly impenetrable to the editors of *Esquire*, who promptly rejected the article. It does reveal, however, an uncanny anticipation of our present era of even greater divorce and promiscuity, massive participation of women in the work force, single-parent families, surrogate motherhood, universal day care, and other post-monogamy phenomena.

Other articles of social criticism that he wrote during the war years included "Dale Carnegie: America's Machiavelli" (begun in 1939 and never published), a scathing analysis of *How to Win Friends and Influence People* as a textbook in the cynical manipulation of human beings, and "Out of the Castle into the Countinghouse," a continuation of his argument that commercial and industrial life were destroying monogamy and traditional family and community life. The latter was published in 1946 in a now defunct journal, *Politics*, edited by the social critic Dwight MacDonald.

McLuhan's greatest successes were the articles that combined the culture and environment strain of social criticism with the literary perspective of his trivium studies. The first of these was "Edgar Poe's Tradition," published in the *Sewanee Review* in 1944. In this essay, McLuhan placed Poe in the tradition of rhetoric, of Cicero and the Renaissance humanists, which he believed took root in the American South. The result of that tradition, according to McLuhan, was Poe's naturally cosmopolitan and aristocratic outlook, his high standards of literature, and his ideal of a society informed and led by men of wide learning, political savvy, and eloquence.

He expanded this theme in a series of essays for the *Sewanee Review* over the next few years. In "The Southern Quality" he elaborated on the southern tradition, associating it not only with Renaissance humanism and the study of rhetoric but with a way of life that was based on agriculture, communal living, and aristocratic values. Opposed to this was the tradition of New England, associated with dialectics — utilitarian logic — and a way of life that was commercial and industrial, based on aggressive individualism. The southern tradition obviously encompassed many of the qualities he would later describe as "acoustic" and the northern tradition qualities he would describe as "visual," although he was not yet thinking in those terms.

As always in his writings, there was no doubt which tradition he favored. McLuhan's admiration for the American South was a product not only of living with a wife who had an inextinguishable east Texas accent but of a genuine regard for the civilization that Jefferson had attempted to nurture. McLuhan was impressed by Jefferson, the aristocrat who was a scholar and a gentleman. Wholly free of any snobbery himself, he found immensely attractive anyone who was courtly, respectful of tradition, self-possessed, and also deeply responsible.

In "Footprints in the Sands of Crime," McLuhan cited Edgar Allan Poe's Dupin, Sherlock Holmes, Lord Peter Wimsey, and other notable sleuths as belonging to the gentlemanly and erudite tradition of Cicero and the Renaissance rhetoricians and humanists. The direct lineage of the sleuth was through aristocratic dandies such as Byron and Baudelaire, with a bit of the man-hunting noble savage of James Fenimore Cooper thrown into the gene pool. The significance of the sleuth was his utter contrast, given such a lineage, with the commercial society around him and with its paid functionaries, the police. While McLuhan had no admiration for either the style or the substance of detective fiction, he maintained an ongoing fascination for the sleuth figure and once allowed himself to be photographed in a Sherlock Holmes outfit. He saw his own work with the media as an example of sleuthing, the relentless search for the telltale clue.

In this article he also articulated a theme that would remain his until the end of his career — the theme of Poe's story "The Maelstrom." In that story, a sailor caught in a giant whirlpool eventually saves himself from drowning through detached observation of the vortex. For McLuhan, the sailor's action became a symbol, along with the sleuth, of his own work — of his freeing himself from the vortex of threatening social change through the process of understanding it. (McLuhan never mentioned that the hero of the story was broken in mind and body after the experience.)

These articles are brilliant and stimulating exercises in social and cultural criticism — and they all appeared in the *Sewanee Review*. McLuhan had found in this respected and well-established publication, then edited by Allen Tate, his first sympathetic ear outside his own immediate circle. Under Tate, the magazine reflected many of McLuhan's own interests: New Criticism, the respect for the vanishing agrarian traditions of the South, and opposition to the abstract and utilitarian ethos of modern civilization.

As he published more articles in the *Sewanee Review* and its like-minded cousin the *Kenyon Review*, founded and edited by John Crowe Ransom, McLuhan became known as an important, if junior, member of the Southern Agrarian/ New Criticism axis of Tate, Ransom, Cleanth Brooks, Robert Penn Warren, and others. When Tate resigned as editor of the *Sewanee Review* in 1946, he asked Ransom to recommend a successor, and Ransom wrote back with two names, one of which was McLuhan's. Ransom characterized McLuhan, in a phrase reminiscent of Conrad's *Lord Jim*, as "one of us." (This recommendation was supported by Cleanth Brooks, then teaching English at Yale, who had become personally acquainted with McLuhan.)[30]

Up to this point, McLuhan had written very little about media or technology, other than to mourn the effects of the mechanistic way of modern life. While he was in St. Louis, however, he was exposed to two writers who left him with important suggestions about technology that he would develop in his mature work. One of these writers was the American social critic Lewis Mumford. In his 1934 book *Technics and Civilization*, Mumford posited a radical distinction between the first stage of industrial civilization, based on steam power and highly mechanical in character, and a second stage, based on electricity and organic in nature. This second stage had introduced, through the telegraph and telephone, instantaneous worldwide communication; Mumford did not refer to a "global village," but his description of a world bound by a communication network is not far from the vision behind that famous McLuhan phrase. Mumford hoped that the widespread use of electricity would decentralize society once again, reverse the process of building up huge cities and factories, and restore something resembling a rural, community-based, artisan way of life.

Despite his sympathies with the agrarian and handicrafts mentality, McLuhan suspected Mumford of envisioning a Rousseau-style pastoral utopia, with little wired villages existing in happy anarchy, for which McLuhan had no use. Nevertheless, he picked up many valuable hints from Mumford. On print, for example, Mumford sounded a note strikingly similar to McLuhan's later thoughts: the printing press, he said, was the prototype for the "completely standardized and interchangeable parts" of mechanistic artifacts, and it contributed to the "lost balance between the sensuous and the intellectual, between image and sound, between the concrete and the abstract" that had been achieved, to some extent, by medieval civilization.[31]

The second writer to influence McLuhan's thought was a Swiss architect, Sigfried Giedion, whom McLuhan met in St. Louis at the beginning of the war. Giedion confirmed McLuhan's undergraduate suspicions that advertising had an interest far beyond most "serious" contemporary writing. In a single advertisement or a single item in a department store, Giedion demonstrated, the secrets of a whole society could be discerned.[32]

Giedion's *Mechanization Takes Command*, published after McLuhan had left St. Louis, remained a resource for McLuhan throughout his career. In that book, Giedion examined a wide range of human objects — nineteenth-century bathroom fixtures, Marcel Duchamp's painting *Nude Descending a Staircase*, and a Chicago meat-packing plant — and demonstrated how they all reflected a single process, the increasing mechanization of human life. The book showed McLuhan how fundamental changes in technology affected all aspects of human existence and how any artifact, no matter how humble, could reveal clues to the new patterns of life — the same patterns that were showing up in philosophy, advertising, and the most exalted art forms.

At the same time as he continued to absorb influences, to read, and to argue with friends, McLuhan also continued to come up with ideas for new projects and books. Between 1944 and 1946 he tried to compile a literary anthology with a fellow English professor, Carroll Hollis. The anthology, entitled *Reason and Tradition* and based on McLuhan's trivium studies, was completed after stupendous labor, but it proved no more practical or adaptable for classroom use than its original, McLuhan's textbook squelched by Father Dreyfus. In the same period, McLuhan worked on *Sixty Million Mama's Boys*, a book-length study of Poe based on his *Sewanee Review* article, a revision of his Ph.D. thesis for possible book publication, and a volume that he tentatively called *From Machiavelli to Marx: A Study in the Psychology of Culture*, which was inspired by a reading of Karen Horney's psychoanalytic work *The Neurotic Personality of Our Time*. The theme of this last project was the possibility of using the present "nervous breakdown" that Western culture was experiencing as an opportunity to undo the damage of the Renaissance and the Reformation by returning our civilization to the principles of right reason. McLuhan was also seriously interested in starting a new magazine, and he talked about it and promoted the idea constantly. Finally, he began work on the book that would eventually be known as *The Mechanical Bride* — the only one of his projects at this time ever to be fully realized.

His energy was a marvel. He worked and read and schemed and talked and wrote as if he needed three or four lifetimes to do all that he wanted to do on this earth. Unfortunately, his facility for coming up with projects was not matched by a determination to see them through. This characteristic never changed. An associate who worked with McLuhan toward the end of his career recalls, "Marshall was never interested in finishing anything — certainly in the sense of polishing off something. He always wanted to get on with something else. He was a bit manic in that regard."[33]

McLuhan's last year at St. Louis University, 1943–44, was marked not only by the beginning of his public career as a culture critic but by one of the more remarkable episodes of his personal life: the beginning of his friendship with the English painter, novelist, and critic Wyndham Lewis. In a letter to Lewis a year or so after their first meeting in 1943, he claimed that Lewis had been a significant influence on his thinking for years.[34] It is McLuhan the courtier speaking in that letter, though if he exaggerates it is not by much. Certainly traces of Lewis's great polemic *Time and Western Man* (1927), which McLuhan first read with much excitement as an undergraduate at Cambridge, crop up in McLuhan's prose in the late thirties. That book, an attack on what Lewis called the contemporary "time school," upheld certain classical values of contemplation, of clear forms and precise definitions, and of the autonomy of the individual, which made it very congenial to Catholics reared on Aristotle and Thomas Aquinas.

McLuhan was astounded when he found out from his mother in 1943 that this personage was actually living in Windsor, Ontario, across the river from Detroit. Elsie at the time was separated from Herbert and was living in Detroit, where she worked with the Detroit Drama Study Club. She had attended a lecture given by Lewis in the Christian Culture Series sponsored by Assumption College, Windsor. The organizer of the series, Father J. Stanley Murphy, had rescued Lewis from poverty and isolation in Toronto, where Lewis had been spending the war, by offering him a job at Assumption College. On hearing the news of Lewis's relative proximity, McLuhan and his St. Louis friend Felix Giovanelli, also a Lewis fan, took the train to Windsor to meet the great man in the summer of 1943.

The meeting went very well. Lewis was interested in McLuhan and Giovanelli's account of American education and university life and was talkative about his own work both in literature and in painting. The two then returned to St. Louis with the notion of helping the financially pressed Lewis by rounding up some portrait painting assignments for him. They wasted no time trying to sell the idea to the rich and famous of St. Louis. This campaign was an adventure for the unworldly McLuhan, since he had never really tried before to influence people outside of his colleagues and students.

The exercise proved beneficial for him. He was at a low point in his career, just before his first significant published articles appeared, and at the height of his immersion in the teaching of report writing to nurses. As he wrote Lewis, soliciting commissions was instrumental in helping him recover self-respect.[35] In fact, he discovered he had rather a taste for counseling the mighty. His campaign in St. Louis introduced him to circles in which no respectable academic content with his station in life was supposed to mingle. To his delight, McLuhan found the people in these high social circles rather easy to handle, particularly compared with someone like Dreyfus.[36] He succeeded in arranging a lecture for Lewis at the St. Louis Museum and another at the Wednesday Club, an organization of prominent society women. His topic at the club was to be suitably gossipy: "Famous People I Have Put on Canvas and in Books." McLuhan was immensely relieved when Lewis showed no offense at this indirect disparagement of his stature.

McLuhan found it more difficult to obtain portrait commissions from the monied of St. Louis. "The extreme timidity and shyness of these people astonishes me," he wrote Lewis, perhaps not fully aware that his prospects owed their wealth in part to their ability to say no to people like him. He was amused when they suggested that portrait painting was too grand and expensive a proposition for the likes of them. McLuhan and Giovanelli did manage to win over Ernest Hemingway's mother-in-law, Edna Gellhorn, who commissioned a chalk portrait for forty dollars. Realizing that she had secured a superb portrait at a very low price, she was, McLuhan gleefully reported, eager to see her canniness confirmed by the willingness of others to pay Lewis higher prices. An influential woman and head of the League of Women Voters in St. Louis, she began herself to solicit other commissions for Lewis, eventually securing for him a $1,500 commission for a portrait of a Nobel Prize-winning physicist at Washington University.

With these commissions settled, Lewis arrived in St. Louis in February 1944. He remained, as it turned out, until the following July, staying at local hotels and at the apartment of Giovanelli and his wife, who moved in with the McLuhans for the duration. Lewis received enough portrait commissions to keep him going for six months, and in the process, got to know the McLuhans and Giovanellis very well.

Relations between McLuhan and Lewis, though occasionally warm, were never entirely smooth. Lewis, basically a shy and fearful man, compensated by displaying a good deal of aggression in odd ways. He made a charcoal sketch of McLuhan, for instance, that showed him missing one eye and the top of his head, to the great distress of its subject. In Lewis's 1954 novel, *Self Condemned*, which was based on his wartime experience in Canada, McLuhan thought he recognized himself in the character of Professor Ian Mackenzie, described in the novel as "a little routine teaching hack" possessed "of a very good mind of a routine kind."[38]

Lewis managed to find pretexts to quarrel with McLuhan, who had been almost heroically obliging to him, but McLuhan remained steadfast in his friendship and looked forward to joining Lewis in Windsor to help edit a new version of *The Enemy*, an aptly named periodical that Lewis had put out in the twenties. In the summer of 1944 McLuhan accepted a job as head of the English department at Assumption College, a job immediately offered to him when he applied for it at Lewis's suggestion.

The offer was made by Father Stan Murphy, who knew McLuhan from lectures he had given at the Assumption College summer school before the war and who had been present when McLuhan and Giovanelli first met Lewis in Windsor. With or without the presence of Lewis at Assumption College, the offer was attractive to McLuhan. For one thing, it removed him from the threat of the St. Louis draft board, and his salary at Assumption would be no less than at St. Louis. He expected to be able to rent a house in Windsor with yard space to accommodate his lively son.[39] Not least important, Murphy arranged for McLuhan to enjoy a year with a light teaching load so that he could concentrate on his research and writing. Ever hopeful, McLuhan anticipated that his work might be furthered at Assumption College by "conversation with congenial minds — Catholic minds."[40]

Assumption College at that time was run by the Basilian order of the Catholic Church (then reputed to be the most liberal of the teaching orders of priests) and had an enrollment of about 150 students. Academically and in almost every other respect it was a step down from St. Louis University — except for the presence of Father Murphy. Murphy was an extraordinary individual, evidenced by the fact that he had managed to win the respect of Wyndham Lewis. His chief claim to fame was his Christian Culture Series, an ongoing series of lectures and performances by writers and artists of international repute that he launched in 1934. Working with no budget and no staff, Murphy attracted such luminaries of the Catholic intellectual world as Jacques Maritain and Martin D'Arcy, another indication of his ingenuity and dedication. He was also cheerful, self-effacing, and completely generous.

This exemplary priest made heroic efforts over the years to turn Assumption College into a first-class Catholic college. He never succeeded. Certainly after he arrived in 1944, McLuhan felt that he was sinking into what he called "a little backwater in a stagnant stream" (the "stagnant stream" being Canada). He discovered that students in his day classes were even more lethargic and dull-witted than the students at St. Louis — a harsh assessment considering that he felt he had not had *any* good students in his last years at St. Louis.[41]

Even the physical setting was unfortunate. His heart must have sank when he first walked into the old wooden barracks, once used to house R.C.A.F. trainees, warmed by a coal furnace, where he was to teach. On one occasion, an officious janitor walked into the hut in the middle of a lecture and switched on the motorized hopper, causing a tremendous racket. McLuhan, whose tolerance for noise was limited in the best of circumstances, switched off the motor. The janitor promptly switched it back on. McLuhan switched it off again. The two continued this comedy for a few minutes until McLuhan marched to the president's office, and demanded the janitor be put under some restraint. The president, Father John Guinan, agreed.

Nor was McLuhan free from the usual academic intrigue at Assumption. He particularly felt the need to mollify Edwin Garvey, head of the philosophy department. Lewis would later portray Garvey as Father O'Shea in *Self Condemned*, describing him as "the uncrowned King of the college, who made his views felt to such good effect that no one dared to do anything without his consent."[42] McLuhan perhaps felt that any disapproval he aroused through his unconventional ideas or approaches might be used against Murphy by this powerful figure. To propitiate him, McLuhan went out of his way to discuss a book Garvey was then preparing on the subject of education. McLuhan's campaign was probably successful not because of this discussion but because he had already earned several points in Garvey's estimation simply by virtue of his conversion to Catholicism.

As far as his native land was concerned, McLuhan was harsh. He described it as a "mental vacuum," full of "terrible social cowardice." Perhaps inspired by Karen Horney, he speculated that some "unacknowledged guilt" might be behind the omnipresent Canadian furtiveness.[43] In an unpublished article titled "Canada Needs More Jews," he put forth a notion he had picked up from Wyndham Lewis that the importation of two million or so Jews might liven the place up.

He asserted that the Jews had a few strikes against them: they tended to be egotistical because of their long effort to retain self-respect in the face of hatred and persecution; they were ostentatious in their wealth, clannish, all too eager to join in the American habit of obsequiously fussing over women; and, because of their superior energy and imagination, they were more outrageous than Gentiles in the cultivation of both virtues and vices. Nonetheless, McLuhan felt that this very superiority in energy and imagination, combined with a reverence for art and learning and a rich instinct for community life, made Jews in massive numbers extremely desirable additions to postwar Canada.

McLuhan kept up his relationship with Wyndham Lewis, who interrupted his portrait painting in St. Louis to return to Windsor from December 1944 until early spring 1945. Although the two never did get around to collaborating on a periodical, McLuhan continued to try to help Lewis in any way he could. He wrote to his former student Marius Bewley, who had become the secretary and assistant to the noted New York gallery owner Peggy Guggenheim, to ask

Bewley if Guggenheim would commission work from Lewis.[44] In a letter to Lewis the previous summer, McLuhan maintained that he would always try to interest others in the work of Lewis, particularly when he had become what nature intended him to be, an influential figure on the intellectual landscape. McLuhan acknowledged that, for the present, he had very little influence. Yet he regarded his efforts on behalf of Lewis as a kind of foretaste of a larger mission he would someday undertake, the precise nature of which was still unknown to him. It would have something to do with increasing human awareness; for now, championing the cause of Lewis was the best possible thing he could do in that regard.[45]

Unfortunately Lewis, as was his wont, eventually picked a serious quarrel with McLuhan. In a note written in February 1945, Lewis accused McLuhan of being "sick with unsatisfied vanity" and enumerated a few instances of what he imagined was McLuhan's rudeness to him. He accused McLuhan of leaving his wife in the car when he visited them, which must, of course, have been meant as an insult to either Mr. or Mrs. Lewis. He accused McLuhan of neglecting him and his wife when they returned to Windsor. He added, somewhat cruelly, that behaviour such as this "is probably why you had so few serious friends in the college where you last taught." Neither these accusations nor anything else in Lewis's note is very coherent, and it is safe to say that McLuhan would never have dreamed of behaving offensively to the Lewises. The quarrel was entirely on the part of the man that Giovanelli referred to as "Der Ogre."[46]

McLuhan nonetheless remained faithful to Lewis, continued to correspond with him until his death in 1957, defended him in critical essays, and responded positively when Lewis asked him to help find a New York publisher for *Self Condemned*. McLuhan also continued to read and admire Lewis's works. (It was a sentence from Lewis's 1948 book *America and Cosmic Man* — "the earth has become one big village, with telephones laid on from one end to the other, and air transport, both speedy and safe" — that inspired McLuhan's phrase "global village," for example.)[47]

He could not help but continue to admire Lewis, partly because McLuhan never really minded anyone who was outrageous and even ornery, as long as that person was free of malice, envy, and underhandedness. For all that he could lash out at people in an irrational way, Lewis possessed none of those defects. Conversely, Lewis's way of disrupting, on a grand scale, the routine habits and expectations of people around him rather appealed to McLuhan, who felt himself surrounded in the late forties with people who coped with dissatisfaction by moping. McLuhan hated the *puny*. He himself rather enjoyed looming over people with his six feet two inches — he had a knack for seeming even taller than he was. Lewis's personality loomed over others in an equivalent, psychic manner.

Of course, it was Lewis's intellect that really compelled McLuhan's admiration. Initially, he was attracted to Lewis because they shared the same enemies — the philosophy of sensation seeking, the "revolutionary simpleton," the world of advertisers, vulgarized science — all those forces that tend, as McLuhan put it in an early essay on Lewis, to the "destruction of family life... the flight from adulthood, the obliteration of masculine and feminine." But Lewis did more for McLuhan than reinforce certain views of the world. One sentence from that early essay sums up Lewis's legacy for McLuhan: "Sheer annoyance, at first, led [Lewis] to study a hostile environment to see how he might the better accommodate his creative work to it."[48] In later years, McLuhan tirelessly repeated this theme as the key to his own work on the media, insisting that he had little use for the changes he was describing but felt impelled to study them so as not to be steamrollered by them. The only option for a rational person, more so for an artist, according to McLuhan, was "unremitting observation and analysis" of the environment.[49]

This unremitting observation profoundly suited McLuhan's temperament. He was one of those men who, without any prompting, find observation of the world an excellent strategy for coping with life. Lewis, however, helped him to elevate this strategy into a ruling principle of his work. After meeting Lewis, McLuhan gave full rein to the daily process of studying the environment. There was no longer a single thing in that environment that was not interesting (a lesson McLuhan had also learned from Giedion). "Even if it's some place I don't find congenial, like a dull movie or a nightclub, I'm busy perceiving patterns," he once told a reporter.[50] A street sign, a building, a sports car — what, he would ask himself and others, did these things *mean*? The noted critic George Steiner once recalled a lunch with McLuhan in which "the entire meal consisted of seeing everything in the restaurant, from the lighting to the menu, in a new way."[51]

For McLuhan, the alternative to this kind of observation was a hallucinatory trance, or waking sleep. He took to heart Lewis's comment "The world is in the strictest sense asleep, with rare intervals and spots of awareness. It is almost the sleep of the insect or the animal world."[52] McLuhan tended to annoy his students by telling them that he noticed they were all in hypnotic trances: his job, he felt, was to try to wake them up. (If he managed to wake up one student per year, he felt he was being more than successful.)

In one respect, however, Lewis's influence was not altogether to the good. Lewis's habit of seeing intellectual or artistic antagonists as personal enemies encouraged a similar tendency in McLuhan. At times McLuhan seemed to view schools of thought that he disliked, and those who upheld those schools of thought, as melding together in a vast, malignant, personal conspiracy against the truth. The force behind the incarceration of Ezra Pound in a mental institution, for example, was not the U.S. Justice Department but "Hegelian History."[53] This was an instance where McLuhan went overboard in perceiving patterns; such instances would be repeated many times in the years ahead.

After Lewis left Windsor in 1945, McLuhan continued teaching and writing, occasionally giving public lectures, which he found distasteful but which helped pay his income tax. In late 1944 at Notre Dame College in Ottawa and the Detroit Women's Catholic Study Club, McLuhan gave a lecture entitled "A Catholic Looks at Latin America." He used the subject to make some familiar points: Latin American culture, being Catholic, was of course superior to North American culture and was oriented far more to the pursuit of truth and beauty than to industrial and commercial "progress." Scholars, artists, and poets, not businessmen, were the heroes of Latin America. The culture also was marked by strong family life and strong parental authority. There were few Hispanic Dagwoods.

Another aspect of this theme was elaborated in McLuhan's 1944 Christian Culture Series lecture "Delinquent Adults Behind the Super-Comics." In this talk, McLuhan did for the Windsor and Detroit public what he had been doing for almost a decade with his students. Using slides, he analyzed samples of comic strips. His point was that the decay of philosophy and religion, the pervasive North American Darwinian approach to success, and commercial and materialistic values had reduced adults to a kind of emotional and mental delinquency. In this state they were extremely vulnerable to the crude daydreams fostered by the mass media. The violent sensationalism, the sadism and masochism reflected by cartoon figures like Superman were an essential ingredient of these daydreams. McLuhan repeated his conviction that the educational system was helpless against the powerful onslaught of such daydreams, that children received their real education from the media and not from their schoolteachers, and that the only hope of educators was to bring the media within the classroom and try to encourage students to conduct some sort of rational analysis of it.

The McLuhan family enjoyed an active social life at Windsor, as they had in St. Louis, swimming and picnicking at Peche Island in Lake Saint Clair with some of the priests at Assumption College and attending faculty parties where McLuhan, as always, indulged his immoderate love of talk. One colleague recalls a party at the college librarian's house where the librarian's husband played on their new piano, which was obviously meant to be the proud centrepiece of the gathering. McLuhan, excited about a new breakthrough in his thinking, talked through the playing. Suddenly the man stopped, so that the only sound in the room was that of McLuhan's voice. Everyone except McLuhan stiffened with tension. But the piano player merely turned to McLuhan and politely asked him how he liked the piano. McLuhan quite affably told him it was a splendid piano and returned to his conversation.[54]

As ever, McLuhan's family life remained a dubious source of consolation for him. By four years of age, Eric had acquired an "insolent ingenuity," which he spent in pranks and mischief not always harmless or amusing.[55] One of his milder tricks was putting cats in drawers and shutting them in. If guests sat on the porch of the McLuhan home they were sometimes in danger of getting doused by a garden hose wielded by the young terror. He also showed whose son he was by revealing a precocious glibness, including occasional displays, his father complained, of "snide propositions and concupiscent sophistries."[56]

As his other children were born, McLuhan frankly confessed to friends that he was incapable of understanding them. (The children, in turn, took a kind of symbolic revenge by firing beebees at McLuhan's books from time to time.) It fell to Corinne to act as a buffer between the children and their father, who occasionally remembered that he was no Dagwood and demanded respect from them, even to the point of being called Sir. Fortunately, the children did not take these displays of paternal authority very seriously, although McLuhan occasionally acted the part of the old-fashioned father, even going so far as to use his belt as a strap.

Corinne became even more the dominant parent after she gave birth to twins, Mary and Teresa, in October 1945. Following their birth McLuhan told his little son that they had just lost the sex war in their own household.[57] He was, of course, partly joshing; but the man who resolutely opposed the feminization of America knew for certain at that moment that the gods had rewarded him for his opposition by granting him a household that would become a female preserve. He was always proud of the four daughters he sired (two more daughters were born in 1947 and 1950; a second son was born in 1952); the girls were beautiful creatures. He continued to feel, however, that humans of the opposite gender were a distinct, almost exotic species — a feeling colored sometimes with a neurotically Victorian tint, as when he passed on to a friend the startling generalization that women go downhill after the age of twelve.[58]

His wife ran the household as best she could. In the early years of their marriage her husband did some household tasks, to be sure — ran the diapers through the washing machine (which he considered, as he did most machines, a tyrant converting him into its own human "servo-mechanism"), washed and ironed clothes, and did the dishes, though, with a touch of that male eccentricity that sometimes leads to the divorce court, he believed that no soap was needed for this operation. In general, he was not fond of labor-saving devices in the home and purchased a vacuum cleaner only after his wife borrowed a neighbor's, in the early fifties.

Managing the household was made more difficult by one otherwise admirable trait of McLuhan's, a sense of hospitality almost endless in scope. "He loved to entertain," Pauline Bondy, a family friend in Windsor, recalls. "He would invite anybody to stay with them. He would invite the dean of Ely Cathedral, who happened to be in town, to stay with them, when they had no spare room at all for him. They would put the dean of Ely on the chesterfield to sleep. As long as there was a hunk of bread in the kitchen he would invite people to stay. Sometimes it was hard on Corinne's pride."[59] This habit of impromptu hospitality stayed with McLuhan until virtually the end of his life and was certainly not limited to celebrities or important people. In one respect, Corinne was an ideal faculty wife. No colleague or friend of Marshall's, no matter how disagreeable, ever received a frosty reception from her, as long as the individual was on good terms with her husband. According to Bondy, "If Corinne set herself to get along with somebody, she got along."

One person who became caught up in the McLuhan vortex in Windsor actually proved to be a boon for the household. Amy Dunaway was an Englishwoman living with relatives in Windsor who seems to have met McLuhan at one of his public lectures sometime in 1945. She ended up living with the family for a year as a friend and unpaid assistant housekeeper and nursemaid. She was, in all respects, a godsend.

Near the end of his second year at Windsor, in the spring of 1946, McLuhan received an offer to teach at St. Michael's College at the University of Toronto. The superior of the Basilians at the college at that time, and its de facto president, was an austere priest named Father Louis Bondy, brother of the McLuhans' friend Pauline Bondy. "He had met Marshall often at Windsor at my place," Pauline Bondy recalls. "He was perceptive enough to know that, as he put it, McLuhan was too big for Windsor." Bondy suggested to the head of the English department at St. Michael's, Father Lawrence Shook, that he invite McLuhan to join his department.

McLuhan, Bondy pointed out, was relatively young; a good Catholic; a scholar with impressive training; and an expert in modern poetry and criticism — the only such expert, probably, in the province of Ontario at that time. In that respect, he would fill an important gap in the expertise of the English department at the University of Toronto. McLuhan, who had lit no fires under the students at Assumption College and who was eager, in any case, to escape that backwater, accepted Shook's subsequent invitation. The move to Toronto finally put an end to his days as a wandering scholar. With the exception of one year at Fordham University in New York City, McLuhan would spend the rest of his life living and teaching in Toronto.

6. Twilight of the Mechanical Bride (1946–1951)

Bosh! Stephen said rudely. A man of genius makes no mistakes. His errors are
volitional and are the portals of discovery. — James Joyce, *Ulysses*

St. Michael's College in 1946 was an institution similar in many ways to Assumption College and St. Louis University: it possessed the trappings and the spirit of a Catholic educational system now vanished as thoroughly as lectures in Latin. At St. Michael's, professors began each class with a prayer. They needed permission from the college librarian to see volumes that had been placed on the Vatican Index of Forbidden Books (which were kept in a locked closet), and it was understood that they would refrain from teaching works like Pope's *Eloisa to Abelard*, with its indelicate treatment of clerical celibacy. More important, the college still had its air of a *petit séminaire*, a place run by and for Basilian priests. Mediocre instructors who wore the right kind of collar found shelter there from time to time, often in the form of reinforcements called up from the Basilian high school in Rochester, New York. The result was that students of the English novel at St. Michael's sometimes found themselves listening to lengthy plot summaries of *Tom Jones* or *Vanity Fair*.

It was therefore an innovation when the college hired two superbly qualified laymen after the war, a philosophy professor named Lawrence Lynch and Marshall McLuhan. McLuhan eventually tired of the role of "civilian Basilian" and felt the same frustration he had felt at St. Louis and Assumption. In 1952, he wrote to his former mentor Gerald Phelan voicing his by now almost ritual complaint about students. In five years, he claimed, he had not had one good student from among the undergraduates at St. Michael's College.[1]

His frustrations were mitigated, however, by the fact that St. Michael's College was not wholly autonomous. The college was only one of the federated colleges within the University of Toronto. The university as a whole, a secular institution, established the curricula of the English courses, graded the final exams, granted the degrees, and also controlled the graduate school. In many ways, this was a fortunate situation for McLuhan. In the college he had his circle of coreligionists who could respond to his ideas from an impeccably orthodox point of view. Yet that circle was also challenged, enriched, and to some extent held accountable by a larger institution, more broadly based and at least as formidable intellectually. This institution also supplied McLuhan with some decent graduate students.

St. Michael's College in the late 1940s labored under a sense of inferiority to the other colleges of the university: Trinity College, an Anglican institution; Victoria College, a United Church (largely Methodist) institution; and, above all, University College, the predominant, entirely secular college. Partly this sense of inferiority was the result of a widespread belief that Catholic education was, in the nature of the case, bound to be inferior, and partly it was a reflection of certain realities. One reality that McLuhan felt keenly was that his colleagues at University College made about a thousand dollars more than he did — a significant sum of money at that time.

On an entirely different plane, McLuhan felt that Catholic education was stymied because of its remoteness from twentieth-century achievements in the arts and sciences, particularly the arts. As he wrote to Phelan, Catholic thinking could be nourished only by attaining a rapport with the artistic techniques of the moderns — and by concentrating on the problem of communication. It was this problem, first encountered by McLuhan in the writings of Richards and Empson and in his own studies of rhetoric, that offered the single most fruitful area of investigation for contemporary thinkers.[2]

In the first years of his career at St. Michael's, McLuhan promoted this goal of expanding Catholic thinking with more diligence than tact. He was perpetually buttonholing his colleagues in the philosophy department with questions about their discipline — questions meant partly to satisfy his own curiosity and partly to prod these men into working in areas he felt needed investigation. (One of McLuhan's favorite openers was "For your information, let me ask you a question.") He was particularly dogged in pursuing the philosophers since their department at St. Michael's, with the Pontifical Institute of Medieval Studies at its center, carried real intellectual weight. As at St. Louis University, the stronghold of scholarship at St. Michael's was the work of the Thomists.

The chief Thomist in residence at the time was Anton Pegis, director of the Pontifical Institute and the kind of painstaking scholar who was always a thorn in McLuhan's side. A rigorously logical thinker, Pegis was shocked by many of the comments McLuhan made in his presence. He would shake his head and ask, "What are we going to do about Marshall?" on innumerable occasions when McLuhan had lobbed some extraordinary philosophical proposition his way. (McLuhan's colleagues at St. Louis University had labeled some of his ideas "bloodcurdling.")[3]

McLuhan was impervious to Pegis's reactions, however, consoling himself with the thought that Pegis was, after all, far inferior to his old friend Bernard Muller-Thym as a metaphysician. An altogether more significant figure than Pegis was a visiting professor at the institute, Étienne Gilson, the greatest medieval scholar of the twentieth century and a major philosopher, who had founded the institute in 1929. To have won the respect of a mind of Gilson's caliber, to have communicated with him in a fruitful and collaborative sense, would have been an achievement, in McLuhan's eyes, that would have made up for a great deal.

It never happened, however. Aside from a certain shared knowledge of detective fiction, the two men had little in common. Gilson, for one thing, possessed a kind of hardheaded, French bourgeois practicality that deflated many of McLuhan's cherished schemes. He tended to regard McLuhan as a bit of an oddity. Gilson was primarily interested in what Aquinas or Matthew of Aquasparta or any of the other historical figures he wrote about had actually *said*; McLuhan was interested in them for what ideas they might spark in his own head. Sometimes Gilson lost patience with what he regarded as McLuhan's more outrageous remarks. On one occasion in the St. Michael's faculty lunchroom Gilson said that it was a tragedy that Bernard Muller-Thym had ever abandoned metaphysics. McLuhan, who admired Muller-Thym's successful career as a business consultant, replied that, in his opinion, Muller-Thym was more deeply into metaphysics than ever. This was too much for Gilson. He threw his napkin on the table and stormed out of the room.[4]

McLuhan gave up the idea of having any real influence on the St. Michael's philosophers. "He called us the most boring group on campus," one of them recalls.[5] McLuhan maintained until the end that his theories of communication were "Thomistic"; certainly he would have felt uncomfortable severing connections entirely with the official philosophy of his Church. Nonetheless, he always remained rooted in the tradition he admired, that of the Sophists and rhetoricians, rather than in the more logical and dialectical tradition of Aquinas.

If McLuhan received limited understanding from his colleagues at St. Michael's, he obtained even less sympathy from professors in other colleges at the university. For years he had to contend with the disapproval of one of the most formidable figures on campus, A.S.P. Woodhouse. Woodhouse headed the English department of University College and the graduate English department. He had been one of the two men chiefly responsible for setting up the English undergraduate curriculum at the university — a curriculum that remained virtually unchanged for twenty years after McLuhan arrived at Toronto — and he retained throughout his career an absolute dominance over English studies in Canada. His protégés, trained at the university graduate

English department — Toronto was the only Canadian university to offer a doctorate in English for several years after the war — filled the English departments of other universities throughout Canada.

Woodhouse was the embodiment of a type once familiar on campuses but now something of a dying breed: the eccentric bachelor, with a fussy disdain for women and the biology of procreation. He could sleep only on his side, facing south, and with feet pointed east; and he barely recovered from the trauma of having to serve, on one occasion, as an examiner at the Ph.D. orals of a pregnant woman. As a scholar, his forte was the "history of ideas"; his approach to literature was the kind that makes the theology of Milton the key to understanding *Paradise Lost*. Under Woodhouse and this historical scholarship approach, the teaching of English at the University of Toronto attained a stature in North America in the 1950s not far beneath that of Harvard.

It was an approach utterly antipathetic to the New Criticism McLuhan brought with him from Cambridge. The standing joke about Woodhouse was that he had enjoyed his last aesthetic experience at the age of four. In his maturity, he sat at a desk in front of a portrait of Dr. Johnson, conveying the suggestion that he might have been one of Johnson's inner circle, perhaps a spiritual descendant of the great man himself. McLuhan, on the other hand, possessed a curious humility regarding the writers he taught. He might have viewed himself as a collaborator of, say, Joyce and Pound — in the sense that any good, creative reader of those men becomes an artistic collaborator in their works — but he never felt himself their peer.

Woodhouse saw in McLuhan the death knell for his world. And the politics of literary criticism aside, McLuhan, possessor of a Cambridge Ph.D., recognized that the prewar university in which a British or a Toronto M.A. was sufficient for an academic career was finished. (Woodhouse himself had never obtained a doctorate, a fact that was carefully buried at Toronto.) Seeing the coming dominance of the Ph.D. before anyone else in Canada, McLuhan insisted that his brightest graduates obtain one. It was no wonder that Woodhouse sensed danger. Even before McLuhan's arrival, Woodhouse, who had of course heard every rumor circulating on the Toronto campus, told Father Shook that he hoped the good father was not intending to bring McLuhan to St. Michael's. Later he told Father Bondy, with the forcefulness of one who is used to having his suggestions treated as commandments, that McLuhan "was not the sort of person we want at the University of Toronto."[6]

When McLuhan arrived at Toronto, then, he found himself having to refight the battles that Leavis had initiated at Cambridge — even to the extent of having to obtain recognition for the work of T.S. Eliot and other moderns. That work was regarded by his colleagues as something that could take care of itself without professors having to fuss over it. McLuhan was almost the only member of the graduate department to teach twentieth-century English literature throughout the 1950s.

To defend this specialty in the faculty lounges and at committee meetings, McLuhan relied on his ace in the hole — his Cambridge Ph.D. — and on his lethal debating skills. He could always throw opponents off base by saying something incomprehensible or something that seemed completely irrelevant but that just might, somehow, be relevant. He possessed the consistent unflappability of a man who can discover, in the course of an argument, what he wants to say and who then knows that he will always be able, afterward, to find reasons for having said it. McLuhan could be merciless in his evaluation of colleagues who thwarted him, but he seems to have been singularly free of malice or resentment toward Woodhouse; he was almost fond of him, perhaps seeing in him a not ignoble specimen of a dying academic tradition.

F.E.L. Priestley, another University College professor, was a Woodhouse ally. Priestley's view of McLuhan was typical of that of many colleagues teaching in colleges other than St. Michael's. He recalls a long chat with McLuhan in the late forties, during which McLuhan complained of his finances and his career prospects.

> He was limited in the chances open to him by the very fact that he was in a Roman Catholic college. He said to me, "You know, going to Assumption locked me into the Roman Catholic network because nobody in the outside world would ever look to Assumption for a scholar." That was true. There was a prejudice pretty well in every university in Canada against employing Roman Catholics — and especially Catholics who had announced their Catholicism by being employed in a Catholic institution. So McLuhan felt trapped. He had come from Assumption to the top Catholic college in Canada — and where could he go from there?[7]

According to Priestley, McLuhan eventually realized that the way out of this trap would not lie in pursuing conventional English studies. It would lie in pursuing some subject that would not only get him out of Catholic higher education but win him larger audiences than just his fellow academics.

This view of McLuhan as a man desperate to get out of poverty and obscurity — a view increasingly held by his colleagues as McLuhan began to receive more and more invitations to speak from other universities, in the period of his growing fame — is not without its taste of sour grapes. It is also not without some truth. As an undergraduate at the University of Manitoba, living in threadbare circumstances with an amiable but uninspired father, McLuhan had vowed to eventually become both solvent and prominent in some good work. In all the changes he had experienced since that time, he had never forgotten that vow.

But this view of McLuhan held by many of his colleagues was overaccentuated by personal antipathy toward him — an antipathy that McLuhan, to some extent, reciprocated. The English department in the late forties was, as he put it, "a ghastly crew."[8] In the mid-fifties, he complained to a friend at Columbia University that his colleagues frankly thought he was a "nut."[9] In later years, he made the best of the situation by citing Socrates's famous answer to the question of why he had married the shrewish Xanthippe: if he could handle her, he could handle anybody. In like manner, McLuhan felt that his marriage to the University of Toronto was irritating but at least kept him on his toes.

But McLuhan did worry that the hostility of his colleagues would affect his graduate students. As a teaching assistant at Wisconsin, he had learned that Ph.D. candidates might be penalized by professors feuding with their thesis supervisors. Now he was in the position of being one of those supervisors and he realized that, while his enemies could do nothing about his own work, they could get back at him through his students.

As the years passed, it became impossible for McLuhan not to notice that graduate students were being discouraged by other professors from taking courses with him. It was not necessary for the professors to mention McLuhan by name; all they had to do was drop a few subtle remarks about the subjects with which he was associated: "Is this the kind of serious subject you want to pursue?" or "Wouldn't it be better for your professional development to do such and such?" The hints were taken, especially after it became obvious that graduate students associated with McLuhan tended to have a more difficult time going through the system. That phenomenon explains, to some extent, why only seven Ph.D. theses were completed under McLuhan's supervision in his more than thirty years at the University of Toronto — a relatively low number given his popularity among many graduate students and his fertile imagination for good thesis ideas.

Even when it came to being asked to sit on committees examining other Ph.D. candidates, McLuhan often felt he was excluded. In later years, it is true, he could be cranky; but in most respects he was a brilliant examiner, even though he rarely seemed to have prepared questions before the examination. "McLuhan could steal the show when he wanted to, by asking some crazy question which had a real depth to it," Father Shook recalls. "This could have caused a certain jealousy among other examiners. Certainly the students talked about him and quoted him a great deal."[10]

He was also a favorite topic of conversation among undergraduates. When he first began teaching at St. Michael's — under the impression the students might really be good — McLuhan resolved to give teaching his full attention, which he had not done since his early years at St. Louis. His resolution soon evaporated. Between McLuhan-style lectures and the model university lectures of A.S.P. Woodhouse loomed a great gulf, into which students sometimes peered with dizzying effect. Woodhouse polished and perfected his lecture notes year after year and bequeathed to his auditors, in turn, a wealth of material for their own notebooks. McLuhan never gave a lecture remotely resembling any previous lecture he had ever given. Nothing was predictable. He would enter the classroom with a huge pile of books under his arm; sometimes he would refer to them, sometimes not. And like his favorite Cambridge lecturer, Mansfield Forbes, he refused to "cover" a course: he would only point out the most fertile areas for the students to dig around in. This meant that students often heard detailed references to Batman and Johnnie Ray in lectures ostensibly dealing with Francis Bacon.

Many students were dismayed, since their marks were almost wholly dependent on final exams, which were set and graded by the university English department as a whole, and lecture notes were supposed to be the key to success on exams. McLuhan seemed incapable of comprehending that students were anxious to have neat, point by point summaries of who influenced Spenser and what the "theme" of Milton's *Comus* was and how Shakespeare utilized the Petrarchan sonnet form.

Insecure students complained that in the modern drama course, for example, Professor McLuhan spent ninety percent of his time talking about T.S. Eliot, one of his favorites, and one percent on George Bernard Shaw, a playwright he had little use for — despite the fact that both authors carried equal weight in final exams. "He announced that since Shaw had only three original ideas in his life, we could prepare material on Shaw ourselves," one former student recalls. "He never even hinted at what the three original ideas were."[11]

Like most professors, McLuhan also disliked marking; at Wisconsin he had complained that it made him feel like a hangman.[12] He skimmed essays and exams for any insights they might contain, just as he did any other potential intellectual resource, but he had little patience for actually evaluating students' work and certainly took no interest in pointing out gaps in their knowledge. With graduate essays, he simply marked them by noting "one point," "two points," and so on. A well-researched, well-written, well-documented paper that was basically a rehash of current literature on the subject — par for the course in graduate essays — received the notation "zero points."

With undergraduate exams, however, he could be extremely indulgent. He once told a colleague who was failing a number of students on an exam to be more generous on the grounds that the students were suffering from anxiety. He gave them generous marks, he said, if they just knew what the novels were about. He certainly failed very few. McLuhan's rationale was that it was the job of the admissions department to keep out poor students, not his to flunk them.

In this and other ways, McLuhan was far from being the conventional pedagogue. He rarely told students what to do; it was up to those who were serious to do something themselves, McLuhan believed, and it was perhaps up to him to nudge, tickle, or provoke them in the right direction. In a way, McLuhan had profound respect for these students. He sometimes terrified them by assuming they knew things they did not know or by taking their comments and questions more seriously than they themselves did. "If you made a remark you might find yourself in the situation where he'd say, 'Well, maybe we should do a book on this,'" one of his former students, Father Robert Madden, recalls.[13]

But McLuhan really did tend to see his students as, in a sense, de facto collaborators. "His relationship to his classes was very different from that of any other teacher I had ever encountered," another former student, and later a colleague in the English department at St. Michael's, Frederick Flahiff, maintains. "At the deepest level, aside from all the authoritarianism of McLuhan — and there was a good deal of it in him — one always felt that what he enjoyed and thrived on was a sense of common participation. He didn't feel himself leading a class so much as being involved with it in a common pursuit."[14]

To assuage his students' worries, McLuhan usually spent the last few class meetings talking about the rhetoric of exams. When a student mastered this rhetoric, he had no need to study for exams, McLuhan assured them. The key was to play with exam questions. One could talk about the question, break it down into subquestions, respond to it by inventing two contrary opinions and then comparing and contrasting them, and so on. One could make up one's own authorities, citing nonexistent books and authors if need be. One could, in short, organize one's ignorance and give it all the solidity of knowledge. These lectures were meant both to satirize exams — McLuhan believed that they paralyzed independent thinking — and to reassure his students that they did not need to fear them. The lectures, however, did nothing to ease the suspicion and hostility of other professors.

It was a pity, because the last thing McLuhan wanted to be was a troublemaker — except in a controversy where genuinely important issues were at stake. He had no real interest in academic politics. As far as the academic chores of his department were concerned, moreover, McLuhan usually proved cooperative. One of the heads of his department at St. Michael's called him "absolutely docile" in accepting assignments.[15]

McLuhan's habit of exaggerating the hostility of his environment — which certainly existed — somewhat obscures the fact that he really did enjoy many aspects of his life at St. Michael's. He and his family settled into a comfortable house on campus when they arrived, for which they paid the college $65 a month rent. In 1950 they moved into a second campus house, which had once served as the college infirmary. They stayed there for six years, the longest period McLuhan had lived at any one address since Gertrude Avenue in Winnipeg. The Toronto house was typical of the places he had lived in since he'd begun raising a family: it was slightly rundown (the front door didn't close properly and the doorbell didn't always work), and it was utterly lived in. Books were piled everywhere, prints of works by the French Catholic artist Georges Rouault were taped to the walls, a pad of paper was fastened to the kitchen wall with the saying of the week lettered in felt-tip pen for contemplation by the family. The children even scribbled on the walls from time to time, apparently with impunity. The McLuhans were not house proud. A friend who helped them move from a later house recalls, "There was nothing left of the house. They damn near destroyed it."[16]

The atmosphere in the McLuhan home in the old infirmary was quite pleasant, however. The living room had the coziness of comfortable spaces that are poorly lit and, above all, it had a fireplace. McLuhan, who enjoyed roaring fires even in the summer, doted on the familial hearth. "Fire is thermal, visceral and auditory," he once explained. "With a fire burning you don't have to worry what to say next, which in a highly visual culture like ours is a frequent concern to some people. Fire is a kind of dialogue."[17] In later years, he was a keen advocate of artificial fire logs.

Within this domain, Corinne McLuhan did her best to raise her children. A third daughter, Stephanie, had been born to the McLuhans in October 1947, and a fourth, Elizabeth, was born in August 1950. Their sixth and last child, Michael, was born in October 1952. Caring for these children was, in a way, a heroic endeavor on her part, given her husband's limited assistance, the small amount of time he devoted to the needs of his children, and his inability to understand and sympathize with their emotional lives.

Because he could be impatient if things were not done properly or if the children seemed to engage in willfully bad behavior, Corinne ended up simply not telling her husband about many of the things the children did. She herself became virtually submerged in the role of housewife. Seldom seen without her apron, she continued to provide excellent meals for her brood, despite the family's limited budget, and to iron her way through baskets of laundry, determined that her four daughters would, at all costs, look "nice." She tried to control these children as best she could; a priest at St. Michael's remembers one vivid scene of this "tall, long-limbed Texan chasing after the two little twins who were running ahead of her across the baseball diamond on campus."[18] Mostly, she was concerned with making sure the children were quiet during the time their father was working at home.

Eric, sometimes referred to as Ear Ache, continued to be the most difficult of the children. A quirky, almost disturbingly serious child — his father thought that he lived much more intensely than he himself ever had[19] — Eric seemed perpetually in trouble. He would try to sneak out the back door, for instance, when his father insisted that the family say evening prayers for an hour around the parental bed. (His children were not the only ones dragooned into prayer. As Pauline Bondy recalls, "After dinner, it didn't matter who was there, everybody had to get down on their knees and say the rosary. Nobody ever thought of contradicting Marshall. If he said we'll do it, then we did it.")[20]

McLuhan seemed to churn up the atmosphere wherever he went, to increase the electrical charge in the air by his very movements. As one colleague put it, "There's a machine inside him that never stops. He never seems to tire. After three or four hours of conversation, he's still fresh while others are limp with exhaustion."[21] To maintain this formidable energy, McLuhan frequently had steak and potatoes for breakfast, occasionally even consuming his meat raw. He received other dietary tips from his mother, who moved to Toronto in the 1950s. By then a Rosicrucian and health faddist, Elsie McLuhan — who was still quite capable of creating her own electrically charged atmosphere — introduced her son to the virtues of distilled water and organic bread.

All the McLuhans had a weakness for this kind of holistic fare and the home remedies that usually go along with it. In the late forties McLuhan hit upon a coal oil and turpentine mixture as a cure for the indigestion that had long been the bane of his life. Later he employed an electric gadget called the Tension Master to massage the top of his spinal column. He claimed that this device, which he fondly dubbed Thumper, stimulated his trapezius muscle and his cerebral cortex and also prevented lung congestion and bronchial coughs.[22]

McLuhan's salary as an associate professor of English at St. Michael's in 1946 was $4,200. As more children arrived, that salary became increasingly inadequate. To save on grocery bills, McLuhan occasionally bargained with the manager of the nearby Pickering Farms supermarket for a discount on bulk purchases and latched on to bargains with a childlike enthusiasm, even if the bargains were for commodities, such as a case of ripe olives, that were not exactly foremost on his shopping list. Although he disliked automobiles on principle, McLuhan was forced to obtain one, choosing a twenty-year-old Plymouth that he habitually patted on the hood as a kind of fetishistic gesture to prolong its life. He proved to be a menace on the highway, and it was a great relief to his friends when he stopped driving altogether in later years.

Even when he became wealthy in the late 1960s, McLuhan retained a charmingly naive attitude toward finances. He gave his wife all his funds (he thought this was the "traditional European" way)[23] and she in turn doled out his pocket money. Rich or poor, he continued to watch his nickels and dimes and keep a sharp eye out for bargains, however trifling. He was also one of those who automatically pocket restaurant receipts, for "tax purposes."

At the same time, he remained astonishingly generous. In innumerable instances he lent and sometimes gave away money to friends and to worthy causes even in his most poverty-stricken years. When an appeal was made to the Christian ethic of charity, McLuhan was particularly susceptible, and not merely in monetary matters. He made a point of appearing every Sunday at a nearby nursing home, for example, to read to some of the patients.

To escape from poverty, McLuhan dreamed of various projects. One of the more exciting resulted when Cleanth Brooks asked him to visit the University of Chicago, where Brooks had just spent six months as a guest professor. Brooks had left that university disappointed in its rigorously "logical" approach to literary criticism. Robert Hutchins, then president of the university and ally of Mortimer Adler in the Great Books approach to education, had asked Brooks, whose reputation as a literary critic was then rising steadily, to speak to him personally about what might need to be changed. Brooks agreed on condition that McLuhan accompany him. Brooks had been enormously impressed by McLuhan's Cambridge thesis and its historical review of the trivium. That account, he felt, had a direct bearing on what was wrong with the University of Chicago.

In June 1946 McLuhan went to Chicago with Brooks and argued the case for change. Basically he told Hutchins that if he wanted to produce graduates who were capable of "learned eloquence" instead of mere technical expertise — as Hutchins had told the world he was going to do — then he would have to do something about his deans, who were all complete dialecticians, strangers to rhetoric and grammar. "Marshall spoke brilliantly," Brooks recalls. "He put into what seemed to be very cogent, lucid English, in concise form, what I was thinking in inchoate form after six months of teaching there. He who had never set foot on that campus before had it all beautifully diagrammed."[24]

McLuhan returned from Chicago with more than his usual share of optimism, sat down, and wrote a 10,000-word brief expanding on what he had said at the meeting. Ernest Sirluck, a colleague at the University of Toronto who saw that brief, was amused many years later to note that it "was an enormous piece of special pleading, the most blatant preaching for a call."[25] McLuhan, in a not so subtle fashion, urged Hutchins to hire him and a few friends to take over his program in the humanities. As McLuhan wrote to one of those old friends from St. Louis, Felix Giovanelli, Hutchins was being offered a plan to revamp his university that would allow McLuhan, Brooks, Muller-Thym, Giovanelli, and others to form a cadre of New Critics and like-minded metaphysicians on his campus. The "circle of fine intellects" McLuhan had been dreaming of since his undergraduate days in Manitoba would be given an institutional power base and institutional prestige.[26]

For a man whose mode of originating and transmitting insights was not through the writing of books but through conversation — for a man who also worked instinctively through coteries — it was vital to seize whatever platforms might be available, even if it meant that colleagues accused him of being an opportunist or a charlatan. In this case, McLuhan's personal self-assurance was augmented by his conviction that Catholics, with their coherent outlook and internal spiritual supports, could assume easy intellectual mastery over non-Catholic colleagues.[27]

This time, however, like many others, McLuhan's assurance got him nowhere. Hutchins ultimately resisted his proposal — he was planning to leave the university shortly anyway and had no intention of overhauling it before he departed. McLuhan was left to take revenge by writing an essay, never published, denouncing the University of Chicago's instruction in literature. In that essay, McLuhan cited, with devastating effect, some examples of multiple-choice questions on *Hamlet* and *Tom Jones* from University of Chicago exams. These examples, he claimed, were symptoms of the university's "naive rationalism," the kind of rationalism that reduced literature to categories and concepts instead of training students to sharpen their perceptions of literature through the analysis of the effects sought by writers.

The argument in this essay, like that of his oral presentation to Hutchins, was a variant of a theme McLuhan had pursued since his Cambridge thesis. According to this theme, the rational, or dialectical, mind must be controlled by the other two disciplines of the trivium: by rhetoric, which emphasized the relationship between the author and his audience, and by grammatica, which emphasized verbal analysis. This argument also reflected the ongoing war between the New Criticism and other forms of literary scholarship, since verbal analysis, in McLuhan's mind, meant analysis along the lines of Empson and Richards. As he wrote in another context, in words that echoed T.S. Eliot's definition of "auditory imagination," it was analysis that treated "words not merely as signs but as things with a mysterious life of their own which could be controlled and released by establishing exact relationships, rhythmic and harmonic, with other words."[28]

As part of his campaign to promote his insights, McLuhan continued to dream of starting a new magazine. Writing to Giovanelli in 1946, McLuhan promised that his magazine would focus "on the hatred of being manifest everywhere."[29] The word "being" was a tip-off that the magazine would be an organ for analogical thinkers and Thomists, who talk about Being the way Marxists talk about History. In pursuit of his dream, McLuhan attempted to see no less a personage than Henry Luce, the man whose publications — *Time, Fortune, Life* — McLuhan occasionally subjected to scathing analyses. That is, at least, the story that circulated at the time. If McLuhan and Luce ever did get together in the forties, there is no record of such a meeting — although later, in the early fifties, McLuhan did write to Henry Luce, demanding $50,000 in return for his communicating to Luce information on the trends that were transforming society.[30]

This kind of audacity was beginning to turn McLuhan into a legend as early as 1946, particularly among a certain circle in New York. Seon Manley, then an editor with Vanguard Press in New York and a friend of Felix Giovanelli's, recalls that a McLuhan coterie was already beginning to form, consisting of literary intellectuals with mainly Catholic backgrounds interested in James Joyce and the complexities of human language. Members included Giovanelli, an old St. Louis associate and contributor to various periodicals named John Farrelly, and the author James T. Farrell. The group was a counterpart of the much more celebrated *Partisan Review* clique. "It was a group of people who always had a good time together," Manley comments. "There was no sort of rivalry or over-intellectualization. It was just very spontaneous — much more of a European café group than the *Partisan Review* group. Nobody enjoyed themselves in the *Partisan Review* group. That was very bad form."[31]

As an editor at one of the more literary New York publishing houses, Manley saw members of both groups from time to time. Occasionally representatives from these two distant intellectual planets even ran into each other in the corridors of Vanguard. "I remember one unbelievable meeting with Paul Goodman, who was also an author I worked with, and McLuhan. Of course, neither one of them understood *anything* the other was talking about."[32]

The gulf between the two groups was highlighted in 1949 when the controversy broke out over the awarding of the first Bollingen Prize to Ezra Pound; in the words of Leslie Fiedler, "everybody's deep politics began to show." McLuhan's "deep politics," and those of his friends in New York, were not altogether unsympathetic to Pound's. In any case, McLuhan aligned himself with the Southern Agrarians, people such as Cleanth Brooks and Allen Tate (who also tended to be "antiprogressive" in their thinking) in their defense of Pound's poetry. McLuhan viewed the New Criticism of Brooks, Robert Penn Warren, and others as a high school version of the criticism of Richards,

Empson, and Leavis, but he shared their basic attitudes to politics and literature, and he was cheered by Yale's hiring of Brooks, which he regarded as one up for his side in English studies. He certainly stood with the New Critics against the left-liberal polemics of *Partisan Review* critics such as Irving Howe and William Barrett in the case of Ezra Pound.

It was not surprising, then, that Giovanelli, in a letter to McLuhan in 1949, termed *Partisan Review* "the most pestilential menace to autonomous art and free intelligence in this country," after its comments on the Bollingen Prize. He added, "Those guys have displayed about as complete a contempt for literature as it is possible for a bunch of political lonely hearts to display" and suggested that a counterattack be launched not only in the Southern Agrarian/New Critical journals like *Sewanee Review* and *Kenyon Review* but in *New York City*, a magazine edited by a former *Partisan Review* editor, Dwight MacDonald.[33] Giovanelli, who acted in the late forties as McLuhan's tireless (and unpaid) agent and promoter, seems to have known anyone of any intellectual importance in the city — even MacDonald, who was far from sympathetic to the ideas of the McLuhan circle. Nonetheless, Giovanelli did occasionally show MacDonald the work of his Canadian friend.

Giovanelli continued, in general, to encourage McLuhan's restless ambitions. He urged that they continue exchanging ideas, thinking up new projects, knocking on more doors, harassing more people. *Something* would happen as a result, surely.[34] McLuhan did not need much encouragement; so eager was he to pursue his projects that he occasionally felt the urge to move to New York City, despite his oft-stated preference for the vantage point on Western civilization provided by the Canadian boondocks. He would have liked to be in the fray directly and to be within easy reach of the offices of publishers and the editorial departments of magazines. He felt toward the city of Toronto much the same way he felt toward the University of Toronto — it was an irritant that removed any temptation to complacency. In Toronto, McLuhan felt, culture was considered to be "basically an unpleasant moral duty."[35] The locals made up for their sense of being outside any important intellectual life by maintaining a kind of surly dogmatism.

Although McLuhan had a habit of exaggerating the hostility of his environment, he was not far wrong in his estimate of Toronto. In the postwar years, the city was characterized by charmless architecture, omnipotent banks and life insurance companies, Sundays in which no life ever stirred, and a citizenry who regarded not only culture but life itself as a grim duty.

Nonetheless, McLuhan cast about for a "circle of fine intellects" in this unpromising milieu, to duplicate the one Giovanelli was involved with in New

York City. One of his first and most promising recruits was a graduate student named Hugh Kenner. Kenner had just finished his M.A. at the University of Toronto in 1946 when he met McLuhan. During that summer they saw a great deal of each other, and over the next two years they continued to meet, although Kenner was based in Windsor, having taken McLuhan's old job at Assumption College. When they got together they often spent ten hours a day simply talking. "He was, of course, unlike any mentor I'd had," Kenner recalls.

> He pushed at me T.S. Eliot, who'd been the type of unintelligibility to my Toronto profs. And he had me read Richards' *Practical Criticism*, Leavis's *New Bearings in English Poetry*, and (eventually) the entire file of *Scrutiny*. He kept mentioning Wyndham Lewis, whom I'd never heard of, notwithstanding that for two years I'd lived half a mile away from him. . . . So many windows opened![36]

Kenner became a McLuhan disciple; and McLuhan, through his connections with Brooks, got Kenner into a Ph.D. program at Yale. Writing to Giovanelli, McLuhan expressed his fears for Kenner among that group of students and professors who were much less bold and insightful than his young friend. He knew that Kenner was brilliant and that Kenner knew he was brilliant, and he was afraid Kenner would not have sense enough to keep his mouth shut and get his "union card" (his Ph.D.) by giving the impression that he was actually learning from his professors. He was afraid Kenner would try to educate those professors. Behind all this concern was McLuhan's feeling that he had been isolated in his own academic career because he had never learned to keep his own mouth shut and to betray no surprise or frustration at bumping up against endless and incurable mediocrity. He had a genuine sympathy for what he felt was Kenner's loneliness and lack of contact with people who could smooth his way — a state similar to what his own had been.[37]

Kenner, meanwhile, continued to correspond with McLuhan from Yale, providing caustic accounts of the intellectual and temperamental follies of his professors (some of the biggest names in English studies) and trying to reassure McLuhan that he was not divulging any of the material from his Cambridge Ph.D. thesis, of which McLuhan had loaned him a copy. (Ironically, Brooks, who too had been loaned a copy by McLuhan, was also showing the work to all of his colleagues at Yale.) McLuhan hoped someday, when his intellectual prestige was sufficiently established, to get his thesis published, and he lived in mortal fear that its insights would be purloined before that could take place. The thesis took on the aura of a treasure map he had to keep from the clutches of pirates.

If McLuhan opened windows for Kenner, Kenner may have led him to at least one rather large window, too. That was the window labeled JAMES JOYCE. As Kenner recalls, "McLuhan despised [Joyce] as merely 'mechanical,' a 'contriver.' McLuhan in those days took the Leavis line on nearly everything, though he did smuggle in Wyndham Lewis. He told a mutual friend (Pauline Bondy) that I was 'wasting my time' on Joyce."[38]

McLuhan had, in fact, read Joyce at Cambridge with great interest, and when *Finnegans Wake* was published he made sure he obtained a copy. The latter work seems not to have made a great impression on him, however, before his arrival at the University of Toronto. Whether it was through Kenner's influence — Kenner had decided the year before meeting McLuhan that Joyce was "the supreme nut to be cracked"[39] — or whether his curiosity was piqued for some other reason, McLuhan began to read Joyce very seriously around 1950, particularly *Finnegans Wake*. His method, he wrote one former student, was to go through three or four pages a day, reading aloud in an Irish brogue (McLuhan fancied he had a talent for mimicking accents) and using Robinson and Campbell's *Skeleton Key to Finnegans Wake* as a guide.

Very soon McLuhan, characteristically, became obsessed with Joyce. "The atmosphere was Joyce inundated," a friend recalls of McLuhan in the early 1950s.[40] McLuhan, who loved mysteries and puzzles and thought of himself as a sleuth perpetually on the track of breakthroughs (toward the end of his career he kept a chart of his major breakthroughs on the wall of his office), found in the impenetrable text of *Finnegans Wake* the ultimate mystery. He claimed that his investigation of it yielded some of his most important discoveries in media studies — but whether he really made new intellectual discoveries there or whether he read into Joyce notions and themes that were at least already half formed in his mind from other sources is very difficult to say. Almost anything can be inferred from the multileveled puns of *Finnegans Wake*, and no particular interpretation of that text can easily be shown to be nonsensical or misleading.

Whether Joyce supports the notion, for example, that all human artifacts are extensions of the human body (a notion articulated in McLuhan's *Understanding Media* that probably owes more to McLuhan's reading of the anthropologist Edward Hall); whether *Finnegans Wake* is, in fact, devoted to a monumental history of human technology;[41] whether McLuhan learned the importance of what he later termed the "resonant interval," the space of intervals, or the space characteristic of the sense of touch from *Finnegans Wake* — none of these conjectures is, from the nature of the case, easily verifiable. (It is beyond dispute that Joyce deals, in his own way, with the whole of human history in the *Wake* and with the various ways in which human beings have communicated, including radio and television. To that extent, certainly, McLuhan was not wrong in seeing Joyce as an explorer of the media.)

Finnegans Wake also reinforced McLuhan's suspicion that a logical, sequential approach to reality grew out of a primarily visual imagination, as opposed to T.S. Eliot's "auditory imagination." In the auditory imagination, words resonated with levels of meaning linking, as Eliot said, the most modern with the most primitive; and whatever else it may be, *Finnegans Wake* does present a view of language as affording, in McLuhan's words, "an unbroken line of communication with the totality of the human past."[42] In that sense, the book certainly shatters the habit of seeing human history as a linear and sequential process by showing both past and present as present in the medium of human speech — not speech as nailed down in print, with one or two dictionary meanings attached to each word, but speech the way Eliot and the ancient grammarians saw it, echoing with a rich and elaborate and meaningful history.

Joyce's work as a whole also reinforces the psychology of sensory activity that McLuhan had already absorbed from Thomas Aquinas via Muller-Thym. Joyce, perhaps the greatest Thomist of this century, reminded McLuhan that the senses are a form of reason; they recapture, by analogy as it were, the forms and movements of reality in the human mind. Therefore McLuhan felt that to leave *any* of the senses out of one's perception of something was literally irrational. And the sensory preferences of an individual were not a mere idiosyncrasy but a clue to his very character.

As a result, McLuhan's studies of media tended to emphasize the sensory effects, but not sensory effects as measured, say, by a clinical psychologist. His repeated and baffling insistence that television is "tactile" rather than "visual" begins to make some sense when his Aristotelian and medieval theory of sensation is recalled. In that theory, the senses are agents operating in the mind as a kind of artistic collective, working together in a very delicate balance to recapture reality. When the media upset that balance, disaster strikes.

During their years of intellectual collaboration, Kenner and McLuhan discussed many more figures than Joyce. McLuhan was a great fan of Al Capp, creator of the comic strip "L'il Abner." At one point, when Kenner was at Yale, Capp had the characters L'il Abner and Daisy Mae on the brink of marriage. McLuhan thought this was an excellent opportunity for Capp to satirize Dagwood and his mate the way he had satirized Dick Tracy with the creation of Fearless Fosdick. McLuhan urged Kenner to grab some impressive Yale stationery and write to Capp with the idea.

In June 1948 McLuhan and Kenner visited Ezra Pound, then incarcerated in St. Elizabeths Hospital in Washington, D.C. McLuhan never saw Pound again but corresponded with him for several years afterward. The visit and correspondence inspired him with yet another idea for a book, a critical introduction to Pound that would present his work "to the lazier readers of our time," as

McLuhan explained to the poet in a letter.[43] He also continued to be impressed by Pound's critical books such as *ABC of Reading* and *Guide to Kulchur*, in which Pound promoted works he believed would provide a student with crucial standards of taste and perception. As McLuhan wrote to Giovanelli at that time, every sentence in those books sent him "scurrying to fifteen books. To reading twenty poems *aloud*." Under the inspiration of Pound's critical prose, McLuhan conceived the idea of a *Guide to Kulchur* of his own. Such a work would list important books for the student, works to whet the appetite for, and enable the mind to recognize, the truly "relevant" in arts and sciences. To that end, he asked Giovanelli to ask every expert he met for the one or two "indispensable" books in his area of expertise.[44]

Of all the twentieth-century literary giants, Pound was closest to being a soul mate of McLuhan's. Not that the two men felt close to each other or that Pound ever recognized McLuhan's intellectual contributions as being parallel to his own — indeed, he rather brusquely dismissed McLuhan's magazine *Explorations* when copies were shown to him in the mid-fifties, and he had no use for McLuhan's later obsession with "acoustic space," and "visual space." Both men had a certain horror of popular culture but defended themselves against it in very different ways: McLuhan by studying that culture, Pound by resolutely ignoring it.

Nonetheless, the two men functioned in the same way. They were both rather isolated figures who desperately attempted to circulate information and insight among a network of colleagues, who saw themselves as sleuths on the track of the real lowdown on vital subjects (ranging from the Elizabethans to the workings of high finance), who complained bitterly that important clues had been kept from them because of general indifference and obtuseness, who were both convinced that if they could get the ear of the right people they could change the world, and who occasionally succumbed to paranoiac interpretations of that world. Both men had a tendency to jump quickly from point to point without visible connection and to toss off huge ideas in a few brilliant phrases. They were both born talkers who ended their careers in silence — in McLuhan's case, involuntary silence.

McLuhan's admiration for Pound never ceased. As in *Finnegans Wake*, the very obscurity, the constant gnomic references in Pound's work appealed to McLuhan's sense of sleuthing in arcane lore. When Brooks mentioned to McLuhan that he thought Pound was crazy, McLuhan replied cryptically, "Crazy? Crazy like a fox."[45] He also prayed constantly for Pound's conversion to Catholicism and was particularly gratified to hear of Pound's desire, expressed shortly before the poet's death, to be buried in a Benedictine church. (McLuhan could hardly bear to think of his other idol, Joyce, dying unreconciled to the

Church.) It was an unspoken article of faith with McLuhan that all great artists were really Catholic, either overtly or in their secret sympathies. Those who could not, by any stretch of imagination or conjecture, be termed Catholic, like Milton or Samuel Beckett, were hopeless cases.

For a year or so after Kenner began studying at Yale, he and McLuhan continued to discuss joint projects. The two thought of themselves as "champions" of the twentieth-century greats — Lewis, Pound, Eliot, Joyce — against the modern enemies of human awareness — the aesthetes, the Marxists, the dialecticians, and so on. They planned to do a book on Eliot together and an anthology of some of their critical essays. Neither project was realized. Writing books was never really McLuhan's forte. Kenner, on the other hand, began publishing books and articles even before he left Windsor for Yale. By 1971, when he published *The Pound Era*, he had firmly established himself as the leading critic of the "men of the Vortex" — Ezra Pound and his friends.

McLuhan became increasingly upset as he saw more and more of what he considered his original insights in Kenner's prose without adequate acknowledgment of their source. "I have fed Kenner too much off my plate," he told Brooks.[46] At the end of 1948, he wrote Kenner a sharp letter on the subject. Kenner replied in his defense, although not to McLuhan's satisfaction. Giovanelli, who by this time knew both Kenner and McLuhan, sided with his old friend, referring to Kenner as a "phonograph record" of McLuhanisms, but adding, in a letter to McLuhan, that the loss of Kenner as a collaborator would be a powerful blow to McLuhan. No one else understood so well, Giovanelli commented, the mental world of McLuhan. Kenner's comprehension of that world, in fact, was "uncanny."[47]

Giovanelli was right; McLuhan needed somebody like Kenner badly. McLuhan's accusations of plagiarism against Kenner were the kind that would have poisoned any hopes of real collaboration between the two men, however, even if the accusations had been entirely just. Kenner possessed the ability to absorb thoroughly, retain to the point of total recall, and simmer in the juices of his own brain anything he had heard or read; to suggest that this was plagiarism was almost petty. McLuhan was very much like him, in fact. When he quoted reams of *Scrutiny* material to his classes in St. Louis without credit to the source, he was no more plagiarizing than Kenner was stealing from him. McLuhan's touchiness on the score of Kenner's "thefts" might owe something to the unfortunate influence of Wyndham Lewis, who felt that his contemporaries were continually purloining his ideas.

Later McLuhan displayed a certain generosity to Kenner, giving his first book on Pound (*The Poetry of Ezra Pound*) a friendly review in 1952 and praising his "spectacular intellectual performance" in a letter to Kenner's employer, the University of California, in 1967.[48] He recommended Kenner's books to his students without mentioning anything about Kenner's indebtedness to him; in private, he sometimes suggested to friends that it would have been nice to have more acknowledgment from Kenner but that he was glad, anyway, that somebody had developed his ideas and written the books he had never gotten around to writing. (Kenner did in fact acknowledge McLuhan's influence, more than once, as in the preface to Kenner's 1959 book on Eliot, *The Invisible Poet*.)[49]

Kenner's defection did nothing to alter McLuhan's instinct for working with coteries. The problem, he wrote to Pound, was rounding up "10 competent people" in one city, such as Toronto.[50] In reality, a person did not have to display particular competence to join a McLuhan coterie — only a willingness to take phone calls at 4:00 a.m. to listen to McLuhan hold forth over some insight he had just arrived at or even to tell a joke he had just heard. (This disconcerting habit, which increased with age, was a function of his irregular sleeping patterns. He made up for lost sleep at night with catnaps during the day. Like many extremely active people, he could fall asleep with no trouble and frequently interrupted conversations with friends by lying down on his office couch and nodding off for ten minutes. Faculty meetings were prime occasions for these catnaps, as were movies or plays.)

McLuhan's unappeasable appetite for an audience was the chief motive behind his seeking collaborators. "Wherever I am, there's a discussion group," he once told a reporter.[51] He was particularly attached to the faculty lunchroom at St. Michael's. "Very often, if he knew something would pique you he would bring it up just to start a fight," one colleague recalls. "He couldn't stand a silent table, and he would hold forth just to get something going."[52] This ability to create a perpetual conversational stir around him explains why he once claimed, "I have never been lonely in my life for one minute." In the same breath, however, he would add that he was a "loner" — the endless discussions never penetrated a certain august personal isolation.

McLuhan's graduate students formed a natural pool of potential collaborators. Many found his courses a refreshing break from the usual graduate seminars, in which students took turns reading papers on subjects no one else was interested in and the professor used the occasion to relax and suspend all activity except judgment. On the other hand, some graduate students were repelled by McLuhan's monologues. Ann Bolgan probably speaks for many of McLuhan's graduate students in the early fifties:

> [He] was never dull. He would establish a connection which you
> hadn't thought of yourself and that was stimulating, but it was never
> anything you could get your teeth into and there was never sufficient

pause to say, well, let's think of the implications of that, or where
would that lead, or how would that have come about. So I got stimulus
from him, but no training in anything, and I was constantly haunted
by a sense of chicanery.[53]

Toward the end of his career, McLuhan sometimes became very quick to
dislike certain graduate students. He failed some of them despite his earlier generosity as a marker and despite the fact that outright failures were virtually
unheard of in University of Toronto graduate courses. Students who knew how
to handle him fared well, however. Regarding oral exams, one student recalls,
"He gave us a reading list six to eight weeks before the exam, consisting of
about a thousand books — so you had to be clever about the exam. The trick
was to get him to do most of the talking."[54]

This was not, of course, difficult, especially as McLuhan had heard quite a
bit about those thousand books by the end of his career. Occasionally he
described the oral exam to a student as a "casual chat in which you will force
upon me your knowledge of the texts."[55] Given McLuhan's intolerance for
being told what he already knew, he was barely kidding.

From 1946 onward, McLuhan began to surround himself with a coterie of
graduate students — dubbed by outsiders as "McLuhanatics" — at the University of Toronto. The existence of this coterie doubtless consoled McLuhan for
the fact that, as Ann Bolgan noted, he was an "outcast" among his colleagues.
As time went by, members of the coterie also became useful as filters for, or substitute readers of, the growing numbers of periodicals and books sent McLuhan
by his correspondents. Occasionally one of these graduate students, like Kenner,
really did open a major window for McLuhan. In the late forties, Marianna
Ryan, a graduate student in comparative literature, gave McLuhan a guided tour
of the French Symbolist poets, the subject of her thesis.

McLuhan, who was too impatient to master the French language, thereafter never ceased citing Mallarmé and the Symbolists as key influences in his
thinking. McLuhan had known since Cambridge, in a general way, of the significance of the Symbolists from his readings of Pound and Eliot, but Ryan's
work heavily reinforced the basic notions McLuhan associated with them — for
example, the notion that the content of a work of art was really its technique or
that studying the effects of things on the mind was more important than studying ideas.

McLuhan's knack for creating discussion groups wherever he happened to
be and his hunger for information from people meant there was a lot of traffic
in his home. "His house on St. Joseph Street was an ongoing seminar room,"
Lawrence Lynch recalls. "Everybody who knew Marshall and Corinne felt that

he could just drop in."[56] Not satisfied with the haphazard or individual exchanges he enjoyed with colleagues and students, McLuhan organized on the St. Michael's campus in the late 1940s semiformal discussion groups on various subjects. One group was formed to study Greek, another (after 1950) to read *Finnegans Wake*. Still another was organized to discuss specific books that Lynch and McLuhan felt were particularly influential on modern thought. Many of these books were by contemporary psychologists and anthropologists.

McLuhan insisted the Church had to study such works because they lay at the heart of twentieth-century thought.[57] Just as Aquinas had used Aristotle, so contemporary Thomists ought to use explorers of the emotional life, such as Karen Horney and Carl Jung, to help educate "emotional illiterates" — the Mortimer Adlers and Robert Hutchinses of the world.[58]

Occasionally, McLuhan used coteries for purposes other than strictly intellectual. As part of his plan to escape poverty, he joined with two of his non-academic friends in 1955 to form a company called Idea Consultants. The closer of the two friends was a neighbor and public relations man named William Hagon. McLuhan had formed a habit of dropping in on Hagon nearly every Saturday morning, partly to get away from his kids and partly to toss around ideas; Hagon was a good listener and a good storyteller, two qualities McLuhan prized in his friends.[59]

McLuhan and Hagon's firm offered creative ideas to businesses and was based on McLuhan's cherished notion that an outsider to a business could often come up with solutions to problems that had evaded the "experts." One of their slogans, which has an authentic McLuhanesque ring, was "A headache is a million-dollar idea trying to get born. Idea Consultants are obstetricians for these ideas." The partners rented an office in downtown Toronto, designed a logo for their firm, and sent off proposals to a variety of companies. They submitted ideas for transparent potties for use in toilet training children; for illuminated panels in buses and subway cars that would flash the names of stops with a sponsor's name or advertisement attached; for sealed airline dinners made available for general consumers at a moment's notice; for package tours to pollen-free areas for hay fever sufferers. One startlingly prescient suggestion was for the manufacture and sale of taped movies for replay on television sets. (McLuhan called these tapes "television platters.")

The idea most characteristic of McLuhan's thinking was a television program that would select a business problem, dramatize it, and offer viewers a reward for the best solution. His notion was that the odds were better that someone in the millions of ordinary viewers would come up with the right solution faster than the most illustrious think tank.

Unfortunately, Idea Consultants never sold an idea. The firm was abandoned after a year or two. But its demise in no way diminished McLuhan's appetite for projects, which continued to absorb his time and attention.

Such absorption left him with little or no time for normal recreational pursuits. Perhaps the only nonintellectual form of recreation McLuhan ever pursued in his early years at St. Michael's was his old love of sailing. In Toronto he bought himself a not very seaworthy boat, which he once capsized in the Toronto harbor. (He and the two other occupants of the boat had to be rescued by the police.)

He also enjoyed going to the movies from time to time, although it was rare for him to sit through one from beginning to end. "There's only one movie I knew Marshall to have seen right through — Laurence Olivier's *Henry V*," Hugh Kenner recalls. "He would say, 'Let's check this out,' and we'd go into a movie theater and after twenty minutes he would say, 'That's enough.'"[60]

In fact, he seemed to have a good grasp of the language of film. Lou Forsdale, a friend he made at Columbia University Teachers College in the fifties, recalls, "I went to one movie with him, at which he fell asleep after twenty or thirty minutes. When he awoke after it was over, he talked with some accuracy about the structure of the movie, the organization and rhythm and sequence of shots and so on. He had picked it all up at the start of the movie. It was almost uncanny."[61] (In St. Louis he had introduced his students to the writings of the Soviet filmmakers Pudovkin and Eisenstein and had given them exercises in adapting novels to film scenarios.)

When television appeared in the fifties, the McLuhans were the last on their block to purchase a set. Even after the purchase, McLuhan used the device very sparingly. He was careful to put the set in the basement so that it would not dominate the living room. Each week he marked on a copy of *TV Guide* the programs he wanted to watch and stuck to those choices. He had a soft spot for lighter fare, such as "Hogan's Heroes" and "Bonanza." In the seventies, he was captivated by the "Bob Newhart Show," claiming that the program, in which characters frequently spoke to each other as they were getting into or out of the elevators outside Newhart's office, was the only one that understood the nature of the elevator and the spaces it created.

Talk, however, remained his paramount recreation. He planned his parties to maximize opportunities for good talk, often starting off a soiree by playing a comedy record (Tom Lehrer was a McLuhan favorite at the time). Once things got going, McLuhan would then direct the conversation along the lines of some theme he had chosen beforehand. "He always had an agenda in his mind and virtually set out the topics of conversation," Joan Theall, wife of one of his graduate students in the fifties, comments.

The entertainment was based on his self-dialogue of the week. It could
be anything from the latest phenomenon of popular culture to what
the recent literary reviews had been saying about some poet or painter.
He almost resented nonintellectual conversation. If he got to know
you very well he might speak about personal matters, but you could
always tell that he was making a very great sacrifice in doing it.[62]

McLuhan's conversational agenda was based on the themes that obsessed
him. In 1948–49 Mallarmé and the Symbolists were such a theme. Shortly thereafter and throughout the early fifties, he became fascinated with the epyllion, the
little epic. McLuhan considered the essence of this literary form to consist of the
interplay between plot and subplot and was convinced that in that interplay lay
the secret to interpreting Western literature. He started, of course, to write a
book on the subject — a book he was still working on a dozen or so years later.
In the meantime, he began to spread the word about the epyllion through his
personal network. At Victoria College, Northrop Frye, then beginning to
acquire his considerable reputation as a critic, recalls,

> His enthusiasms were so contagious, he would go off the deep end
> with some word like "epyllion," and for the next six months all you
> would hear around St. Michael's students was the word "epyllion." But
> he left it to other people to work out the scholarly implications. I think
> some of his students who did spend a long time trying to work out
> these implications felt they were getting out of that academic trough,
> when they were really falling behind the world that we're living in.[63]

McLuhan did not apologize for his attitude toward scholarship or for the
effect of his enthusiasms on his students. He was hot on the trail of clues to reality, and anyone who wanted to follow was welcome. He kept discovering so many
clues that he felt it impossible, in any case, to stop and elaborate on each one.

One vital discovery he thought he had made was the influence of secret
societies, particularly the vast and occult powers of the Masonic order.
McLuhan became obsessed with this discovery in the academic year 1951–52,
after he had studied Pope's use of Rosicrucian material and some of the similar
esoteric rituals and lore used in the writings of Yeats, Pound, Eliot, and Joyce.
McLuhan made the leap from noting that such lore persisted from the Renaissance to the twentieth century to positing that actual elite, secret societies promoted it in every area of contemporary life.

He was strongly attracted to occult lore himself and was also deeply super-
stitious, regarding the numbers three and seven with a religious awe. "He…

does everything in threes," his daughter Teri once commented. "If he has two letters to mail, he'll wait until there's a third."[64] He also nursed a traditional abhorrence of the number thirteen. He once balked at sitting down with twelve other people at a dinner party, and a secretary had to be rustled up to add one more to their number. McLuhan also called himself "a Cancerian, a moonchild, very affected by the behaviour of the moon."[65] During periods of full moon, he tended to brace himself for the worst, convinced that this lunar phase inevitably meant a frantically busy time for him.

As an undergraduate, McLuhan had attended a lecture on psychic phenomena such as telekinesis and ectoplasm and had noted that these subjects had a perilous fascination for someone with the kind of imagination he possessed. Such occult lore had to be shunned at all costs.[66] The observation was an astute one. Just how susceptible he was to this lore is demonstrated by the completeness with which he indulged his obsessions with secret societies and the occult after 1952. To Ezra Pound, he wrote that he had spent the year 1952 investigating the rituals of organizations such as the Masons and the Rosicrucians. To his surprise and overwhelming disgust he had discovered that these organizations had spread their tentacles almost everywhere in the arts and sciences. He expressed some disappointment that Pound himself (who must have been slightly bewildered by this outburst) had used occult rituals in his own poetry.[67]

Of course, it was emotionally as well as imaginatively satisfying for McLuhan to indulge this phobia. As one of his colleagues remarks, "McLuhan enjoyed constructing the world as a conspiracy. It was childlike fun, putting one over on the sneaks and the Masons."[68] Even opposition to him from colleagues at the University of Toronto could be explained as centered in a secret society or two flourishing in other colleges. (Since McLuhan tended to operate with a coterie, itself characterized by a mildly conspiratorial air, he assumed his antagonists did likewise, only more so.) He also began to feel that he had been a fool in his earlier writings, expressing his views without an awareness that those views were contrary to the interests of the secret societies. Now he was sure the adherents of those societies had become alerted to his presence; as a result, he was finding it more and more difficult to find anyone to publish him. In a letter to Pound, he quoted a remark Wyndham Lewis had once made to him: "The secret of success is secrecy." He finally understood the significance of that remark, he told Pound. (Pound wrote back with the sardonic reply "No harm in McL/succeding [sic] by secrecy if it don't degeneete [sic] into mere non-being.")[69]

The secret societies, McLuhan claimed, were characterized by a use of ritual and liturgy designed to put adherents directly in touch with occult spiritual forces existing in timeless patterns. For the societies, those forces and patterns were alone real, not the world of appearances. McLuhan characterized the philosophy behind these societies with varying labels: "gnostic," "hermetic," "Buddhist," "Neo-Platonic," and so on. They posed a deadly threat to the Catholic Church, which, along with Aristotle, insisted that the senses did not deceive the mind and that the material world was dependably real.[70]

The Church insisted that the means of salvation were not artistic or occult rituals, but its sacraments. The rituals of the secret societies tended, in fact, to be parodies of the sacraments, most notably in the case of the black mass. McLuhan was convinced that black masses were being celebrated all around him and that the "personal" columns in the Toronto newspapers contained coded messages regarding the times and places of these masses. (For some reason, McLuhan believed that they were frequently celebrated at Casa Loma, a huge, empty Edwardian mansion and Toronto tourist attraction not far from the university.) He became so vocal about the subject that two graduate students decided to pull his leg by making up a story about black masses being celebrated in the Egyptology section of the Royal Ontario Museum at midnight. McLuhan, in the simplicity of his heart, believed them absolutely.

The foremost of the secret societies, in McLuhan's view, was the Masonic order, historic foe of the Catholic Church. McLuhan began to see the history of the West as having been shaped in unknown ways by the works of the Masons. He assured one of the daughters of Bernard Muller-Thym that the American Civil War had really been a struggle between the northern and southern branches of the Masonic order. Watching John F. Kennedy's inauguration in 1961, he observed to a friend that the delay in taking the oath (most likely occasioned by a severe snowstorm) was really the Catholic Kennedy's way of not taking the oath at high noon, a time of great significance in Masonic ritual.[71]

The influence of the Masonic order in the "arts and sciences" was even more pervasive and was probably even responsible, McLuhan surmised, for the fact that many scientific names were in Greek rather than Latin. Needless to say, he held the workings of the Masonic conspiracy responsible for many of the setbacks that he endured. He felt the Masons were ultimately responsible, for example, for his failure to launch his cherished magazine and a private Catholic school (based on Stonyhurst, the English Benedictine institution) he had dreamed of establishing in the early fifties. They also controlled book reviews in important periodicals, he felt.[72]

McLuhan had sense enough not to express these views in public, and by the 1960s he pretty well kept quiet about the Masons even in conversation with friends and colleagues at St. Michael's, although, appalled at the post-Vatican II changes in the Catholic Church, he occasionally voiced the suspicion that some prelates of the Church, including Toronto Archbishop Philip Pocock, were secretly Masons. He seems never really to have abandoned his thesis about a

Masonic conspiracy and even wondered occasionally whether some of his own employees at the Centre for Culture and Technology (established by the University of Toronto as a McLuhan research center in 1963) might be Masons.[73]

He certainly never abandoned his belief that his great rival in the English department of the University of Toronto, Northrop Frye, was a Mason at heart, if not in fact. Frye's theories of literature, his solemn dedication to myth, symbol, archetype, and so on, gave McLuhan a good deal of ammunition for this view. The omnipresence of great overarching categories in the Frye school of literary criticism, aside from its tendency to put literature into a series of intellectual straitjackets, is closely linked to the timeless, unearthly visions of gnosticism and Neo-Platonism that McLuhan detected in the secret societies.

Frye's critical ideology also tended to make a religion of literature, a religion that McLuhan, who already had one that he considered perfectly satisfying, most heartily detested. Of course, McLuhan himself was sometimes suspected of ruthlessly imposing intellectual patterns on the real world and, in a philosophical sense, of being more impressed by ideas and forms than by reality. But however much he may have strayed in practice from the notion that all the particulars of God's creation were unique and radiantly existent, he paid a certain undying allegiance to that notion.

Understandably, McLuhan's personal relations with Northrop Frye became rather strained as the years passed. In 1961, in a letter to a close friend, McLuhan mentioned Frye's leaving Toronto for a conference and added that he hoped Frye would not bother to return. He was only half joking.[74] When Frye's *Anatomy of Criticism* appeared in 1957, McLuhan longed to reply to it but was afraid that it would look like personal rivalry or jealousy on his part and that a reply from him would indirectly boost Frye's prestige.

A panel of graduate English students was organized by the Graduate English Association at the University of Toronto to discuss Frye's book shortly after its publication. One of the panelists, Frederick Flahiff, recalls, "One morning after the announcement of the panel had gone out, Marshall appeared in my room carrying a copy of [an] essay entitled 'Have with You to Madison Avenue; or, The Flush Profile of Literature.'" The essay, written by McLuhan, was an attack on Frye's criticism, maintaining that it represented a switch from the old notion of criticism as the formation, via literature, of a perceptive mind to a pseudo-scientific charting of the features of literature vaguely analogous to Madison Avenue profiles of consumer groups. ("Flush profile" is a reference to a method of measuring viewer response to radio and television programs by gauging the incidence of toilet flushing.)

McLuhan was not at his best in this essay. His argument, studded with tortured metaphors, was extremely convoluted and would have succeeded in confusing any audience, no matter how well versed in Frye's book. One thing was clear, though: no one but McLuhan could have written it. Nonetheless, McLuhan asked Flahiff if he would read the essay on the panel as if it were his own response to Frye. "We went out and walked around and around Queen's Park [a small park near the St. Michael's campus]," Flahiff recalls.

> McLuhan was at his most obsessive. I don't mean that he was hammer-
> ing away at me to do this thing, but he was obsessive about Frye and the
> implications of Frye's position in the same way that he had talked about
> black masses. It was the first time I had seen this in Marshall — or the
> first time I had seen it so extravagantly. As gently as possible, I indicated
> that I could not do this and that I was going to write my own thing.…
> Later, on the night of the panel, he phoned me before my appearance
> and asked me to read to him what I had written. I indicated that he could
> come to the session if he wanted, but he said, "Oh, no, no."[75]

Doubtless there was at least a hint of simple jealousy in McLuhan's attitude toward Frye. They were, after all, contemporaries who had both gone to England to study in the thirties, Frye at Cambridge's rival. It must not have been altogether easy for McLuhan, ideology aside, to see this colleague win increasing acclaim in the fifties with books that were written in a magisterial style, congenial to the academic mind, that could not have been further removed from McLuhan's aphoristic mode.

Yet McLuhan himself had not been wholly out of the literary arena since coming to St. Michael's. Despite his complaining in 1953 that there was a virtual boycott of his work among periodicals, he had appeared in a number of magazines since 1946. Chief among them was *Renascence*, then edited by his old friend and coreligionist at Wisconsin, John Pick. Pick always claimed, with the indulgent air of an artistic soul confronted with a faulty carburetor, that he could never understand McLuhan. Nonetheless, he seemed happy to publish McLuhan's articles, which were almost entirely on purely literary subjects.

McLuhan's social criticism found its way to a somewhat broader audience. In 1947 he published an article entitled "American Advertising" in *Horizon*, the well-known English cultural journal edited by Cyril Connolly. The article, written with wit and the unmistakable F.R. Leavis tone of moral seriousness, repeated McLuhan's plea for a rigorous study of the effects and techniques of advertising, a subject that continued to fascinate him. Articles such as "American Advertising" demonstrate clearly that no man ever hated the effects of advertising more than McLuhan. For this very reason, he had a genuine, if qualified, respect for the artistry of advertising — and he refused to put quotation marks around the word *artistry* in that phrase. Like great artists, advertisers knew that their job was to achieve certain effects in the minds of their audience. Therefore, they often used the same means as artists, including the techniques of the Symbolist poets. They were also as skilled as any of the ancient rhetoricians in swaying the minds of their listeners.

Although in the sixties McLuhan was often accused of glamorizing the advertising industry when he pointed out these realities, he never lost sight of his central thesis: that the power of ads could be defeated only when their victims stopped ignoring them and started paying serious attention to them. It is true, as McLuhan once commented, that this very process turned ads into something "highbrow,"[76] but being highbrow, in McLuhan's eyes, never conferred the slightest moral value on anything.

The culmination of McLuhan's own examination of advertising occurred in 1951 with the publication of his first book, *The Mechanical Bride*. For years prior to that publication, McLuhan had been conducting lectures with slides of advertisements for students and the public. As early as 1945, in a series of these lectures at Windsor, Ontario, McLuhan characterized the present era as the Age of the Mechanical Bride. McLuhan later claimed that the enthusiastic reception of one of these lectures by a nun, Mother St. Michael (Winnifred Guinan), was the reason he decided to attempt a book on similar lines. Mother St. Michael was indeed excited by the lecture and may have given McLuhan special encouragement, but his plans for publishing these lectures had been hatched well before 1945. As Walter Ong notes, "As long as I knew McLuhan he had been talking about publishing *The Mechanical Bride*."[77]

The theme of the book was essentially the theme of all of McLuhan's social criticism in the 1940s. He succinctly restated that theme in an essay on John Dos Passos in 1951, in which he observed that our technological society vitiated family life and the free, human expression of thought and feelings.[78] The plan of the book, which was to reproduce an ad or comic strip as an exhibit and then attach a short essay analyzing it, was a printed form of his slide-and-lecture series on the same subject; ultimately the plan was inspired by Leavis's *Culture and Environment* and reinforced by the use of similar exhibits in Wyndham Lewis's 1932 book *The Doom of Youth*.

For years before *The Mechanical Bride* was published, McLuhan — who had been assured by Wyndham Lewis that advertisements were not subject to copyright laws and could be reproduced freely — was in the habit of clipping ads (as well as cartoons and articles from newspapers and magazines), putting them in envelopes, and stuffing the envelopes into grocery boxes. Shortly after the war's end, one of those boxes, filled with clippings plus a manuscript, found its way to the editorial offices of Vanguard Press in New York, steered there by Felix Giovanelli, who had done some reading and translating for the publisher. As Seon Manley recalls, the box was, to a busy editor, "a terrifying thing, the last thing that you're going to open." It is possible, Manley thinks, that the box was not even recognized as a manuscript and was shoved away in some "back office" until McLuhan called to inquire about it.

Manley was appalled when she finally opened the box. Inside was a five-hundred-page manuscript plus hundreds of yellowed newspaper and magazine clippings attached with paper clips to various pages. These clippings, McLuhan later calculated, represented only about .01 percent of his collection at home.[79] "I was even afraid to show the manuscript to anybody, because it seemed to me that it was going to require so much work," Manley comments.

> But there was something intriguing about it. No matter where you picked it up, one of the paragraphs was fascinating. And we finally decided that it was just a question of finding some way to do it, of just making a good selection.... Jim [Henle, then president of Vanguard Press] always thought of the book as a kind of exposé of advertising because he did enjoy exposé books and had a history of publishing them. I knew it certainly wasn't that, but I went along with the treatment of it as that because it was getting at something which I thought was tremendously valuable.[80]

The thirties and forties had seen a number of "exposés" that raked advertising over the coals for perpetuating misleading information and using hokey psychology. Manley was quite right in supposing that McLuhan's book was different. It was, if anything, a critique of an entire culture, an exhilarating tour of the illusions behind John Wayne westerns, deodorants, and Buick ads. The tone of McLuhan's essays was not without an occasional hint of admiration for the skill of advertisers in capturing the anxieties and appetites of that culture. Although McLuhan was still very much under the influence of Leavis, he was also beginning to find his own voice in *The Mechanical Bride*, a voice considerably more playful than Leavis ever dreamed of being. As he wrote to his mother while working on the book, he had found a certain tone of "flippancy" in writing that seemed "right." There was quite enough solemnity in the air without his adding to it. In 1949, in a letter to Giovanelli, he expressed a sort of formula that would characterize his writing from then on, a formula of sharpening insight while maintaining a tone of perfect emotional equanimity.[81]

McLuhan signed a contract with Vanguard Press in June 1948 for publication of a book tentatively entitled *The Folklore of Industrial Man*, which became the subtitle of the published book. He received an advance of $250, not an outrageously low sum at the time for a "highbrow" book. Vanguard's mistaken notion that it was buying an exposé, however, was a portent of trouble in the editorial offices of the press. In 1950, McLuhan found himself rewriting the book, although he was incapable of rewriting in the sense of refining or clarifying prose he had already written. Whenever he sat down to rewrite, he inevitably found himself launching off in new directions or bringing in new material.

He could work very quickly: when the editors suggested that he come up with four or five descriptive sentences for each of the thirty-nine essays in the book, to be set beside them in caption form, McLuhan polished them all off in one day. Of course, one-liners always were his forte. (In this case, they were pointed queries addressed to the reader, many of them very witty.) But for the wearisome work of making the text more readily understandable to the reader, McLuhan had no patience. He disliked making cuts; he also disliked expanding on a subject in the sense of giving examples or otherwise underlining his point. "It bored him," Manley comments. "He wrote in the style of a man who refused to bore himself."[82]

From Toronto McLuhan wrote to Giovanelli, who had been pressed into service as McLuhan's personal representative in New York, suggesting various schemes for getting the outline, selection, and arrangements of exhibits accepted by Manley. McLuhan felt that Manley was basically unsympathetic to the book and that, while she would not present arguments against it, she would not fight for it either. The real villains, in his mind, were Henle and the woman who did most of the actual editing of the book, Evelyn Shrifte. While Manley at least understood his material, McLuhan wrote Giovanelli, Henle and Shrifte did not. He felt that Henle and Shrifte wanted to see in the book a clear argument that was capable of being easily summarized and with all the logical links made explicit. In other words, they were still breathing the air of Thomas Babington Macaulay, while he was living in the universe of Ezra Pound and *Guide to Kulchur*.[83]

After the book's publication, McLuhan complained of some vague homosexual influence in the publishing world that was horrified by the masculine vigor of his prose and was trying to castrate his text. McLuhan abused those who frustrated him at Vanguard as he had once abused Father Dreyfus at St. Louis, assigning the most malign motives to them. He even felt Henle and Shrifte were deliberately trying to wear him down physically with their editorial demands so that he would no longer resist them.

In fact, the unusually difficult editorial process *The Mechanical Bride* underwent was partly because of the nature of the material and partly because there was, no doubt, an absence of full comprehension of McLuhan's intentions on the part of the senior people at Vanguard Press. McLuhan transformed this difficult process, however, into a melodrama. What was definitely not a product of his imagination was the rage and frustration he felt. He began to suffer from severe headaches at the time of the publication of the book, psychosomatic expressions of his rage and an ominous foreshadowing of more serious health problems in the coming decade.[84]

The actual publication of the book in the fall of 1951 was something of an anticlimax and brought no particular satisfaction to McLuhan after his six years of struggle with Vanguard Press. Years later he felt this book "appeared just as television was making all its major points irrelevant." He published it, he thought, "just under the wire," just when the Mechanical Bride was being replaced by the Electronic Bride.[85] It was, in any case, his last protest against the ravages of capitalism, industrialism, dialectical thinking, and mechanistic automatism in general. He was soon to discover that the automatism portrayed in *The Mechanical Bride* was yielding to a new tribalism. The study of this new tribalism would strip the last traces of moral earnestness from his prose and immerse him completely in the role of explorer, the relentless seeker of insights unhindered by the striking of moral attitudes.

Despite McLuhan's disclaimer, *The Mechanical Bride* remains the most delightful of his books. The dissection of North American culture in the late 1940s is accurate, merciless, and curiously high-spirited; the energy of the author comes through in the prose, despite his fear of being castrated by his editors. Many of McLuhan's favorite themes — the Dagwoodian American male, the origins of the contemporary sleuth, the desiccated rationalism of the Great Books program, the delinquent adult behind the comic strips, the crudity and cynicism behind the Dale Carnegie type of self-improvement — these are all gathered in a final, mordant farewell to Machine Age civilization. It is also a farewell to McLuhan's own long effort to oppose that civilization.

The book received respectful reviews but sold no more than a few hundred copies. It is, of course, impossible to say what might have become of McLuhan's career if the book had been a success. As it was, he was certainly prodded by its failure to seek new approaches to the study of society and technology. In the meantime, he bought up a thousand copies of the book when it was remaindered, stored some in the basement of the Muller-Thyms' apartment in New York, personally distributed others to bookstores, and gave copies away to friends. (A colleague received a copy studded with bee bees fired by the McLuhan children; McLuhan evidently failed to notice the book's condition.) For years after its publication he acted as a salesman for the book, sometimes selling copies to students who came to see him about their problems with his courses.

7. The Discovery of Communications (1951-1958)

In our time the wireless telegraphy has produced a new outbreak of antient
speculations. — Ezra Pound, *Guide to Kulchur*[1]

In the years immediately following the publication of *The Mechanical Bride*,
McLuhan was like a high-strung thoroughbred who had been exercised almost
to death but had not yet entered a race. He had his intellectual bearings, he
had ferocious energy to call upon, he had a craving to be heard and to make his
ideas felt. In a letter to Pound in 1951, he called himself "an intellectual thug,"
more than ready to clobber others with his insights.[2] But he was no longer interested in fighting on the battleground of literary criticism. He had instead discovered Technology.

Of course, McLuhan had never been one of those literary critics who concentrate on technique or the inner dynamics of a given work of literature. He
had almost always, despite his regard for the New Criticism and its emphasis on
the text, related a poem or novel to something in history or the contemporary
milieu. Now he wanted to stray even farther from the examination of literature.
In a letter to Pound, he lamented his bad luck in having settled on English literature as a scholarly profession.[3] His interest in technology was due not just to a
disgust with English studies or to his discovery of the shadow of the Masons
lying heavy upon contemporary literature but to his exposure to the works of a
fellow University of Toronto professor, Harold Adams Innis.

Innis was a professor of political economy who, prior to World War II, had
written two important books on Canadian economic history, *The Fur Trade in
Canada* and *The Cod Fisheries*. To the economic dynamics of beaver pelts and
dried codfish he had brought a poetic sensitivity and a historical imagination
that enabled him to perceive how the pursuit of even these humble commodities
had transformed a society. That same sensitivity and imagination Innis then
applied to the history of the world. In the process he asked himself which commodities were the most far-reaching in their effects. McLuhan described this
process, with typically aphoristic flair, as Innis's turning "from the trade-routes

of the mind."[4] Along the way, Innis discovered — perhaps through his study of another great Canadian commodity, pulp and paper, and its connection with newsprint and the circulation of opinion — that the greatest staples of all were the various media of communication.

When McLuhan and Innis first met at the University of Toronto in the late forties, Innis was still grappling with the implications of that insight. In April 1949, at one of McLuhan's informal discussion groups, Innis spoke about the characteristics of press and radio that these media necessarily possessed apart from what they actually printed or broadcast. He said, for example, that newspapers fostered a concern with the immediate at the expense of any sense of continuity. Radio, with its semipublic character and consequent need for government regulation, fostered government centralization.

Before and after this meeting, McLuhan and Innis maintained sporadic contact, although they never became close as friends or as colleagues. Innis occasionally joined the McLuhan coterie that met almost daily at 4:00 p.m. in the coffee shop of the Royal Ontario Museum. McLuhan was enormously impressed by the extent of his colleague's knowledge of political economy and history, especially in those aspects dealing with communications media. As a man of encyclopedic lore, Innis was a rhetorician's dream. He was in the process of cranking out a thousand or more pages on the history of communications in Canada, expressing his seminal reflections on pulp and paper, newsprint and public opinion. He did, in effect, a great deal of McLuhan's research for him.

In 1950 Innis published *Empire and Communications*, an epic study of world empires since ancient Egypt and the modes of communication that shaped them. "The significance of a basic medium to its civilization is difficult to appraise since the means of appraisal appears to be peculiar to certain types of media, and indeed the fact of appraisal appears to be peculiar to certain types of media," Innis wrote in his introduction to the book. "A change in the type of medium implies a change in the types of appraisal and hence makes it difficult for one civilization to understand another."[5]

In 1951, a year before his death, Innis published *The Bias of Communication*, a series of essays in which he articulated his major themes. He explained that he had attempted to trace, throughout the history of the West, "the implications of the media of communication for the character of knowledge."[6] He discerned a basic division in these media between those that favored extension in space and those that favored continuity in time. Stone clay tablets, for example, because of their durability, imparted the character of timelessness to the messages impressed on them. But they could not be transported very far or reproduced very easily. Papyrus and paper, because of their lightweight disposable qualities, could spread messages over great distances but imparted a character of imme-diacy and transience to those messages.

A civilization employing clay tablets and cuneiform, therefore, such as the ancient Sumerian culture, was apt to be more limited in area and preoccu-pied with religious and moral themes that by their nature endure relatively unchanged over centuries. Papyrus, on the other hand, encouraged the growth of empires of vast extent and a preoccupation with more secular problems of law, administration, and politics, as in the Roman empire.

Underlying this split was an even more basic one. The ultimate time-biased medium of communication, in Innis's terms, was speech, since it could hardly be transported at all and since it encourage the memorization of ancestral lore — poetry, proverbs, and so on — unchanged over countless generations. The basic conflict between time and space biases was introduced with the advent of writing. With writing came the first blow against the magic, incantatory quality of the spoken word, against the tradition of the elders, against sacred authority. With writing came the first step toward science, secularism, and mastery over space instead of mastery over time.

Innis did not carry these reflections very much further. Occasionally a sen-tence crops up in *The Bias of Communication* that might have been written by McLuhan: "In oral intercourse the eye, ear, and brain, these senses and the faculties acted together in busy co-operation and rivalry, each eliciting, stimu-lating and supplementing the other."[7] That sounds the true McLuhan note. For the most part, however, Innis did not view the various media of communi-cation as extensions of specific senses. He offered no suggestion that these media might subliminally distort human perception because they favored one sense over another. He spoke rather of how institutions in history had seized upon various media of communication and used them to build up monopolies of knowledge — as the Egyptian priesthood had used hieroglyphs on stone, for example, or the medieval Church had used parchment books. He saw the major conflicts in history as the struggle of institutions to ride on the backs of these media of communication until they were unhorsed by rival institutions using rival media of communication.

Innis, then, left McLuhan with a great deal of filling in to do, but with a most valuable hint of how to construct a new theory of culture. Not since F.R. Leavis had any professor stimulated his thought so much. Innis was the "extra boost" that got him into media study, McLuhan wrote to a friend years later.[8] "Harold Innis is the real freak," McLuhan once said when someone commented on his career as media guru. "How did that hick Baptist ever come up with this amazing method of studying the effects of technology?"[9] Of course, it was a method for which McLuhan had been preparing since Cambridge. Was not the greatest technology, the greatest medium of communication, human language? Had not Richards and Emperson and Leavis taught him to look to the *effects* of this particular technology? Had not his study of the ancient trivium taught him to analyze the way people experienced this technology, not what its "content" was supposed to be? By emphasizing communication, Innis had set up a perfect arena for McLuhan's work — although McLuhan entered it via Cicero and T.S. Eliot instead of through fur trappers and wood pulp manufacturers.

To prepare himself for the great work ahead, McLuhan wrote Pound in 1951 that he wanted to absorb the basic structure and approach of twenty major disciplines.[10] By absorbing these structures, he would master not only the technology of language but the whole of modern science and technology as well. Aided by Innis's focus on communications media, he would repeat — but in a much more sweeping fashion — Giedion's great attempt at showing how mechanization altered all human sensibility.

McLuhan would have loved to collaborate with Innis in this attempt. Developing the idea he had mentioned in his letter to Pound, McLuhan wrote Innis shortly afterward proposing a newsletter to be sent to a few dozen people in various fields. The newsletter, to be entitled "Network," would point out the underlying "grammar and general language" of their fields and would facilitate far-reaching dialogue. It would bypass the verbose and technical academic quarterlies in establishing communication among people of different specialties.[11]

Innis was interested in the idea. He had, in some measure, also been influenced by McLuhan, despite his initial abhorrence of McLuhan's conservatism. (Innis was a classic liberal who could hardly abide McLuhan's support of such figures as General Franco.) He was not insensible to McLuhan's claim that artists offered the best insight into the effect of technology and communications media. Among these insights was the observation of the French Symbolist poet Mallarmé regarding the form of the newspaper: that the newspaper's juxtaposition of news items placed a weird symbolic vista before the reader. McLuhan urged Innis to read Wyndham Lewis, although Innis never quite understood what Lewis was getting at.

Innis's death in 1952 cut short any further intellectual collaboration. In later years, some of McLuhan's Canadian colleagues accused him of distorting Innis's insights for the sake of fame and fortune. (Canadian Marxists, in particular, were always eager to reclaim Innis from McLuhan.) McLuhan's response was plaintive: of the innumerable students Innis had taught, he would ask rhetorically, how many *listened* to him? How many actually *read* him? McLuhan could take credit at least for that. McLuhan was even overgenerous in calling his book *The Gutenberg Galaxy* a "footnote" to Innis. If Innis is read in the future it will be as a footnote to McLuhan and not vice versa.

Not long after Innis's death, McLuhan found an opportunity to explore the new intellectual landscape opened up by his work. It was to be the turning point of his career. From a close friend in the anthropology department, Edmund Carpenter, McLuhan learned that the Behavioral Sciences Division of the Ford Foundation was awarding up to $50,000 for a two-year interdisciplinary research project. Under pressure from Canadian academics to spread some of its largesse north of the forty-ninth parallel, the foundation had sent application forms directly to the president of the university, Sidney Smith, who in turn had passed them on to a professor of political economy, Vincent Bladen. Bladen, one of McLuhan's most detested enemies on the faculty of the University of Toronto, was already at work on a proposal.

McLuhan immediately went to his friend and colleague Claude Bissell, then a professor of English and an emerging force in the little society of the university. (He later came its president, where he functioned as an important ally for McLuhan.) Bissell confirmed the news and supplied McLuhan and Carpenter with appropriate forms. The criteria for proposals included few specifics other than that the project involve scholars from the behavioral sciences as well as related disciplines and that it organize an ongoing seminar to deal not only with a particular problem of research but with the general problems of interdisciplinary collaboration.

Edmund Carpenter was then a young anthropologist and an "intellectual thug" after McLuhan's own heart. Following their first meeting in 1948, they became fast friends and saw each other almost daily. In many ways, Carpenter had the kind of vivid personality that McLuhan appreciated. He was a born raconteur and possessed an abrasive and sometimes bawdy wit. His lectures, like McLuhan's, became notorious on campus. During one particularly graphic lecture on Polynesian sexual mores, a female student stalked out in disgust. Carpenter called out after her, "You don't have to rush, my dear, the boat doesn't leave for two days."

It was almost inevitable that the two would end up as fellow outcasts in league against the rest of the university. McLuhan did his best to give his younger colleague the benefit of his long experience in academic politics. When Carpenter was embroiled in a dispute with the administration, McLuhan's advice to him was invariably pragmatic: give the university what it wants, he would say, and then go your own way and forget it. But Carpenter almost courted notoriety. There was a whiff of wickedness about him. "I made life as difficult for them," he says of his enemies on the faculty, "as they did for me. A good many faces fell when I walked into the faculty lounge."[12] (If one of Carpenter's enemies came into that lounge looking for a place to sit and there was an empty chair beside Carpenter, he would ostentatiously tip the chair over.) At St. Michael's, it was rumored that Carpenter possessed the largest collection of books on the devil and diabolism in Canada. The rumor turned out to be only a slight exaggeration: Carpenter had an excellent library on the black arts. It was almost as if he deliberately cultivated the reverse image of McLuhan's deep-seated piety.

Carpenter supplied an additional element of energy and élan to McLuhan's lonely struggle for recognition in the face of his colleagues' disapproval. "He said to me once, 'We've got it made here, there's so little talent,'" Carpenter recalls. "He would say, 'Come on, Ted, we'll win kudos from people we respect.'" McLuhan undoubtedly got a welcome charge from the fun he and Carpenter had together. The two would go about the city indulging in McLuhan's practice of perceiving patterns in whatever he saw. In the early fifties, Carpenter recalls, they were on a streetcar full of the pinched and lifeless forms of their fellow Torontonians.

> There were all the little widows in black, all the mean faces — it was a
> mean life then, and the faces reflected it — and Marshall began to talk
> about the ads on the streetcar in a very loud voice. He pointed to an ad
> of a girl drinking a Coke with some man watching her, and he said,
> "Coke sucker." I said, "Marshall, lower your voice." He was totally
> oblivious to the sound of his voice. And he went down the line analyz-
> ing one ad after another. Everybody just looked straight ahead and
> pretended they didn't hear him.[13]

In Carpenter's class the two sometimes engaged in dialogues with a loud tape recording of African war chants as accompaniment. "Our exchanges were like Lenny Bruce and jazzmen doing rap sessions, where Bruce would tell five jokes at once," Carpenter recalls. "Sometimes it was like a comedy act we had going, with the two of us making up stories as we went along. People listening to it were *appalled*." They would refer in public, for example, to a mysterious entity they named Hugo. This turned out to be a reference to their common enemy Northrop Frye, who at that time usually signed himself H. Northrop Frye (the *H* standing, in fact, for Herman).

In 1949 Carpenter confronted their enemy in a debate at Hart House on the campus over the nature of archetypes. (Carpenter attacked Jung and Frazer's understanding of the term, while Frye defended their use of it from a literary point of view.) McLuhan stood by, in Carpenter's words, "egging me on." Anything to strike a blow at Frye's mistaken theology. "Carpenter was quite an aggressive, belligerent person," Frye recalls. "He could be a real son of a bitch. It was all quite impersonal, but he tended to tackle people on their own ground, and of course he would always win. Although I don't think either of us won that debate, particularly."[14]

When they heard of the Ford Foundation grant, McLuhan and Carpenter seized the opportunity. McLuhan had previously referred, accurately, to foundations as institutions dedicated to tax loopholes and public relations.[15] Fifty thousand dollars, however, was serious money. If any cabal at the University of Toronto was going to get its hands on the loot, it might as well be their cabal. Ultimately, three groups from the university applied for the grant. The Bladen group submitted a proposal for research into "problems of social learning and cooperation in an industrial society." Another group, including the distinguished Canadian historian Donald Creighton, submitted a proposal to study the origins of radical and conservative behavior. McLuhan and Carpenter submitted a proposal entitled "Changing Patterns of Language and Behavior and the New Media of Communication."

The proposal cited the work of Innis as demonstrating that changes in the media of communication resulted in vast social, political, and economic change. With the new media of television, radio, and movies, society was being radically reshaped; a "new language" was being created, since the media of communication were themselves languages, or art forms. (McLuhan had recently noticed the work of the linguists Edward Sapir and Benjamin Whorf, who theorized that human beings learn to perceive reality through language and that language shapes the way we experience the world — not a startling thesis for someone exposed to the work of I.A. Richards on language and communications or to the belief of the ancient grammarians that the order of language was analogous to the order of reality.) The McLuhan-Carpenter proposal went on to state that their view of language and the media as art forms — that is, as instruments of focusing perception — gave them an advantage over all other academic groups studying human communication since those groups tended to view communication as a problem of information engineering.

On May 19, 1953, the Ford Foundation announced that the McLuhan-Carpenter group had hit the interdisciplinary competition jackpot — a grant of $44,250. It was a moment of triumph McLuhan was more than ready for. He assembled his group that summer. Besides McLuhan and Carpenter, it consisted of McLuhan's old sparring partner from University of Manitoba days, Tom Easterbrook, now a professor of economics at the University of Toronto; Carl Williams, a professor of psychology; and Jacqueline Tyrwhitt, a professor of architecture and town planning and a friend of Sigfried Giedion.

That first summer each member summarized the developments of the past century in his or her own field. McLuhan interpreted these summaries as confirmation that there was, in the present era, an underlying unity of approach in most major intellectual disciplines. No longer was there a split between the scientific and the aesthetic approaches, since both were based on the drama of human cognition. McLuhan hoped the group's findings would signal to the world that the era of the specialist was dead and that what would now be termed a holistic view of things was possible.

In fact, from the very beginning it was apparent that there were irreconcilable intellectual differences in the group. Carl Williams stood as the champion of the traditional scientific reliance on quantitative measurement. McLuhan, of course, has a rather different approach to reality. Some of his colleagues noted sourly that on any given fact he was likely to be wrong. The truth was, McLuhan was constitutionally unable to allow an inconvenient fact to dim the luster of a good story. "It never disturbed him in the least when he got his facts wrong," one colleague, former chancellor of the University of Toronto Omar Solandt, comments. "He was never intimidated by facts." On one social occasion, Solandt recalls, the two argued about McLuhan's latest insight, which in this case had to do with airplanes.

> He had figures for the speed and the carrying capacity of airplanes that were wildly wrong. His argument would have held quite well anyway if he had used the right facts. But it used to bother me that he didn't think facts mattered enough to get them right. I don't think he often reached conclusions that would have been different if he had the right facts, it's just that he never bothered with them.[16]

As Carpenter recalls, "He didn't really care what the reality was, in a way, if it fit into his system. If I told him about some scientific report, he would usually put a twist on it to make it more interesting."[17]

When challenged by a colleague, McLuhan never for a moment betrayed a hint of uncertainty. "He would usually just carry on with his idea as if he had had a happy thought or was sort of recounting a dream," a fellow English professor comments.[18] McLuhans' notions, however careless with facts, released a great deal of intellectual energy; sometimes they enjoyed, like the insights of G.K. Chesterton, an intuitive and happy relation to the truth; but they were a poor basis for genuine collaboration with a scientist like Carl Williams.

At the start of the project in 1953, McLuhan made one key decision that turned out be immensely valuable. He decided to realize his old dream by launching a periodical to publish the results of the group's efforts. Such a periodical might alert other researchers to their work, sparking further collaboration. It might also provide a forum for individual egos that became mildly claustrophobic from constantly working with a group. Above all, it just might inform the world that something interesting was happening in Toronto. "Canadians are all a very humble bunch," McLuhan commented years afterward. "They take it for granted that everything they do must be second rate. Carpenter and I just blithely assumed that, since nearly everything in the world is second rate at best, there was no reason why we couldn't do something that was first rate right here."[19]

To fund the publication, the group dispensed with the services of a secretary and finagled free accommodations and telephone service, saving six thousand dollars. The key element in starting the magazine, however, was not seed money but the talent and energy of Ted Carpenter. "He was a brilliant editor," one of the graduate students attached to the group recalls. "Everything about the magazine that eventually came out was impeccable — the copy, layout, design. It would have been a mediocre production without Ted."[20]

To start with, Carpenter sent off queries to about thirty possible contributors in North America and Europe and received positive responses from almost every one. "Carpenter had an extraordinary gift for prying things loose from people," Claude Bissell comments. He was a more worldly character than McLuhan, for one thing. McLuhan, for all his hustle and his talent at getting his way, was most at home in his office with books stacked dangerously high on every flat surface, thousands of sheets of paper stuck in each book, and his circle of loyal students doing chores for him. The dry odor of that scholarly den clung to him in a way it never clung to Carpenter. In the end, Carpenter obtained articles from an amazing array of individuals: scholars and scientists like David Riesman, Sigfried Giedion, Hans Selye (who contributed a highly technical piece on the physiology of stress), Jacques Maritain, Ashley Montagu, Jean Piaget, and even H. Northrop Frye. Robert Graves, Jorge Luis Borges, and e.e. cummings were some of the literary luminaries who also appeared in the pages of the magazine, which McLuhan and Carpenter dubbed *Explorations*.

In the fall of 1953 the group started the weekly seminar that was to be the heart of the two-year project. Eleven graduate students from various disciplines also attended the seminar, which began with a consideration of the work of Harold Innis, although the work of Whorf and Sapir also made an appearance very quickly. McLuhan immediately went beyond Innis in insisting that each medium of communication, at the same time as it affected all the senses, involved a specialized emphasis on either sight or sound. That emphasis, in turn, profoundly affected the entire outlook of those who used the medium. This twist on Innis was, in a way, a natural development from a man who had long studied poets like Eliot, Yeats, and Hopkins with their pronounced views on the differing effects of ear and eye in poetry. In McLuhan's view, only Innis's ignorance of such artists had stood in the way of his recognizing this obviousreality. The analogy between a poem and a medium of communication was, according to McLuhan, extremely tight. Just as a poem imposed its own assumptions on the listener, created its own world, so too did a medium of communication.

The first issue of *Explorations* was published in December 1953.[21] It contained an article on comic books by Gershon Legman, editor of an immensely vital "lay psychiatric journal" called *Neurotica*. The journal, which lasted from 1948 to 1951, had published excerpts from *The Mechanical Bride* as well as McLuhan's article "The Psychopathology of *Time* and *Life*." Legman at that time was a legendary New York intellectual, a "wild man" according to Carpenter, who indulged in such proto-hippie eccentricities as wearing a rope instead of a belt to hold up his pants. He collected a huge library on folklore and erotic literature and wrote books such as *No Laughing Matter: The Rationale of the Dirty Joke* and *The Limerick: 1700 Examples, with Notes*, which enjoyed a small but devoted readership (including McLuhan, who was particularly delighted by the latter volume).

It was Giovanelli in New York who brought McLuhan and Legman together; in a letter to McLuhan he described the New Yorker Legman as a "walking encyclopedia of anecdotes, scandals, inside stuff." Legman was too Freudian for their taste, but he redeemed himself with his general disdain for North American psychiatrists and psychoanalysts and his criticism of the baneful influence of homosexuals.[22]

McLuhan was particularly intrigued when Legman showed him a rare two-volume dictionary of sexual slang, derived partly from material compiled by the editors of the *Oxford English Dictionary* that never made it into the *O.E.D.* For years afterward McLuhan marveled at the existence of this work, which he referred to, inaccurately, as "the pornographic supplement to the *O.E.D.*" He thought its existence one of the best kept literary secrets in history.[23] The work was an example of precisely the kind of hidden lore, with ramifications both literary and otherwise, that was pure gold to McLuhan.

Legman's article in *Explorations* was a denunciation of comic books, written with all the moral earnestness McLuhan had abandoned and with much of the insight into the role of sadism in American life and the Dagwoodian fall of the father that had characterized McLuhan's own writings. McLuhan's article in that issue was very different. Entitled "Culture Without Literacy," it was a brilliant summation of the intellectual yardage he gained since his exposure to Innis. The prose was, as usual, compressed, but it was never more charged with meaning and the excitement of intellectual discovery. Every sentence radiated the energy of the preceding few years during which he had made, as he wrote to his mother, so many discoveries that he was "dizzy."[24] Not since his Ph.D.

thesis had McLuhan attended so strictly to the elucidation of a new intellectual world or laid aside so completely previous attitudes.

He called the approach exemplified by his article "observation minus ideas,"[25] meaning that he attempted to discard as thoroughly as possible mental structures that would interfere with pure insight. He described the process in a *Playboy* interview in 1969 as akin to cracking a safe:

> I don't know what's inside [the safe]; maybe it's nothing. I just sit down and start to work. I grope, I listen, I test, I accept and discard; I try out different sequences — until the tumblers fall and doors spring open.[26]

This was the method of the "probe" — the word McLuhan consistently applied to his intellectual investigations throughout the latter part of his career. Of course, even in "Culture Without Literacy," where, if at any place in McLuhan's work, the tumblers fell and the doors sprang open, he did not proceed without certain preconceptions or assumptions. Nonetheless, that article did mark a turning point, in which McLuhan positively exulted in the lack of moral "tone" now manifest in his work.

McLuhan being McLuhan, he never stopped explaining this lack. Somewhat absurdly, he claimed that bringing in personal feelings was too time-consuming and distracting: by the time one had finished emoting about a situation, the situation had changed entirely. On other occasions, he deemed it "impertinent" for a "private individual" to express approval or disapproval for technological changes, which had almost the force of nature. More credibly, he asserted that moral and emotional indignation was simply an indulgence on the part of those powerless either to act or to understand.[27] If there really was any human villainy or stupidity involved in the changes McLuhan studied, it would be better exposed, he felt, through understanding than through the expression of futile rage.

Sometimes McLuhan called his nonmoral approach "satire." He meant that simply putting the spotlight on the features of a situation that most people ignore tends to bring out the latent ridiculousness of the situation — in the way that advertisements, when studied very objectively, often become ridiculous. There was nothing particularly moral about the process.[28] The eye of the satirist did not always conduce to peace of mind, but then McLuhan felt that anxiety was sometimes necessary for the survival of the species in this age. For the most part, the adoption of a satirical rather than a moral frame of mind in his investigations gave him "great joy."[29]

Some of that joy comes through in "Culture Without Literacy." McLuhan takes the Thomistic position that the world is basically coherent; the instantaneous character of modern communications media, however, has made the world seem irrational. Like tribal peoples, we are bombarded with impressions that don't add up in a logical or literate fashion. To discern again the underlying coherence of the world we must learn the "grammar and general language" of both science and art.[30] We must also, McLuhan argued, master the art forms of the new media. In doing so, we would not only recapture the coherence of the world but enrich our culture unimaginably.

In this article McLuhan summarized briefly the history of the West in terms of media. With the invention of writing, and particularly the phonetic alphabet, which translated the spoken word into abstract, meaningless visual symbols of abstract, meaningless sounds, the primitive magic of the spoken word, conjuring up an absent reality in the very tones of the speaker, was broken forever. The word was transformed into a mere sign or label. Human beings achieved a kind of psychic distancing from their sensate perception. Ultimately the phonetic alphabet made possible such phenomena as logical analysis and the activities of bureaucracies.

In the Middle Ages, this effect of the phonetic alphabet was, to some extent, counterbalanced by the manuscript format. Monks did not read their manuscripts silently, as we do printed books; rather they mumbled over them, word by curious word, and memorized their contents because the manuscripts were scarce and not easy to refer to. Johann Gutenberg's invention changed all this. As McLuhan later commented, "If the phonetic alphabet fell like a bombshell on tribal man, the printing press hit him like a 100-megaton H-bomb."[31] Everything about this invention — its process, which was the first application of an assembly-line, mass production technique to a handicraft, and its product, which was uniform, highly sequential, and even further divorced from the audible than manuscripts, since books were read much faster and in silence — was a tremendous victory for the abstract, the mechanistic, the visual. McLuhan concluded in "Culture Without Literacy" that the print form still had a kind of hypnotic sway over us: we were unable, because of that influence, to recognize the effects or the "language" of the new media of communication.

In the second issue of *Explorations*, published in April 1954, McLuhan extended this thesis in an article entitled "Notes on the Media as Art Forms." He repeated his notion that any medium of communication radically alters the experience being communicated. So potent are these media that it is true to say, as he later expressed it, that nothing ever printed is as important as the medium of print. Nothing in Plato is as important as the fact that in a given classroom all copies of *The Republic* have the same word at the same place on the same numbered page. Anthropological and historical research has allowed us to see how societies are transformed by their media of communication and to recognize in the process that the study of individuals and society must be based on the study of communication.

In this article McLuhan for the first time speculated on the nature of the newest medium of communication, television. It was a medium fundamentally different, he suggested, from another visual medium, the movies. Somehow we did not look at the television screen the way we looked at a movie screen. In fact, he asserted, "in some sense the spectator is himself the [television] screen," A year later, in the fifth issue of *Explorations* (June 1955), he elaborated this curious perception in an article entitled "Radio and TV vs. the ABCED-Minded." The electrons of the television screen, McLuhan argued, penetrated the head of the television watcher in a way that the photographic images of the movie screen never did.

It is useless to search for some physiological underpinnings for this state-ment. McLuhan was groping for an explanation of something he no doubt felt intuitively: that the effects of the television medium on viewers were drastically different from the effects of anything else experienced by human beings. Later he offered the hypothesis that the mosaic of light and dark spots that made up the television screen presented a low-definition image, which the viewer was forced to complete in his or her own head. (McLuhan insisted on this point so strongly that his colleagues began to joke that McLuhan owned a poor television set.) These pseudoscientific explanations were really attempts to substantiate in some way the insight Mumford had initially offered in *Technics and Civilization* — that the new civilization based on electricity was organic in character and therefore radically opposed to the mechanistic civilization that preceded it. McLuhan was trying to say that, if Mumford's insight was true, no electric device was more organic than the one that brought us "I Love Lucy."

McLuhan was not then or at any other time ready to jettison literacy and call for complete immersion in the new world of pulsating electrons. The awareness fostered by books and writing, he felt, could still play its part within the larger awareness generated by the new media. There was no need to obliterate two or three thousand years of civilization. Nonetheless, McLuhan's intellectual energy was now harnessed to explaining — with a zeal that could almost be mis-taken for glee — the new power of television to transform every aspect of con-temporary life.

The next step in the elaboration of his thesis was the articulation of the notion of sensory spaces. In November 1954, Carl Williams read an uncharac-teristically wide-ranging paper entitled "Acoustic Space" to the seminar. The paper, which had the stamp of McLuhan all over it, was published the follow-ing February in *Explorations*. (It was later slightly revised and reprinted in an anthology of *Explorations* articles under the byline of McLuhan and Carpenter.)

In it, Williams — or McLuhan, speaking through Williams — explained that our concept of "space" is almost entirely visual. We think of it as something that separates visible objects. Hence a barrel full of gaseous fumes or a wind-swept tundra is spoken of as "empty."

If space is regarded as the world created by sound, however, its character-istics are completely different from those of visual space. It has no fixed bound-aries, no center, and very little sense of direction. (You can hear, but not see, around a corner. Also, if blindfolded, you cannot tell whether a sound comes from in front of or behind you, above or below you.) It is more immediately connected to the nervous system that anything visual: a warning sound, such as an ambulance siren, jolts the system far more than the sight of the object mak-ing the sound, in this case an ambulance.

According to McLuhan, the notion of acoustic space was first broached when Jacqueline Tyrwhitt discussed Sigfried Giedion's book *The Beginnings of Architecture* in the seminar. Tyrwhitt explained that Giedion cited the Romans as the first ancient civilization more or less to create visual space by enclosing an arch within a rectangle. Before this, no one had ever though of space as some-thing enclosed within a structure, as a static quantity able to be visualized.

The notion of acoustic space provided McLuhan with an all-purpose tool that he used to the end of his career. "Visual space" took its place along with "linear," "sequential," "print-oriented," and so on as a term to characterize almost everything post-phonetic alphabet and pre-electric media. Visual space was the only space where logic ("If A is greater than B, then it must follow that...") applied. Visual space was the space of continuous vistas — of conti-nuity itself — of areas enclosed in three-dimensional rooms and therefore of such unlikely phenomena as central heating. Acoustic space, on the other hand, was two-dimensional, according to McLuhan — and two-dimensional space he regarded, for some reason, as eliciting the richest interplay of the senses. It was the space of the electric world, in which people are hit with almost random bursts of information from all sides. It was the space of "the most eminent form of rational awareness, the analogical," since analogies cannot be visualized or measured any more than sounds can be.[32]

McLuhan tended to attribute any intellectual approach he favored to the acoustic world — or to its close cousin, the "audile-tactile." Thus, the study of "structures" or the study of effects in general was acoustic. Art was acoustic.[33] Visual space was ultimately a human creation, farther from the pulsating heart of reality than acoustic space.[34] The crucial discovery, from McLuhan's stand-point, was that Western culture had returned to this resonating world of acoustic space three thousand years after the phonetic alphabet had detoured it into a long spell of literacy and emphasis on the visual.

134 MARSHALL McLUHAN

From 1955 onward, McLuhan ruthlessly categorized objects, philosophies, and attitudes in terms of the sensory bias he detected in them. In theory, there were "spaces" or biases or forms of energy created by each of the five senses. In reality, McLuhan claimed there were two basic sensory spaces or biases, the visual and the audile-tactile or acoustic. The most interesting fact about an individual was not what set of ideas or what philosophical system he embraced, but what sense he favored. If you knew that, you knew your man. You knew which of the two spaces he inhabited. Much of the rest, including his philosophical outlook, could be deduced.

McLuhan used the weekly seminars to develop these and other themes. One of the more interesting experiments the seminar group conducted was an attempt, orchestrated by Carpenter, to demonstrate that different media of communication did indeed have an effect quite apart from the content of information they conveyed. In the spring of 1954 more than one hundred students were divided into four groups. At the local studios of the Canadian Broadcasting Corporation, one group watched a lecture delivered on television, a second attended the same lecture delivered in a television studio, a third listened to it over the radio, and a fourth read it in printed form. All groups then took an exam to test their comprehension and retention of the contents of the lecture. As Carpenter later wrote in *Explorations*, "About twenty of us in the seminar placed bets on the outcome. Academics all, we each seriously thought print would win and merely selected other media as sporting bet." It was the group watching the lecture on television that scored highest in the test, however. The print group scored lower than even the radio listeners.[35]

The experiment was like many other academic attempts to investigate McLuhan's ideas "scientifically": it was intriguing but not very conclusive and ultimately not very helpful. The experiment did, however, draw attention to the work of the seminar. The March 4, 1954, edition of the *New York Times* ran a small story on the experiment headlined VIDEO BEST TEACHER, RESEARCHERS FIND, with the lead "Television is a first-class teacher, easily surpassing books and its elder cousin, radio." Already McLuhan and his associates were being seen as apostles of the tube.

In October 1954, the five principals of the seminar attended a conference called "Culture and Communications," organized by the University of Louisville and attended by such academic luminaries as Margaret Mead and S.I. Hayakawa. At this conference McLuhan gave a talk reiterating some of his major themes. Writing was a metaphor, he told his audience: it translated sound into sight. Printing was the mass production of writing. Every mode of communication in our era had been transformed by technology until there was no possibility of examining ourselves and the new world we had created except in the manner of primitives confronting *their* world — that is, as participants in a collective, living, holistic environment. In such a world, the classroom had become obsolete. The young, like their elders, needed to become competent critics of radio, television, and other media. They had to deal with a total environment, an environment where everything was art, since everything was shaped by humans.

In April 1955 the seminar held its final meeting. McLuhan was not sorry to see it end. After a visit from McLuhan, an officer from the Ford Foundation noted that McLuhan was "very glad the project is over because the interdisciplinary efforts involved gave rise to violent controversies and political jockeying for the support of graduate students."[36] (The worst controversy, it seems, was between Edmund Carpenter and Carl Williams, whom Carpenter ever afterward termed a "fink for the administration.")[37]

In his report to the foundation, McLuhan concluded that the more members of the seminar understood each other's disciplines, the more they feared a kind of mutual espionage.[38] The more specialized an academic became, in McLuhan's view, the more desperately he clung to his expertise and the less he wanted to understand others or wanted others to understand him. Eventually McLuhan concluded, for these and other reasons, that "all ignorance is motivated." Most people simply do not *want* to expand their awareness. This being so, it was easy for McLuhan to arrive at the belief, not always deluded, that he saw more significance in almost any given subject than the professionals who made it their life's work to study it.[39]

Whatever the cause, it is clear that the members of the seminar never really communicated with one another. Each of the five professors tended to treat intellectual approaches held by the other four with "benign neglect," according to one graduate student involved in the seminar. "Usually someone would read a paper in his own field that no one else understood except the one or two others in that same field," the student recalls. "The participants were talking past each other. And no one took the bull by the horns and tried to deal with the situation."[40] The soul of the seminar, when it possessed one, always remained the property of McLuhan. "If McLuhan didn't get involved in a meeting of the seminar, somebody would go off on some *dreary* academic thing," Carpenter maintains. "But if he did get involved, he would tend to dominate."[41] In the dying months of the seminar, McLuhan complained in a letter to Wyndham Lewis that the other members simply refused to understand or give the benefit of the doubt to his approach.[42]

As far as other University of Toronto faculty members were concerned, the seminar rather confirmed, if anything, McLuhan's outcast status. Graduate students were warned away from it. Easterbrook's colleagues in the economics department asked themselves why this respectable figure involved himself with it. (Easterbrook, in turn, took pleasure in taking McLuhan to lunch at the Faculty Club as a kind of superb pet, whose debating performances could be counted on to leave adversaries numb.) Although hurt by this kind of attitude, McLuhan had his usual answer to it: a better audience awaited him outside the university.

McLuhan realized that *Explorations* was a key to reaching that larger audience. When the initial funding from the Ford Foundation had been exhausted, he desperately sought funds to keep the magazine alive. He and Carpenter applied to the usual academic sources, such as the Canadian Social Science Research Council, but McLuhan, using the experience gained from rounding up customers among the St. Louis gentry for Wyndham Lewis, also approached people outside the university.

One important backer turned out to be the Toronto newspaper publisher John Bassett. Bassett, who prided himself on being a "simple businessman," seemed an unlikely candidate for sponsor of a moderately esoteric intellectual journal. But McLuhan worked his charm on him. "McLuhan was a terrific fellow," Bassett recalls,

> enormously stimulating, with a great sense of humor. Funny man. Brimming over with energy. He couldn't sit still for very long. He'd jump up and down enthusiastically — oh, he was a hell of a fellow. The only problem was, I couldn't understand a goddamn thing he said. I used to laugh at his ideas and say, "For Christ's sake, Marsh, what are you talking about?" And he'd just go on and on.[43]

Bassett ended up giving McLuhan $7,000 for his magazine — not a grand sum, but enough. It was the kind of coup that confirmed Seon Manley's assessment of McLuhan: "One part of Mac was like an absolute, glorious con man. I mean, there was no lying, no cheating, he was just so sure of his own ability and of his own thoughts that anything was possible."[44]

Eventually nine issues of *Explorations,* with one thousand copies of each issue, were published. The last issue appeared in 1959, by which time Carpenter felt the magazine had run out of steam. Many of the nine issues contained articles that were patently filler. Certainly not all of the articles, even the best, illustrated any "common language and grammar" among the sciences. But they did provide McLuhan with wonderful material for his themes. The English scholar H.J. Chaytor, for example, gave McLuhan insight into the manuscript culture of the Middle Ages in an article entitled "Reading and Writing."

Even more important was an article by McLuhan's former student Walter

Ong entitled "Space and Intellect in Renaissance Symbolism." The title promised a rather abstruse and technical work; yet this short essay — the fruit of McLuhan's suggestion, fifteen years earlier, that Ong look into Peter Ramus — provided the key to McLuhan's history of Western civilization. In the article, Ong set out to explain why mathematics and physics emerged at only one time and in one place, western Europe at the end of the Middle Ages. His answer was that math and physics were the result of an emphasis on the visual and that "the central operation in visualizing knowledge was the exploitation of letterpress printing."[45]

Explorations produced an embryonic McLuhan cult, which grew as the magazine came to the attention of more and more people. An English professor at Concordia University in Montreal spoke for many who discovered McLuhan in those years when he wrote in *Commentary* in January 1965, "I have been shamelessly pilfering his work for years, and others have been doing it too: it is easy for a practiced eye to discern little bits of McLuhan nestling like fossils in the gritty prose of many a literary critic or sociologist."[46]

In New York City, recruits to the cult included a Jesuit scholar at Fordham University named John Culkin, who would function as a key promoter of McLuhan in the sixties. After 1958, Culkin became what he describes as a "closet scholar" of McLuhan's work. He recalls academic conferences of communications specialists during the late fifties and early sixties: "In those days we few card-carrying McLuhanites would seek each other out at such gatherings to swap anecdotes, ideas and self-congratulation."[47]

McLuhan recycled some of his writings from *Explorations* in a curious publication he dubbed *Counterblast. Counterblast* was a response to a report issued by a Canadian Royal Commission in 1951 — the Massey Report — outlining recommendations for a national policy on the arts and sciences. The satirist in McLuhan was aroused by the ponderous, high-toned report. *Counterblast,* the vehicle of his satire, was intended to do for Canada what Wyndham Lewis's *Blast* had done for England. Published in London on the eve of the First World War, Lewis's journal consisted of wickedly phrased headlines "blasting" facets of the London scene that Lewis detested and "blessing" those he found tolerable.

For *Counterblast* Carpenter typed the various headlines McLuhan concocted. A typical headline read "B L A S T The cringing, flunkey spirit of canadian culture, its servant-quarter snobbishness resentments ignorance penury." To each headline McLuhan added some observations lifted from his *Explorations* articles. A few hundred copies of the document were run off (only one issue was ever published), and Carpenter and McLuhan hauled them by sled over the snow-covered streets of Toronto to a cigar store, where the manager agreed to sell them for twenty-five cents a copy, all proceeds going to himself. McLuhan had the satisfaction of reporting to Lewis that the copies sold quite quickly. No doubt the venture caused more pitying smiles on the University of Toronto campus.

After Carpenter left Toronto to take an academic position in California, he and McLuhan continued to correspond two or three times a week, relaying to each other fascinating tidbits of information. McLuhan also continued a far-flung correspondence with other academics, businessmen, casual acquaintances — any interesting people he had run into or whose works he had read. He never ceased asking his correspondents if there was anything in their field he should read, convinced that only by the continued operation of this personal network could he keep his head above the flow of information. He always showed an innocent and happy surprise, even at the height of his fame, when someone managed to tell him something he hadn't yet thought of or heard about.

Books remained McLuhan's chief source of information, and he devoured them with a sense of urgency undiminished since his days as an undergraduate at the University of Manitoba. Works of sociology, history, literary criticism and all sorts of oddball topics claimed his interest. As the years went by, however, he read less and less imaginative literature. For all his reverence for twentieth-century art, he seemed oblivious of poets and novelists arriving on the scene after World War II. The appearance of Allen Ginsberg alone was probably sufficient to kill his appetite for contemporary poetry.

In a typical week McLuhan might run through thirty-five books, not including synopses of books prepared by his cronies. To determine whether a book was worth reading, he usually looked at page 69 of the work, plus the adjacent page and the table of contents. If the author gave no promise or insight or worthwhile information on page 69, McLuhan reasoned, the book was probably not worth reading. If he decided the book did merit his attention, he started by reading only the right-hand pages. He claimed he didn't miss much with this method, since there were so many redundancies in most books.

In 1967 he became caught up in the fad for speed-reading and took an Evelyn Wood Reading Dynamics course. (His son Eric became an instructor in the Evelyn Wood program.) McLuhan thought that speed-reading was suited to grasping the essence of printed works. It revealed patterns, not data, he insisted. Occasionally he astonished colleagues with claims that he could read 1,500 words a minute or go through Milton's *Paradise Lost* in five minutes. In a more candid frame of mind, he later confessed to a friend that, although he took speed-reading seriously when he first attempted it, in the end he found it useful mainly for reading junk mail.[48]

Speed-reading was merely the extreme of McLuhan's book-reading habits,

habits he recommended to his students when he advised them to be "intelligent browsers." Occasionally, this "browsing" got McLuhan into trouble, as when he grandly announced to Ph.D. candidates during their thesis examinations that he had no questions for them since it was evident to him, from reading the first three pages of their theses, that they indeed knew their stuff (another reason his colleagues might have avoided including him on thesis examination committees).

Nonetheless it was surprising, sometimes, how much McLuhan retained from his "browsing." Robert Logan, a physics professor and friend of McLuhan's in the latter part of McLuhan's career, when his powers were by no means at their height, recalls, "I would read the same book cover to cover and yet he always got more out of the book than I ever did. He knew the content of the book. It was uncanny."[49]

The more McLuhan processed enormous amounts of material relating to the twentieth century, the more he resisted stating, in any straightforward manner, his personal feelings about the major developments of the century. Partly this was because — in addition to all his reservations about moral indignation — he felt that expressions of personal feelings had become devalued. In fact, he resented people who greeted him by asking how he was "feeling." Clearly, such people did not care to know how he was feeling; they merely wanted to seem well disposed. On a political or social level, expressions of personal feeling were even more fatuous, he thought. It was fatally easy, he insisted, to don the appearance of virtue by inveighing against human corruption.[50] As a serious Catholic who went to confession, McLuhan felt he had no need to demonstrate to the world high moral seriousness.

When McLuhan did display feelings over a public issue, it was not South African apartheid or the war in Vietnam that roused him to indignation but the issue of environmental degradation, especially of the campus and the city he lived in. When a dirt path on the campus — cherished by McLuhan as the last country path in downtown Toronto — was paved over, McLuhan wrote an incensed letter on the subject to a newspaper columnist in Toronto and questioned everybody in the college administration about it. (To a man, they denied prior knowledge of the act, including the guilty party, the college treasurer.) He strenuously opposed the extension of a major expressway through downtown Toronto in the early 1970s and the widening of a city street bisecting the St. Michael's campus.

McLuhan often went out of his way to pick up litter on the lawns of the campus. In 1971, he wrote his old supporter Bassett suggesting that the publisher drum up support for voluntary citizens' groups to help pay the costs of fighting civic pollution. On a global scale, he appealed to Canadian Prime Minister John

Diefenbaker, elected in 1957, to commit Canada to a scientific quest for ways to remove radioactive particles from the atmosphere.[51]

In all of these instances, McLuhan's personal feelings were quite evident. Less evident, from the mid-1950s onward, were his feelings with respect to the McLuhan Subject: media. Did he, many readers wondered, *really* think television and electrotechnology in general were terrific advances, as he seemed to be saying? As the years went by, McLuhan defended himself with increasing vehemence from any imputation of partisan feeling for the new media. The effect of the new media on human society had "never aroused the slightest enthusiasm" in him, he wrote one correspondent in 1970.[52] Such disclaimers were like the earnest protestations of "antiwar" moviemakers who produce films with thrilling battle scenes. Somehow they never convinced his critics.

They did not convince because they were not entirely true. Partly true, perhaps — McLuhan did have an emotional aversion to change and was never awed by gadgets of any kind. "I am not, by temperament or conviction, a revolutionary," he told *Playboy*, in what was decidedly an understatement. "I would prefer a stable, changeless environment of modest services and human scale."[53] Only by understanding change could he ease the burden of experiencing it, he claimed — and therefore the only "extension of man" he desired was that of awareness. Powerful gadgets like television were all the more dangerous, in McLuhan's view, because they fascinated those who used them and often turned those users into dependents. Such gadgets became idols.[54] "I wish none of these technologies ever happened," he told the American critic Richard Kostelanetz.

> They impress me as nothing but a disaster. They are for dissatisfied people. Why is man so unhappy he wants to change his world?
> I would never attempt an improvement — my personal preference, I suppose, would be a preliterate milieu; but I want to study change to gain power over it.[55]

All this was sincerely meant; but in another mood, and for a period of time from the mid-1950s to the mid-1960s, McLuhan felt much more ambivalence — even hopefulness — about the media and the changes he studied. "I honestly don't know if the changes will be good or bad," he confessed to Richard Schickel, who wrote one of the first articles introducing McLuhan to an American audience, in *Harper's* magazine in November 1965. He certainly would never have maintained that there was anything intrinsically evil about technology or media — just as, in the Catholic view, there is nothing intrinsically evil about any aspect of nature. One simply had to be very careful about using the media, and therefore one had to understand how the media really worked. The problem was that the new media were so powerful and all-embracing that they swamped what was left of nature, making it even more urgent to understand them. (When McLuhan insisted that nature was dead, he meant that a certain *concept* of nature was dead, just as the phrase "God is dead" meant that a concept of God was dead. The world was now a kind of universal Disneyland for us to fashion as we wished.)

From the mid-1950s, when he emerged from his longstanding opposition to the Mechanical Bride, until the mid-1960s, when he became more and more appalled at the darker aspects of retribalized society, McLuhan, for the only time in his adult career, remained hopeful that we might manage quite a superb Disneyland. Enriched by the interplay of the old visual print culture and the new acoustic electric culture, we might enjoy a renaissance greater than the sixteenth century's, which had been sparked by the interplay between the old manuscript culture and the emerging print culture. The electronic media, because they created "a total field of instant awareness," could possibly even free us from our old habits of reacting automatically and involuntarily to our human interventions.[56]

McLuhan was all the more hopeful, of course, because of his high estimate of the literary output of those who had already graced the twentieth century — Joyce, Pound, and so on. Although his understanding was much more limited concerning such disparate phenomena as cubism, the new physics, symbolic logic, and analytical psychology, he felt that these too were manifestations of a great new global culture, the culture more or less foretold in 1948 in Lewis's *America and Cosmic Man*. Our era, he told one correspondent in 1963, was "the greatest in human history," an assertion he repeated many times in the early sixties, both privately and in his published work. Art itself might merge with the fabric of daily life, in the manner of tribal societies or of the Balinese he quoted so frequently: "We have no art. We do everything as well as possible." With such exciting possibilities, his own temperamental preference for a changeless world was a meaningless "private whim."[57] It is hardly surprising, then, that his critics accused him of being a cheerleader for television, despite his frequent disclaimers.

McLuhan's first important opportunity, after the close of the seminar, to articulate the possibilities of the new era before an American audience came in November 1955. Louis Forsdale, a young instructor at Columbia University Teachers College and a reader of *Explorations*, invited McLuhan to speak on the topic of communications at a seminar at Columbia. Forsdale felt rather daring inviting this relatively unknown professor to speak at a gathering that included academic heavyweights such as Robert Merton, perhaps the most distinguished

American sociologist at that time. His feeling of risk-taking was fully justified by the event.

The first paragraph of McLuhan's paper stunned the audience. It began with a reference to Freud, included a complex analogy between psychoanalysis and X-ray photography, and ended with a capsule history of the effects of the ancient Roman road. McLuhan then launched into a précis of his recent media discoveries, citing Innis's insight that any changes in the media of communication are inevitably followed by enormous social change and elaborating on the effects of print, the telegraph, newspapers, radio, and television. He ended by warning his listeners that they were living in an "age of paratroopers" and that any attempt to counteract the effect of the new media in the classroom by a chaste concentration on the good old monuments of literature or culture was entirely futile.[58]

When McLuhan finished, Forsdale asked if there were any questions. Robert Merton, his face flushed with emotion, was the first to speak. "Well, Professor McLuhan," he said, "there were many things about your paper that need cross-examination. It's so chaotic, I don't know where to begin... with your title or your first paragraph." A light glimmered in McLuhan's eye. "Let's begin with the first paragraph," Merton continued, vibrant with the resolution of an umpire about to eject a manager who'd gone too far. "You don't like those ideas?" McLuhan interrupted with a shrug. "I got others."

Forsdale is not sure anybody laughed at McLuhan's remark. "You don't laugh at Robert Merton," he points out. McLuhan, in any case, could hardly have done more to win forever the label of "unsound" not only at the University of Toronto but in the highest circles of American university life. It is the kind of remark that is repeated and relished for a long time afterward in faculty lounges. "McLuhan's response was really outside the academic pale," Forsdale comments.

> What you do in academia is debate. You go over points and you
> describe things carefully, you define and you come to an agreement or
> you lock horns and you talk about the research that you can bring to
> bear on this point of view or the research that you can bring to bear on
> that point of view, and McLuhan wasn't doing it. He was just saying,
> "This is my idea."[59]

At that point, McLuhan was not deliberately trying to violate the rules of academic debate. His particular response was the one that happened to occur to him. Later, of course, he cultivated the technique of the outrageous brush-off in encounters of this sort. He could not bear to have his thought cross-examined.

McLuhan delivered a paper on a similar theme to the National Council of Teachers of English (NCTE) that same month, informing his audience that the leading role of the book in our culture had been taken over by the new media and that the job of teachers was to train their charges in media literacy. For years after this address, the NCTE remained one of McLuhan's chief platforms. Neil Postman, later an influential writer on education, recalls sharing a hotel room with McLuhan and another friend at a conference of English teachers in Cincinnati, where McLuhan was to speak, in the late fifties. "We were trying to get to sleep," Postman comments. "It was about 2:00 a.m. and the lights were out and all we could see was the light of McLuhan's cigar. He was generating these incredible ideas, most of them crazy, and we were saying 'Enough already, we have to get to sleep,' but we couldn't stop him. He just went on and on."[60] Unlike Merton, Postman and his friend were enthralled. "We never doubted from the first moment we heard him that he was not only making sense but that he was a genius," he maintains. "This was clear to us."

McLuhan's final appearance in *Explorations*, in the October 1957 issue, neatly caps the intellectual progress he had made since the periodical had been launched. In this summation of his ideas, he reiterated that literacy was merely one mode of perception and the gloomy pronouncements of educators concerning the baneful influence of television and the resistance of the young to traditional classroom fare were entirely beside the point. He repeated that we were returning to tribal, primitive "acoustic space," where "extra sensory perception is normal" and "all time is *now*."

Everything McLuhan said or wrote afterward is directly traceable to something he wrote in the first eight issues of *Explorations*. The heart of McLuhan's approach to reality had been articulated — and never more cogently — in the pages of that quirky and marvelous periodical.

McLuhan did other writing during this period, including some work that would have added distinction to any English professor's curriculum vitae. In 1954 he edited and wrote an introduction for a paperback anthology of Tennyson's poetry. Tennyson was not a poet McLuhan greatly admired, but he provided an admirable illustration of some of McLuhan's ideas about the concurrent development of poetic technique and nineteenth-century technology.

McLuhan also produced some offbeat work that was never published, including a sketch for a musical comedy based on the idea that the United States was rediscovering the tribal, collective world of preliterate man while the Soviet Union was attempting to detribalize and enter the world of the literate, industrial nineteenth century. The yen to write something in this genre was not new; at one point, he conceived of a musical based on the work of the nineteenth-century Irish poet Thomas Moore. For many years he thought of doing a

Broadway production in which the various media would appear onstage as dramatis personae, and he even wrote to Tom Wolfe in 1978 asking if Wolfe was interested in working with him on the project. Had it ever occurred, the collaboration would have been one of the strangest in modern literary history.

McLuhan's musical comedy, entitled *The Little Red Schoolhouse*, featured the secretary of state, John Foster Dulles, and other Washington bigwigs inviting the Russians to try their hand at running the United States. The Russians, seduced by their love of Elvis Presley, agree.

At least McLuhan did not have to peddle his musical comedy sketch as a desperate money-making scheme. By the late fifties, with increased university salaries, the McLuhans were beginning to feel more financially secure. In 1956 the Basilians, in a paternalistic but welcome gesture, lent McLuhan money to buy a house at a low rate of interest. That year the McLuhans moved from the old infirmary on campus, which was shortly to be torn down to make room for a new student residence, into a roomy Tudor-style house on Wells Hill Road, in an old, well-established, upper-middle-class residential area north of the university campus.

In finally owning a house McLuhan had accomplished what his father had never managed.[61] McLuhan was glad to move off campus, relieved that his daughters, who were approaching adolescence and giving promise of great beauty, would no longer be there to distract young men from their studies. He tried to take seriously his role as head of the family in a real neighborhood. At his new parish church, Holy Rosary, he invariably stayed after Mass on Sunday for coffee and refreshments at the parish hall, introducing himself to people with a hearty "Hello, I'm Marshall McLuhan. I haven't seen you before. Are you new here?"

If the McLuhans were moving out of poverty, they were still not exactly affluent. In the summer of 1958, McLuhan eagerly accepted an offer from Hugh Kenner to lecture at the University of California at Santa Barbara, while Kenner was away from his teaching duties there for the summer. He packed his wife and six children into an ancient Dodge station wagon and headed west. They stayed in motels along the way, all eight of them in one room — three of the girls using sleeping bags, the rest of the family distributed on double beds.

While his wife did most of the driving, McLuhan relaxed with an occasional beer and read to the children, usually from a modern version of *The Canterbury Tales*, by Neville Coghill. Most English professors would have disdained Coghill's contemporary versification of Chaucer, but McLuhan, who was never afraid to speak up for the unfashionable, found it interesting, as well as effective in keeping the children quiet.

Once in Santa Barbara, McLuhan fell in love with the beaches, the warm seawater, and the magnificent climate. He also was intrigued by California informality. It seemed to him, in his new environs, that even the greatest person in the world would appear as ordinary and unimpressive as the sunbather one stepped over on the beach. Apparently McLuhan still suffered from his "yokel" complex, first aroused in Cambridge. Had it not meant Hugh Kenner would have been his boss, McLuhan would have been tempted to stay in Santa Barbara for good.

In August the McLuhans headed back to Toronto via Fort Worth, Texas, where they visited Corinne's family, and St. Louis, where they dropped in on their old friends the Strohbachs. Addie Strohbach was amazed that the travelers had managed to get as far as they had in their Dodge, which seems to have had the picturesque qualities of the vehicles used by the Okies in *The Grapes of Wrath*. She recalls it as a "godawful station wagon. It was a real clunker. I thought he was extremely brave going in it — but then Marshall always had an ethereal way of thinking he was going to make out okay. He just never thought of opposition."[62]

8. The Electronic Call Girl (1958–1964)

An acute critic tells me I shall never learn to write for the public because
I insist on citing other books. How the deuce is one to avoid it? Several
ideas occurred to humanity before I bought a portable typewriter.
— Ezra Pound, *Jefferson and /or Mussolini*[1]

By 1959 McLuhan's long intellectual preparation, his radical and now fully
developed notions about the media, his growing reputation inside and outside
academia as an odd but stimulating thinker, and something else — a certain
restlessness, the beginnings of profound ferment in the society around him —
were all coming together in a sort of critical mass, preparatory to the sudden
explosion of interest in his work.

A sign of his growing importance was his 1959 appearance at the first Congress of Cultural Leaders, sponsored by the Institute of Contemporary Arts
in Washington, D.C. The congress was intended to be a "summit meeting" of
foreign cultural leaders and their American counterparts. McLuhan, the "cultural leader" from Canada, got a chance to talk with poets such as Robert Lowell
and Stephen Spender and his friend the English novelist and poet John Wain,
with whom he discussed the engaging phenomena of jazz and poetry readings,
popularized at that time by the beats.

A far more important catalyst for the imminent explosion of interest in
his work was McLuhan's introduction to Harry J. Skornia in the later fifties.
Skornia, a professor in the department of speech and theater at the University
of Illinois, was no pretender to great intellectual depth; nonetheless, he was
open-minded and had the ability to recognize and use the talents of others. In
the late fifties he exploited that ability as president of the National Association
of Educational Broadcasters (NAEB), an organization in the forefront of a
lively midwestern avant-garde in education throughout the forties and fifties.

Skornia had followed McLuhan's ideas through *Explorations* but was by
no means a full-fledged convert. He absorbed the message that television was
not simply an aid to instruction but a medium with unique properties and
that all the mass media needed to be studied and analyzed in the classroom if
only because they were educating children far more effectively than any

Herbert Ernest McLuhan and Elsie Naomi Hall, on their wedding day,
December 31, 1909. Photograph courtesy of Stuart Mackay.

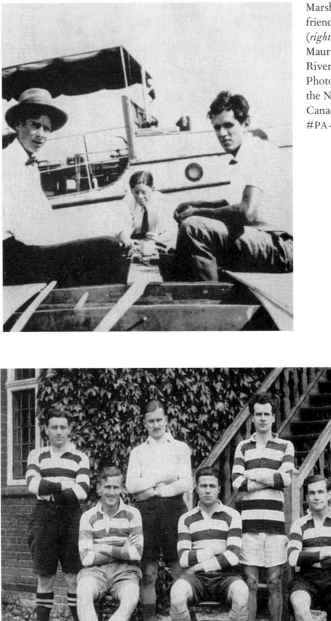

Marshall (*left*) with his
friend Ed Robeneck
(*right*) and his brother,
Maurice, on the Red
River, Manitoba, 1928.
Photograph courtesy of
the National Archives of
Canada/Catalogue
#PA-148461.

Marshall McLuhan (*third from right*) as a member of a rowing team at Cambridge.
Photograph courtesy of Sir John Ellis.

Marshall and Corinne
embarking on their
honeymoon in the
summer of 1939.

Marshall on the snow fields
just under the three peaks
of Mount Olympus in
Washington, August 1936.

Corinne and Marshall at the Hotel Chase
in St. Louis in the fall of 1940.

Wyndham Lewis's 1944 portrait of Marshall McLuhan. Courtesy of the Wyndham Lewis Memorial Trust.

BELOW LEFT:
Herbert, Corinne, and Eric McLuhan in Winnipeg, around 1945. Photograph courtesy of Stuart Mackay.

BELOW RIGHT:
McLuhan and Buckminster Fuller at the Delos Conference on the Aegean, September 1963.

McLuhan recording *The Medium Is the Massage* for CBS Records.
Left to right: Jerome Agel, Quentin Fiore, McLuhan, and John Simon of
CBS Records. Photograph courtesy of CBS Records, New York.

John Culkin, McLuhan, and advertising wizard Tony Schwartz, who
frequently appeared together during McLuhan's stay at Fordham in
1967–68. Photograph by Conrad Waldinger.

McLuhan's Monday night seminar at the coach house, in the mid-seventies. Photograph courtesy of Thomas William Cooper.

Marshall in Toronto. Photograph courtesy of the Department of Information Services, University of Toronto.

Marshall, Corinne, and their beloved dog.

Marshall in his favorite
Hawaiian shirt. Photograph
by Barbara Wilde.

Hugh Kenner, McLuhan, and Leslie Fiedler at the Celtic Arts
lectures at St. Michael's College, December 1979.

Corinne McLuhan, Marshall McLuhan, and McLuhan's Fordham University friend
and devotee John Culkin, September 27, 1980, at the Nova Scotia première of
Teri McLuhan's film *The Third Walker*. Photograph courtesy of John Culkin.

schools. Distinctions such as visual space and acoustic space, however, passed him by completely.

Skornia asked McLuhan to be the keynote speaker at the annual convention of the NAEB in Omaha in 1958. In his address, McLuhan used the phrase "the medium is the message." It was not the first time he had uttered these words — he had used them earlier that year, for example, in an address to a smaller gathering of broadcasters in Vancouver, B.C. But this was the first time he had used them before a truly influential forum, and it was to become a tag identifying him in the public mind.

The phrase was the culmination of a long intellectual process that had begun with his study of Innis — perhaps even before that, in his first year at Cambridge. It was simply one way of stating that all media — including human language, the basis of all subsequent media, according to McLuhan — have effects on the human psyche quite apart from the explicit bits of information they might convey.

McLuhan also told the NAEB members that the "processing and packaging of information" was now the chief business of the age. IBM was eclipsing GM. In this new information age, education was an ongoing necessity for adults as much as for children. Scholars and artists, those most at home with the processing and packaging of information, were moving from the ivory tower to the control tower. It struck McLuhan as perverse that such a development should inspire gloom.[2]

Skornia subsequently enlisted McLuhan for a NAEB project, funded by the National Defense Education Act, to develop a syllabus for the study of media in eleventh-grade classrooms. The objective of this syllabus, as defined by a research committee of the NAEB in September 1959, was basically to impart to students and teachers "familiarity with the various and often contradictory qualities and effects of media." From the start McLuhan made it clear that his focus would be on the "mutational powers" of the media rather than their "content" — on the effects of television in general, for example, regardless of whether it broadcast soap operas or Shakespeare. He wanted, he said, to teach the "grammar" of the new languages of television, radio, and other media.[3]

To carry out the project, McLuhan took a one-year sabbatical from the University of Toronto, beginning in the fall of 1959. As remuneration for that year, he asked the NAEB for $15,000, the equivalent of his university salary plus his lecture fees — a handsome sum of money for that era and an indication of how far McLuhan had come. Eventual funding for the project totaled $30,000. Out of these funds he was able to hire a research assistant and a full-time secretary. It was the first time he had had a full-time secretary, a position that henceforth would become a necessity for him.

To develop his syllabus, McLuhan was expected to consult educators, communications specialists, management consultants, psychologists, and sociologists. He did, in fact, consult many of these people, although in a far from systematic manner. His modus operandi was to "visit on a hit-and-run basis all sorts of places,"[4] pausing to examine whatever took his fancy, consulting anyone who gave promise of some nugget of insight or useful information. In the end, he need hardly have left his office; there was little useful information that anyone could tell him. McLuhan's approach, his basic perceptions, his universe of the media were already implanted in his head — and all of it was terra incognita to anyone who had not followed his intellectual path over the previous eight or nine years. The state of ignorance concerning media was so profound, in McLuhan's view, that he was in the position of somebody who had just discovered subatomic particles: the most useful thing he could do under the circumstances was to dragoon people into peering through his microscope and to frame a few initial hypotheses about his discoveries. Reflecting on his approach to media in these years, McLuhan was fond of quoting from Coleridge's *The Ancient Mariner*: "We were the first that ever burst / Into that silent sea."

McLuhan was also supposed to test his syllabus as he developed it, by bringing it into eleventh-grade classrooms. To that end, he recruited some high school teachers in Toronto and a few staff members of the Ryerson Polytechnical Institute in Toronto. He also enlisted a supervisor of the language education department of the Detroit public school system named Ethel Tincher, who had thoroughly impressed him with her experimental and nonmoralistic classes in mass media. McLuhan thought testing his syllabus — asking students questions about media, the answers to which were by no means certain — would promote his notion of education as dialogue rather than as instruction; he was keen to reverse the revolution Peter Ramus had effected four centuries earlier, when that Renaissance pedagogue had turned the classroom into a processing plant for knowledge, powered by the printed book. McLuhan was sure that the students, who had more experience of the electronic world than their elders, might even provide him with new clues to the nature of that world.

The notion of dialogue as a means of gaining insights and erasing an individual's "fixed point of view" (a mental stance that was itself a product of the visual world of print) was very important to McLuhan. From time to time he would seize upon some bit of information offered by a student with the excitement of an angler hooking a trout in a stream that most other fishermen scorned. But he was not sentimental about students. He looked forward to testing his syllabus with all the dark suspicions of a veteran teacher. He expressed concern to Skornia that teaching students about the effect of the media might lead them to develop a widespread contempt for adults, who were mostly lacking in such knowledge. Students could humiliate their elders with this knowledge.[5]

As his own children entered adolescence, McLuhan's sense of caution regarding the young was doubtless reinforced. His children rebelled, for example, at having to say the rosary every night. The girls in particular troubled him. One family friend recalls witnessing their unruly spirit one night in the late fifties. An invitation to dinner at the McLuhans' had been offered, typically, on the spur of the moment. The guest found herself at the table, crowded with the parents and the six children, all eyes fixed on the head of the family as he carved a roast. A plate was passed to each child with agonizing slowness. More and more of the roast, which was not too large, disappeared with each serving, and Stephanie, last to be served, had not even received her portion when the first child to get a plate was already howling for more. With theatrical flair, Stephanie stood up and proclaimed, "Starve me, if you want" and marched out of the room. The incident was a sort of parable; the children protested continually, in various ways, at the meager subsistence they received from the distracted man who loved them but could not give them a father's attention.[6]

If McLuhan, in his capacity as instructor and with all prudent reservations about the effects of his teaching, approached the young with enthusiasm, there was another class of semibarbarian that he was equally keen to enlighten: the corporate executive. In the late fifties, he began to break out of the academic circuit and speak to businessmen, most notably at the General Electric Management Center at Croton-on-Hudson, New York. He was also eager to try out his syllabus there. McLuhan reasoned that business executives, like children, had less investment in print culture and more desire to keep in touch with social change than did academics and that therefore they would be more receptive to his discoveries.

McLuhan was not merely flattering himself or GE when he wrote to Skornia that the Croton-on-Hudson executives were an "august" assemblage.[7] He had a fondness for men with rich and varied life experience, and the approval of such men, particularly if they were successful and of "august" stature, meant more to him than the approval of his colleagues. "He had a weak spot for Americans," one of his associates recalls. "He liked the kind of cowboy approach to things. Whenever he was drawn to a person, that person tended to be physically robust, heavily built, successful, with a lot of single-mindedness and energy devoted to what he was doing. He was not usually clever." (Conversely, English and European intellectuals made McLuhan nervous. The people he personally made fun of were almost always Canadian.)[8]

Of course, part of the appeal of the successful businessman was the possibility that he might put McLuhan's ideas into action. With this possibility, and the apparent receptivity to his thinking of important groups like the NAEB, McLuhan felt a growing exhilaration as the fifties came to a close. In media, he felt, he was tackling a subject of the first importance, and people were beginning to listen to him. In October 1959 he wrote Skornia that the boldness and historical significance of his original thinking made him shudder.[9]

In January 1960, when he was already well into the NAEB project, McLuhan hit upon what he considered a major development of his ideas about the media. It had to do with his basic notion that within the human sensorium each of the senses reacted with others and translated these other senses into their own mode of perceiving.* If a medium extended a particular sense, the other senses also reacted, more or less subliminally. McLuhan once applied this notion when a disciple of his, a Toronto businessman, complained of problems speaking to his superior over the telephone. In person, the man had no trouble dealing with the superior, but over the phone he felt extremely nervous. His voice shook and his breathing became difficult. McLuhan advised him to try to visualize his superior as he spoke. The telephone, McLuhan suggested, was an extremely intense auditory medium; an increase in the visualizing faculty would serve to water down that auditory intensity. This suggestion doubtless would have sounded far-fetched to many people, but the businessman tried it, and it worked.[10]

At the start of 1960 McLuhan began to elaborate on the basic dynamics of this process. First he postulated that any extension of a sense via a medium — what McLuhan called the "structural impact" of a medium — was not the same as the altered pattern of the senses as a whole that resulted from that extension. The altered pattern was the combination of structural impact and "subjective completion" of that impact within the sensorium.

It made a world of difference, as far as subjective completion was concerned, whether the structural impact was "high definition" or "low definition." The auditory image presented by the radio, for example, was high definition. The listener was not overinvolved in that auditory image, did not have to piece it together or fill it in mentally. His other senses, particularly the visual sense, were free to complement the image with a sort of detached, fantasylike response of their own. Thus it was easy to visualize, say, Fibber McGee or Amos and Andy as they spoke on the radio. Children even imagined that these characters were in the back of the radio, performing among the vacuum tubes.

---

* This idea is partly derived from Francis Bacon's concept of *vestigia communis*, the latency of every sense in every other sense.

The auditory image of the telephone, however — an image created by a speaker who demanded a response from a listener, who demanded the filling in of conversational gaps — was low definition. The other senses were not free to get into the act of perceiving the image in such a detached fashion. They were enlisted, so to speak, in the intense effort to make sense of, to complete, the auditory image. Only by a stratagem such as conscious visualization of the speaker could that effort be relaxed and the other senses distanced from the auditory image. A medium with low-definition structural impact, then, required that the user actively participate both mentally and emotionally. A medium with a high-definition structural impact required low participation. A few years later, McLuhan would characterize the difference as that between "cool" (low-definition) media and "hot" (high-definition) media.

If the members of the NAEB research committee overseeing McLuhan's project entertained the notion that his work would easily fit into any known approaches to the study of media, these latest hypotheses quickly set them straight. In February 1960, the committee took a collective deep breath and suggested to McLuhan that it probably wasn't any use to introduce his syllabus to classrooms right away — better to continue working on his ideas and figure out if there was any way to properly test them.

McLuhan enlisted some of his friends at the University of Toronto to test his hypotheses using computers. By the early sixties a small coterie of science-oriented professors at the university, mostly engineers, had formed around McLuhan. Members included James Ham, professor of electrical engineering; John Abrams, professor of industrial engineering; and the three chief members of the coterie, Arthur Porter, professor of industrial engineering, Daniel Cappon, professor of psychiatry, and E. Llewellyn-Thomas, professor of engineering.

McLuhan, so vulnerable to the charge of being unscientific, exercised a kind of spell over these men of applied science. Of course, such scientists are not always as rigorous in their thinking as popular mythology insists; and McLuhan's highly metaphoric flights of perception sometimes appealed to scientists in a particular mood — a what-the-hell-let's-speculate mood.

McLuhan's appeal to these professors, though, was more complex. Porter recalls:

> Initially, I thought the man was crackers. But he aroused my curiosity. When he made certain assertions, for example, I thought, That's queer. He understands information theory. How can a professor of English understand information theory — which is a highly mathe-matical, technical theory?[11]

Of course, none of these scientists ever claimed that McLuhan's grasp of their fields was incontestably firm. McLuhan's grasp of any given scientific theory, like his grasp of empirical fact in general, was always a bit shaky, and sometimes he committed howlers. Once he used the phrase "exponential increase" in a conversation with his colleague Ernest Sirluck when the context called for "arithmetical progression." When Sirluck pointed this out, McLuhan asked, "What's the difference?"

Often, however, he used scientific terms in a sense that was not wrong so much as it was a kind of poetic appropriation. "He understood them as an artist would understand them," Porter recalls.[12] When he spoke of "resonance" or "field" in physics, for example, Porter assumed he was referring to the precise technical meanings of those terms — until it dawned on him that McLuhan was actually using them in a highly metaphorical fashion. (McLuhan would defend himself by pointing out that the terms themselves were metaphorical in the first place.)

McLuhan deliberately sought out people like Porter in order to have friends who represented different shades of the intellectual spectrum, in much the same way that he was drawn to businessmen. On his part, Porter, who had previously despised literature and the humanities in general — "I never read poetry in my life. I thought, how can anyone read poetry?" — was grateful to McLuhan for showing him that the worlds of the scientist and the artist were not so far apart, after all, especially in their use of metaphor.

Some of the scientists in McLuhan's coterie used computers to measure the degree of the tactile, visual, and audible in the auditory image of telephone conversations or radio broadcasts or whatever. McLuhan was dizzy with the implications of being able to gauge accurately the mix of the senses involved in images. The process would open up the mysteries of the human sensorium for the first time. Cars, clothes, furniture, almost anything could be designed so as to be perfectly adapted to the human sensorium. New methods of teaching would become evident. If the new math had a high auditory component, for example, it might best be taught through a medium that encouraged users to talk to themselves, in the same way that radio encouraged listeners to visualize. Algebra or literature might then be learned with almost unconscious ease. An adult Canadian could learn, say, Chinese with the thoroughness of a two-year-old living in China.

Such engineering of culture and human receptivity McLuhan regarded not as a threat to human freedom but as an enlargement of human awareness, ultimately leading, as he suggested to Skornia, to a state of almost universal ESP.[13] (McLuhan appears to have regarded ESP not as some mystical phenomenon but as a state of total perception in which all the senses were working harmoniously and at full strength.)

In the meantime, McLuhan applied himself to generating insights about the newest and most threatening medium of all: television. He picked up a few clues about the nature of this medium as he went along. A research assistant noted that a class watching a television broadcast inviting them to join in a sing-along responded to instructions automatically — something the class never would have done with a film. McLuhan regarded this as proof that television viewers participated profoundly in what they watched. His conclusion was corroborated when he came across the remark of a doctor at the Kansas City Medical Center to the effect that medical students watching an operation on videotape did not so much see as actually perform the operation themselves.[14]

The participatory quality of television, McLuhan maintained, was due to the relatively indistinct quality of the television image and the resulting low definition of its structural impact. Like the auditory image of the telephone, the television image compelled all of one's senses to get into the act, particularly, McLuhan insisted, the tactile sense. (For McLuhan, "tactile" meant not so much the sense of touch as the interplay of all the senses, an instance of his confusing tendency to invest words with a meaning peculiar to himself.)[15]

McLuhan's theories about television were shaped, characteristically, by a book on a quite unrelated subject. From reading *The Problem of the Form in the Figurative Arts* by the German sculptor and art critic Adolf von Hildebrand, McLuhan concluded that true vision was tactile in its essence. In that book, Hildebrand stated,

> Nature having endowed our eyes so richly, these two functions of seeing and touching exist here in a far more intimate union than they do when performed by different sense organs. An artistic talent consists in having these two functions precisely and harmoniously related.[16]

(McLuhan had already learned from Wyndham Lewis in *Time and Western Man* that the powers of the eye were valuable only when employed in concert with the other senses.) True vision discounted the purely visual as understood by the painters of the Renaissance: it abandoned the use of perspective and three dimensions. Instead, it was two-dimensional. The significant analogue to the television image was the image created by the painter with no-perspective, no-horizon, two-dimensional tactility.

At this point, McLuhan picked up a few hints from his artist acquaintances. André Girard, a painter well known in Catholic academic circles who had done visual experiments with film and television, told McLuhan that the French artist Georges Rouault had been the inspiration for these experiments. Rouault had painted his pictures as if they were stained glass windows. The subsequent quality of his paintings, conceived as if they were illuminated from a light within, reminded Girard of the television image, with its tiny, multitudinous electronic points of light. Both were basically two-dimensional and tactile.

Another friend of McLuhan's, a Toronto artist and museum designer named Harley Parker, pointed McLuhan to the French impressionist Georges Seurat. Seurat's pointillist method struck McLuhan as another painterly anticipation of the television image. His tiny dots of pure color created images that also seemed to be lit from within. (Parker and McLuhan toyed with the idea that the viewer looked at an image of, say, Raphael or Vermeer, but an image of Seurat or Rouault looked at the viewer.) The difference between the television-like art of Seurat and the art of the Renaissance, McLuhan explained, was the difference between "light through" (the stained glass window) and "light on" (Renaissance portraits).

It is difficult to determine whether the example of painters like Rouault and Seurat, any more than technical information he picked up from television engineers, was responsible for the formation of McLuhan's ideas about television. It seems at least as likely that he fastened on all this information as confirmation of ideas he had already formed. Certainly, as far as painting was concerned, McLuhan was no connoisseur. "He was not the slightest bit interested in art except for what it conveyed in terms of ideas," Parker comments — a complaint about McLuhan frequently echoed by literary critics.

Analogies like "light on" and "light through" were intriguing, although they were easy targets for critics. More to the point, McLuhan's investigations into television, while not scientifically rigorous, were obtaining promising results. McLuhan recognized very quickly, for example, that sharply defined, intense, emotional personalities were ill suited to television. Commentator Paul Harvey with his punchy delivery was fine for radio; on television his conservative message was much more effectively conveyed by someone like William Buckley. McLuhan observed that Hitler was made by radio and would have been undone by television. The remark touched a nerve with many left-wing commentators, who seized on it as proof of McLuhan's frivolous approach to things; yet it is an entirely accurate and perceptive remark. Radio excited people because it stimulated the ear, that emotive organ, and encouraged fantasy; television, on the other hand, held people in the grip of a profound involvement with its electronic dots, an involvement that could easily be mistaken for passivity and even mental vacancy. McLuhan went so far as to suggest in 1960 that radio should be shut down for good, while television would be an excellent antidote to restlessness in the "hot spots" of Asia and Africa.[17]

McLuhan was well aware that his theories were going to upset many people. He compared himself to Louis Pasteur, who discomfited his colleagues with talk of germs. McLuhan's theory of change, emphasizing the means of communication rather than the means of production, would supersede Marxism, he confidently predicted. To those who criticized McLuhan (and their numbers would increase as he became better known) for taking attention away from the question of who actually owned and controlled those means of communication, he had a simple answer: it was futile to know who wielded power without knowing what the nature of that power was.[18]

This stance was not uncongenial to the man who had once written articles for his university newspaper denouncing the modern world and urging a return to the pre-industrial paradise of Ruskin and Chesterton. Indeed, McLuhan suggested to Skornia that a conservative (like McLuhan himself) was better prepared than a progressive to understand this era of breathtaking change. Instead of trying to tinker with and adjust to the change like the progressive, the conservative, fortified by his basic distrust of the entire process, could actually *contemplate* change as a process. Of course, this contemplation radically set McLuhan apart from everyone else caught up in change. In a letter to Skornia in March 1960, McLuhan used the word "somnambulists" to describe virtually everyone he was now dealing with. It was a term he would increasingly use to characterize those who singlemindedly pursued goals in near total ignorance of what they — and the media they used — were actually doing. (McLuhan had picked up the term from a comment of Hitler's that had greatly impressed him: "I go my way with the assurance of a sleepwalker.")[19]

McLuhan finished the first draft of his "Report on Project in Understanding New Media" at the end of June. In the preface, he stated the premise that would later inform his book *Understanding Media*: the idea that media are capable of "imposing" their "own assumptions" on the people who use them. McLuhan also warned that, unless people understood the nature of media, including the new electronic media, they were in danger of losing all the traditional values of literacy and Western civilization. This was not, be it noted, the warning of a man besotted with the prospect of the coming global village.

The report was divided into sections devoted to various media: speech, writing, print, prints (engravings, lithographs, and so on), the press, photography, the telegraph, telephone, phonograph, movies, radio, and television. Each section contained an introduction to the medium, "projects and questions" for classroom use, a bibliography with additional comments and questions, and a chart with commentary. The charts were very simply presented. The upper left-hand corner of the rectangle was labeled HD (high definition), the lower left SI (structural impact), the upper right SC (sensory closure or subjective completion), and the lower right LD (low definition). In each corner were a few descriptive words and phrases appropriate to the medium featured in the rectangle.

The charts were intended to show the basic dynamics of the human senso-rium's and the human psyche's reactions to each medium. In fact, they were extremely confusing. It would have taken a very dogged reader to make sense of them, and even then that reader would have been puzzled by inconsistencies and the seeming absence of any rationale behind them.

McLuhan used the charts publicly for the first time in March 1960, when he appeared before a conference of the Department of Audio-Visual Instruction of the National Education Association. He seemed to think that the charts had a "paralyzing effect" on the audience, not dissimilar perhaps to the gestures of a professional magician, which serve to distract the audience until the rabbit is pulled out of the hat. Instead of a rabbit, McLuhan produced his startling con-clusions about the true nature of media.[20]

The "paralyzing effect" of these charts was no doubt due to sheer baffle-ment on the part of his audience, especially at his use of a lecturer's device usu-ally associated with clarity and simplification. The charts were also an attempt to give a scientific aura to his insights: McLuhan was always at his worst when the systematizing fit came over him.

McLuhan maintained that he was only experimenting with the material pre-sented in the charts and no doubt assumed that others would polish them, amplify them, and correct inconsistencies. In the meantime, they underlined his essential point: that the sensory *impression* of a medium was not the same thing as its ultimate sensory *effect*. He was, again, trying to lay down a theoret-ical groundwork for his intuitions about media and artifacts in general — for instance, his intuition that one did not actually have to feel a net stocking in order to have one's tactile sense stimulated.

The whole business of structural impact and subjective completion (ulti-mately derived from Hildebrand's ideas about the unity of vision and other sense impressions and also from Bacon's *vestigia communis*) is one of those areas in McLuhan's thought that will not bear much scrutiny. McLuhan himself said very little publicly about subjective completion in the years after 1960. The dis-tinction that seems to have remained most vital to him was that between high definition and low definition in the volume of information or impressions offered the senses by the media: the distinction, that is, between "hot" and "cool," detachment and involvement.

The NAEB had no idea what to do with McLuhan's report. Skornia put together a final report based on McLuhan's first draft and including oral and written materials McLuhan had passed along. The resulting document was pub-lished late in 1960 under the title "Report on Project in Understanding New Media." Skornia was proud of his part in conceiving and nurturing this final product. Not only had he suggested it to McLuhan and obtained the grant for it, but he had shielded him from irate colleagues in the NAEB who thought the project a complete waste of time.

One of the milder criticisms of McLuhan's syllabus from NAEB members was that the text "was immoderate in its demands on a Grade XI level." That was an understatement. When McLuhan showed Lou Forsdale, his friend at Columbia University Teachers College, his June draft of the report, Forsdale recalls, "I was astonished because in the preface he had stated that this was a high school textbook. That was insane. And that was the first time I noticed McLuhan had no notion of what high school students could do."[21]

McLuhan claimed that the eleventh-grade students he talked to understood his media insights very quickly. We may take McLuhan's word for this; young students, as McLuhan claimed, may have been more receptive to this new approach to the media than their elders. If so, the fact did not make his syllabus any less daunting. Questions such as "Speech as organized stutter is based on time. What does speech do to space?" do seem a touch abstruse.

McLuhan's explanations of the media in the syllabus were not much more helpful to high school students. McLuhan himself dropped some of these explanations or at least de-emphasized them in the years to come. He presented the idea in the report, for instance, that photography inspired an "extrovert generation" because it encouraged active voyeurism. Television, on the other hand, à la Georges Seurat, turned the viewer into the screen rather than the camera eye. McLuhan asked if this would result in an "introvert generation."[22] He would have his answer in a few years — a surprising one. But then all these questions and statements, McLuhan insisted, were wholly "experimental." He was, he always claimed, perfectly willing to drop any statement he made that was disproved by experience.

The final report also included the script of a video McLuhan had made on the Gutenberg era along with the script of a half-hour video on the contemporary teenager inspired, if not wholly scripted, by McLuhan and subsequently aired as part of the Canadian Broadcasting Corporation's series "Explorations" on May 18, 1960. This video opened with a shot of a Lolita figure kissing an older gentleman. The narrator solemnly warned the teenage audience that the show might be too "shocking or controversial" for their parents, who would be better off with an early bedtime. Among the "shocking or controversial" points made by the program was the suggestion that teenagers considered their own subculture to be more mature, worldly-wise, and challenging than the world of grown-ups: in their eyes, adults were mildly pitiable figures, to say the least. McLuhan also emphasized the tribal nature of contemporary adolescents. They happily conformed to tribal mores, performed tribal rituals, and rigorously suppressed the expression of individual attitudes and emotions.[23]

The video, although it plainly did not foresee the emergence of the hippie or the student radical, nonetheless made a few points about the increasing precocity of adolescents that have since become fairly obvious. They were not obvious in 1960. McLuhan received almost no response to the video from the NAEB or the CBC audience. Nor was his syllabus itself ever used in classrooms. The NAEB report remained of interest in the coming years solely to students of McLuhan.

Despite its patent flaws as a textbook, the report nonetheless contained a great deal of value. It tried to impart to teachers and students alike a habit McLuhan himself had carefully cultivated: the habit of observing the effects of things and thereby perceiving these things in fresh ways. (As an undergraduate at Cambridge, McLuhan had shown the germ of this habit when he casually noticed that the effect of central heating was to make English people more interested in windows.)[24] A look at the changes a new photocopier introduced into the life of a school, for example, would tell one much more about that artifact than any description of what the photocopier did or any examination of what was copied on it. Such observation could be made free from the pseudoscientific language of psychological researchers, of the academics who, as McLuhan wittily observed, deem it "indecent to utter high secrets except in technical language." A question such as "Why should the sending or receiving of a telegram seem more dramatic than even the ringing of a telephone?" could have sparked some interesting discussions in a classroom. And a statement such as "War now consists not of the moving of hardware but of information" hit a target, and a very important one.

The report as a whole capped a decade in which McLuhan had moved from his rather grim post–*Mechanical Bride* mood to a new buoyancy and playfulness. It is clear that McLuhan viewed his study of the media almost as an act of revenge for what the media were doing to him, his family, and his world. "It's vital to adopt a posture of arrogant superiority," he was fond of saying.

> Instead of scurrying into a corner and wailing about what media are doing to us, one should charge straight ahead and kick them in the electrodes. They respond beautifully to such resolute treatment and soon become servants rather than masters.[25]

"To the student of media structures," McLuhan wrote in *Understanding Media*, the book that eventually resulted from his NAEB work, "every detail of the total mosaic of the contemporary world is vivid with meaningful life." It was hardly an exaggeration to say of McLuhan, when his spirit was most free and unvexed and when he was rousing the spirits of his friends with his own energies, that he was aware of and enjoyed virtually everything around him. In those moments it seemed that he was in a permanent state of vision, that he was an embodiment of his own theories. If one is not perceptually alert, McLuhan argued, one was as good as dead. And he insisted that everything one perceived was real — even hallucinations, in a manner of speaking.

It is no wonder that in the years immediately following 1960, McLuhan, despite his reflexive disclaimers of partisan feeling for the new media, was optimistic. He articulated the notion that, whereas earlier technologies had extended one sense or one part of the body — the wheel extending the foot, for example — the new electronic technologies extended the entire human nervous system. The very movements of information in these new technologies corresponded to the movements of the human mind.

Previous to this, it had been possible for human beings to numb themselves to the extension of their bodies or senses, to regard these extensions as separate from themselves and not impinging on their psyches. With the extension of the nervous system through the new media, however, this state of unawareness became impossible. Hence the human race was fated to experience a period of intense learning and discovery. The arts and sciences, he assured the Humanities Association of Canada in June 1961, would experience an era of unprecedented accomplishment.[26]

McLuhan was by no means an optimist by temperament. He was, he thought, simply responding to the realities of the new age. It is possible he was also responding to a curious stirring of optimism that gripped the Western world in the early sixties. In the United States this stirring was reflected in the rhetoric of the New Frontier. In a more startling fashion, and with more profound and lasting results, it was reflected in the papacy of John XXIII, who opened the Second Vatican Council in 1962 with a historic address reversing the Catholic Church's long antagonism to the modern world and predicting a benign "new order in human relationships." Whether McLuhan was aware of it or not, his optimism was perfectly in tune with a certain mood spreading, in subtle and mysterious fashion, through the society around him.

When the NAEB published his report in 1960, McLuhan added a note revealing that his health had broken from the strain of overwork on the project. He had been hospitalized, and his doctors insisted on prolonged rest. According to Carpenter, the crisis in McLuhan's health was, in fact, a stroke earlier in that year that put him in such danger that his employer (and confessor) Father John Kelly was summoned to give him the last rites. McLuhan never alluded to the incident again, and to all appearances he recovered completely, in time to teach his classes in the fall of 1960. He did his best, in fact, to pretend that he had never been stricken. But he never regained the appearance of a man enjoying robust good health. His pale coloring, his nervous intensity, and his patent inability to relax revealed ever after a body being driven dangerously beyond the limit.

The NAEB report caused no great stir, although a research paper published in 1960 by the Stanford Institute for Communications Research took note of it. According to the author of the paper, McLuhan was a thinker "whose rate of idea generation is so great that, by chance alone, some pregnant insights are predictable."[27] McLuhan did not mind this ambiguous compliment. He took it for a genuine tribute. As one of his associates comments, "McLuhan thought he just had to get as many ideas out as possible, to utter or 'outer' them, and let the environment decide which were important and which were not."[28] Of course, this scattershot technique also resulted in a certain quota of nonsense.

McLuhan had increasing opportunity in 1960 to employ his technique for generating ideas. (The year also marked the publication of an anthology of articles from *Explorations* entitled *Explorations in Communication*.) Indeed, McLuhan began to assume the stature of a sage, at least in his hometown, where his vast circle of acquaintances, including businessmen and journalists, were spreading his reputation as a man with fascinating comments to make on virtually everything. Newspaper reporters increasingly sought him out for some of these comments on issues of the day. After the U-2 incident in 1960, for example, in which a U.S. spy plane was shot down over the Soviet Union, he told a reporter from the *Toronto Daily Star* that the Russians were upset not so much by espionage as by the fact that the United States used visual means to uncover secrets. The Russians, McLuhan said, were "ear people." The Americans, on the other hand, were "eye people." He then explained why the Russians' love of choral singing and ice hockey put them in the ear camp and also gave a run-down on the general differences between these two types of humanity.

McLuhan also applied his ideas to the 1960 televised Kennedy-Nixon debates. According to McLuhan, Kennedy's image was that of "the shy young sheriff." Nixon's image was that of "the railway lawyer who signs leases that are not in the best interests of the folks in the little town." Kennedy, according to McLuhan, projected a less clearly defined image than Nixon, more nonchalant, slightly blurred — in other words, he was a "cool" personality and therefore ideal for television.[29]

McLuhan's analysis stands up well almost thirty years after the debate — but then, he seemed to have a knack for understanding political drama. Even before he began to study media, he realized that North American politics had become a branch of the entertainment industry. His media studies underlined the point.

The advent of JFK's Camelot intrigued McLuhan and made comment irresistible. In 1962 he observed that "a four-year stint in the White House is no longer easily distinguishable from something arranged by a booking agency."[30] Although deeply shocked by Kennedy's assassination, he was sufficiently collected in his response to write an astute article about the event, entitled "Murder by Television." McLuhan noted that Lee Harvey Oswald's murder in the Dallas police station was made possible by the fact that his guards were wholly distracted by the presence of television cameras. He also noted that the televised funeral of the president was additional evidence that the medium, far from arousing passion or excitement, created a profound, almost tranquilizing communal involvement for viewers across the nation.[31]

McLuhan was advancing on another front by 1960. His talks to the GE executives in their management center in Croton-on-Hudson were an important beachhead in his new career as communications specialist. The talks had been arranged by an old acquaintance of McLuhan's named Ralph Baldwin, a former professor of medieval literature at Catholic University in Washington, D.C. Like Muller-Thym, Baldwin was a Catholic academic who had become frustrated with the low pay and poor conditions of his employment and had found it easy to move from medieval lore to the world of avant-garde corporate management.

In the fifties he joined the staff of the school that GE had established in Croton-on-Hudson for its executives. The school, the first of its kind, had been set up under the influence of the doctrines of still another academic–turned–management theorist, Peter Drucker. Prior to its establishment, promising executive talent had been sent off to regular academic courses at various universities to put the final touches on their intellectual development. Now similar courses would be administered by a school run exclusively by and for the company. GE was prepared to spend whatever money was necessary to bring in the best lectures — academics from Harvard and Yale, high-priced consultants like Drucker, and even Ronald Reagan, then a television spokesman for GE and a communicator of note.

Baldwin had known McLuhan since the forties, when McLuhan had seriously considered teaching at Catholic University. "I invited McLuhan [to the GE school] because communications was part of the course at Croton-on-Hudson," Baldwin recalls. "Corporate 'communications' in those days was very much considered to be a matter of writing letters and memos. I thought that, from top management on down, the Croton-on-Hudson students should be exposed to the latest thinking on the subject."[32]

McLuhan had spoken to the odd group of businessmen before his appearance at Croton-on-Hudson in 1959, but now he was moving into the major leagues. He was delighted at the opportunity and at the indication in yet another arena that the coming age would be one of great cultural enrichment. Executives would become erudite men of ready and eloquent speech, the ideal of the Renaissance that rhetoricians revived in the late twentieth century.

Certain that the intensified application of intellect to business would result in a brave new world, McLuhan perceived nothing but exciting possibilities in his work with the executives of General Electric. For the most part the executives seemed to respond well to his enthusiasm. Of course, those with a penchant for logic sometimes asked McLuhan questions that clearly stymied him — questions, for example, about how his theories applied to particular data. "He wouldn't answer these questions," Baldwin recalls.

> He would fumble and vaporize and go on to something else. It was more the fault of the questioners than anything else — they didn't know how to take what he was saying. They wanted a kind of laboratory plot or plan for McLuhan's insights.

Occasionally, such as in the washroom between lectures, Baldwin heard some of the men grumbling about McLuhan's far-out ideas; during the lectures themselves, however, members of the audience were not about to demonstrate displeasure with someone brought in at considerable expense by their company.[33]

But corporate executives enjoyed bull sessions as much as anyone else, and as one businessman said of a talk McLuhan gave in Montreal, "If nothing else, it was good entertainment." That sentiment was to sum up the feelings of a good many of McLuhan's corporate audiences in the sixties and seventies.[34] For McLuhan, the chance to lecture to businessmen on what their businesses were really all about, at high fees — in 1964 Croton-on-Hudson paid McLuhan $250 plus expenses for one appearance — was, at the very least, sweet recompense for the failure of Idea Consultants in the mid-fifties.

McLuhan's career ascent was interrupted by a moment of personal grief when his mother died on July 10, 1961. Elsie McLuhan had settled permanently in Toronto in 1953 after several years of moving between Toronto and Detroit, where she had worked as a teacher of speech and drama. Elsie was quite proud of her older boy — she occasionally amused and annoyed McLuhan's students and colleagues by insisting that they refer to him as Doctor McLuhan — and she had moved into an apartment on Wells Hill Road to be near him.

In 1956 Elsie suffered a stroke that left her partially paralyzed and unable to speak. Every day McLuhan took the long streetcar ride to the hospital and read detective stories to her. The man who at times could be so oblivious to the emotional states of others was keenly sensitive to his mother's sufferings. He even noted the speechless and helpless woman's chagrin at the hairs growing on her chin and made sure they were continually removed.

When she died, McLuhan was grief-stricken. Elsie had been, at best, a difficult parent, spreading unhappiness about her like an aerosol that made it hard to breathe. Yet if McLuhan had inherited his tender conscience and firm morality from his father (who would die in Winnipeg five years later), he had inherited from his mother a spunkiness, an outrageous independence and indifference to what others thought, that set him apart and marked his approach to every project he pursued. He inherited something else as well: an irrepressible verbal aggressiveness. He was more like his mother than he was like anyone else on earth. When only a handful of people appeared at her funeral, he was deeply hurt. Thereafter he kept her clothes and other possessions locked in a closet in his room. At times he was convinced that her ghost lodged there and could be heard rustling among the dresses she had once worn while performing her monologues.

That summer, McLuhan decided, finally, to put together a book he had been working on since his days in St. Louis, a book that would draw on both his studies of the trivium and his more recent media studies in a major restatement of Western history. Presenting the case for the devastating effects of print on every aspect of life in the West, the book would establish McLuhan as a critic of culture with a new approach to cultural change almost as startling as Freud's approach to personality or Bohr's approach to nuclear physics.

McLuhan had made efforts periodically to polish off this work. In the summer of 1955, he wrote Wyndham Lewis that he was working on a book entitled *The Gutenberg Era* that dealt with the cultural changes involved in the introduction of literacy and the printing press. In the summer of 1958 he wrote to friends, on his way back from Santa Barbara, that he had perceived the "total patterns" necessary to complete *The Gutenberg Era*. The following summer he reported that he was well into the writing of the book. Before starting work on his report for the NAEB he suggested to Skornia that, if he could get a two-year grant, he could preface the report with a completed version of *The Gutenberg Era*, which he said had been more than ten years in the making. In three months, he thought, he might get it all down. He needed only a push. Once he got started, he wrote rapidly, but it was sometimes very difficult for him to make that start.[35]

Like many brilliant talkers, McLuhan found writing tedious and awkward. He had indeed been thinking about this book for a decade or more, jotting down notes, collecting quotations. He had not, however, managed to fully launch into it. The summer of 1961, he determined, was to be his decisive attempt. McLuhan later claimed that he had been spurred finally to write the book after the appearance of an article by J.C. Carothers, entitled "Culture, Psychiatry and the Written Word," in the November 1959 issue of the journal *Psychiatry*.

Carothers maintained that the ear, not the eye, was the main receptive organ of the mind, that thinking and behaviour were governed by the power of words, and that space-time relations and the mechanical notion of causality were dependent on a "habit of visual as opposed to auditory synthesis." The article might have been written by McLuhan himself. It said almost exactly what he had been saying since the early issues of *Explorations*, and it came both as timely encouragement and as a warning to McLuhan that he had better get his ideas between the covers of a book and stake a firm claim to them. McLuhan by this time realized what most writers come to realize sooner or later — that one small book does more for a writer's reputation than a hundred articles.

McLuhan roused his energies with almost heroic determination and took over the entire reading room of the library of St. Michael's College that summer. He sat in front of stacks of index cards, each with a quotation written on it. His massive collection of these cards represented twenty years of reading and hunting for vital clues to the mystery of Western civilization. For three months he flipped through the cards, mentally juggled quotations and his own reflections, and filled pages with elegant handwriting, as crisp and incisive as the writing of medieval scribes. (Afterward, when McLuhan received page proofs for the book, he called in as reinforcements several Basilian seminarians. They located all the books in the library from which McLuhan had quoted and piled them on tables. As McLuhan read out quotations from the page proofs, the seminarians rooted through the piles, located the passages in the books, and checked for accuracy. What would have been an endless task for McLuhan alone was completed in one afternoon.)

The prose in those page proofs was characterized by what McLuhan termed an "oral, noisy brashness."[36] Some readers, like one Toronto reviewer, compared it to the writing of "a mad jackdaw."[37] McLuhan's defense of his prose was an attack on its opposite, the lucid and the smoothly flowing. "Clear prose indicates the absence of thought," he intoned in an interview later that decade. He maintained that his sweeping and startling generalities were intended to poke the somnambulistic reader in the ribs. Highly polished and measured prose merely served to lull such readers into even less alert states of mind.[38] *The Gutenberg Galaxy*, as the book was finally titled, was written to please himself, he told friends. It was as defiant and eccentric as the works of Ezra Pound and Wyndham Lewis and devoid of the "meek and discreet little gestures" McLuhan discerned in the writing of most nineteenth- and twentieth-century authors. (In a letter to a friend, referring to *The Gutenberg Galaxy*, he quoted a sentence from a passage of Wyndham Lewis's autobiography *Rude Assignment*, in which Lewis recounted his defiance of literary and political interest groups: "I have written as if I had the freedom of a man who lived in the 18th century, or at the Ritz.")[39]

THE ELECTRONIC CALL GIRL

The "oral, noisy brashness" of McLuhan's prose would intensify as he developed the habit, in the mid-sixties, of dictating to his wife or secretary instead of sitting down and writing himself. His books took the form of ad-lib comments about random ideas — newspaper articles, ads, passages from "serious" books — dictated to his amanuensis in the manner of some ancient worthy dictating to his faithful scribe. McLuhan would later poke fun at his own style in a phrase in *Understanding Media*. Doubtless the phrase slipped past many readers of that book: "Now that we have considered the subliminal force of the TV image in a redundant scattering of samples...." This "redundant scattering of samples" McLuhan termed a "mosaic."[40]

The Gutenberg Galaxy was such a mosaic — some would say a hodgepodge. More than half the text consisted of long quotations from nearly two hundred authors, interrupted every so often by McLuhan's own ruminations. The quotations sometimes amplified and clarified the ruminations preceding and following them, and sometimes they didn't. Sometimes they pointed in an entirely different direction, like an image in a surrealist painting that seems to have strayed onto the canvas by accident. No wonder McLuhan told a friend who asked about the progress of *The Gutenberg Galaxy* that he was in the process not of writing it, but of "packaging" it.[41]

After completing the bulk of this packaging in the summer of 1961, McLuhan finished the book in the spring of 1962. *The Gutenberg Galaxy*, published that fall by the University of Toronto Press, was by any measure a curious volume. The main text began with an exegesis of *King Lear* and ended with an extended treatment of Pope's *Dunciad* — a reminder, perhaps, to the reader that the author made his living as a professor of English literature and considered himself to be an authoritative literary critic. These beginning and ending sections could have been switched around with no loss of coherence to the text — as, indeed, almost any part of the text could have been placed in almost any other part without interfering with the development of the author's argument.

Odd as the structure of the book was, however, its theme was clear. *The Gutenberg Galaxy* explained why the Western world had become devoted to a visual orientation to reality. It postulated that the tribal state was the normal condition of humanity and that that condition had been disrupted in the West by the invention of the phonetic alphabet, a radical technology unique to the West. Amplifying Walter Ong's thesis, McLuhan argued that the invention of print effected a still more profound transformation in the psyche of Western man, leading to an emphasis on the visualization of knowledge and the subsequent development of rationalism, mechanistic science and industry, capitalism, nationalism, and so on. Laced throughout the text was material from McLuhan's

Ph.D. thesis, used as a kind of subtheme explaining how the printing press eventually de-emphasized traditional studies in rhetoric and grammar and brought logic and dialectics into prominence.

Unlike *The Mechanical Bride*, *The Gutenberg Galaxy* was not ignored. In December 1962, Alfred Alvarez wrote a review of the book in the *New Statesman*; that review established a pervasive theme in subsequent McLuhan criticism: McLuhan as crypto-Catholic. The effect of McLuhan's book, Alvarez wrote, was that of "a lively, ingenious but infinitely perverse *summa* by some medieval logician, who has given up theology in favor of sociology and knows all about the techniques of modern advertising."[42] In February 1963, the noted English critic Frank Kermode wrote a long, intelligent, well-balanced review of the book in *Encounter*. His review ended by saying that McLuhan's book offered a "fresh and coherent account of the state of the modern mind in terms of a congenial myth. In a truly literate society his book would start a long debate."[43]

As if responding to the cue, that venerable organ of the English literary establishment, the *Times Literary Supplement*, published an article by McLuhan on the effects of print in July of the same year. In August 1964, the *TLS* included McLuhan in an issue devoted to the world's current avant-garde thinkers and printed a short manifesto-like statement by him as an indication of work the *TLS* editors believed "genuinely to be breaking new ground."[44] In Canada, meanwhile, McLuhan won the 1962 Governor-General's Award for Non-Fiction, the country's highest literary award. McLuhan and his wife went to Ottawa for a reception for the award winners and were wined and dined in proper style. Toward the end of the evening, Corinne was observed stepping into a limousine and proclaiming, in the rich accents of Fort Worth, Texas, "I just love Canada." This taste of high life was delicious to someone who had endured twenty years of drudgery as a faculty wife.

McLuhan's fame was also spreading in the United States. He was no longer having to hustle the mighty to gain a hearing; now the mighty, or their minions, were coming to him. In 1963, Charles Silberman, a writer for *Fortune* (he later profiled McLuhan in the February 1966 issue of the magazine), was commissioned by the corporation to do a long-range planning study for its top management. He was intrigued by McLuhan's comments on the impact of television on *Time* and *Life* magazines — the Luce publications had long been a subject of interest to McLuhan — and enlisted his help for the study.

McLuhan tactfully refrained from referring Silberman to previous articles he had written on the subject of Luce publications, such as "The Psychopathology of *Time*, *Life*, and *Fortune*," published in *Neurotica* in the late forties. Instead he wrote Silberman a letter expressing his understanding of the unique importance of the Time-Life enterprise, which had much potential for uniting "the cultural and the commercial" in this age when art and science were becoming part of the background of everyday life. (It was in a similar spirit that he recommended the trade journal *Advertising Age* to a friend, assuring him that he would find "every school of aesthetics and art controversy richly illustrated in its columns.")[45] Silberman offered McLuhan a $700 fee for his work on the study.

McLuhan spent considerable time with Silberman in New York City during the summer of 1963 before preparing several written briefs for him, suggesting, among other things, that *Life* bring its features up to the same high standards — the same "iconic inclusiveness and integrality of image" — as its humble advertisements. He suggested that the magazine might feature articles devoted to the "extensions of man" — the various human inventions and their sensory consequences.[46] McLuhan did not say so in these briefs, but it might have occurred to him that, if the magazine followed his advice, no better man could be found to produce those articles than a certain English professor at the University of Toronto.

Silberman had discovered that *Life* magazine, despite the millions of dollars it had spent on market research over the years, had never really investigated the ways in which television affected its readership. He was therefore eager for McLuhan's insights, but he apologized to McLuhan for his inability to swallow them whole by citing the dizzying mental changes involved in becoming an instant "McLuhanite." Silberman ended up including what he refers to as "the first lucid English translation of what McLuhan was saying" in his report to his boss, Hedley Donovan. He also told Donovan, he recalls, that McLuhan himself was probably "half genius, half madman — and that it wasn't always possible to separate the two." (Leslie Fiedler, the noted literary critic who knew McLuhan in these years, had a slightly different estimate. "My sense of him from the beginning," Fiedler comments, "was that he was two-thirds an absolutely fascinating analyst of society and culture and one-third mad. Afterwards the balance tilted.")[47]

To test McLuhan's insights into the nature of print and television, the Time-Life Corporation employed a young psychologist in market research, Daniel Yankelovich. According to Silberman, Yankelovich tested McLuhan's ideas by exposing audiences to both television and print advertisements. "The central insight that emerged from these tests," Silberman recalls,

> was that people acquired far more information from the print form than from television and that the television form conveyed a far more visceral appeal than print advertisement. Viewers were much more

aware of the surge of the salt spray, the foam on the beer — but they had far more knowledge of what the product was from print. So the tentative hypothesis Yankelovich drew was that, for print advertising, one ought to focus on products whose sale required consumers to have more information rather than products whose sale depended on a visceral appeal.[48]

It was an interesting test, coming virtually to the opposite conclusion of the test set up by Carpenter in 1954. Silberman recommended that the corporation launch a large-scale study along the lines of Yankelovich's test, but nothing was done.

The readers of Silberman's report were, by and large, intrigued with the section dealing with McLuhan, but there was little specific response to that section or to Silberman's report as a whole. The management of *Life*, on the defensive (the magazine was already clearly on the road to the demise predicted for it by McLuhan in an issue of *Explorations*), pointedly ignored it. The report did, however, make McLuhan known to increasing numbers of influential media executives in New York. About the same time that Silberman issued his report, a staff member of *Time* who compiled an occasional memo on important cultural trends for circulation among the magazine's staff featured McLuhan in one of her memos. That document also went to several major advertising agencies.

The year 1963 also saw the realization of a dream that McLuhan had nurtured since the days of the Ford Foundation seminar group in the early fifties. When the foundation grant expired in 1955, he had applied for further funding for a Contemporary Institute of Culture, which he felt was a natural outgrowth of the work of the seminar. The foundation summarily rejected the idea. McLuhan nonetheless had continued to hold informal weekly seminars at his home or his office throughout the fifties and early sixties. The seminars, attended by his friends, disciples, curious onlookers, and occasional sparring partners, provided him with the audience he seemed to require to talk out his ideas.

McLuhan had not abandoned the idea of somehow institutionalizing and funding these seminars. The rationale for such an institution was the continuing need for the Innis-like study of the effects of human artifacts. The institution could also confront the emotional resistance that McLuhan believed existed against such studies. (McLuhan puzzled for many years over this resistance — at one point he decided that people resented studying effects in the same way they resented any other invasion of their privacy.)[49]

When McLuhan was finally offered the opportunity to establish an institution of his own making at the University of Toronto, he created the Centre for

Culture and Technology with the stated purpose of investigating "the psychic and social consequences of all technologies." A secondary purpose, setting up dialogue among the different faculties and departments of the university, was a bow to the old hopes of the 1953 seminar, hopes that the new center would have no more success in fulfilling.

An unstated purpose of the university's sponsorship of the center was the desire of Father John Kelly and Claude Bissell to keep McLuhan at Toronto. Kelly and Bissell headed St. Michael's College and the University of Toronto, respectively, and were close personal friends of McLuhan as well. Universities in the United States were offering him "two, three, five times as much money as he was getting at St. Michael's," according to Kelly. Of course, many of McLuhan's colleagues would have been happy to see him go. Kelly, however, had a soft spot for offbeat faculty members whose unorthodoxy did not extend to challenging his authority. In his long years as president of St. Michael's he was always McLuhan's first port of call when McLuhan needed something from the administration.

Bissell was a former neighbor and personal friend almost from McLuhan's arrival in Toronto. Together, these two men had hatched the scheme to provide McLuhan, in Bissell's words, with "the freedom to work easily and effectively and, at the same time, with the support of an institutional base and a certain status within the university."[50]

In 1963, Bissell, Kelly, McLuhan, and Ernest Sirluck, then assistant dean of the School of Graduate Studies, worked out the details of the interdisciplinary research center. McLuhan would be released from at least half his normal teaching duties and would be given a headquarters. The college would pay one-third of an augmented salary, and the rest would come out of the budget for the center, paid by the university. A "draft constitution" for the center was prepared: a simple document for a simple institution.

"Headquarters," for example, was an office in a seedy Victorian house on the St. Michael's campus, with wooden floors that creaked and a door leading to the street on which McLuhan, ever sensitive to noise, hung a sign that read SLAM GENTLY. (He claimed this oxymoron communicated its message very effectively.) Into this office McLuhan piled his six or seven thousand books and a shabby chaise longue with a thin green mattress for his five or six daily naps. On the walls he placed a crucifix, oddities such as a death mask of Keats, and his personal talisman, the oar he won at Cambridge.

The institution itself was supposed to be administered by a council — a sort of board of directors — and an executive committee to oversee the center. McLuhan recruited friends and colleagues of impeccable and, in some cases, luminous academic credentials — people like Claude Bissell, Tom Easterbrook, and Arthur Porter — for these positions. Once constituted, however, the council and the executive committee dropped more or less into limbo. McLuhan ran the center himself, quite happily. And Bissell, as university president, allowed him to do so. "McLuhan couldn't stand bureaucracy," Kelly would later recall. "That was Claude Bissell's genius — not to impose bureaucracy on McLuhan's center."[51]

It fell to Sirluck to ensure that the center was approved by the School of Graduate Studies. The first obstacle was the dean of the school, Andrew Gordon, a man who despised interdisciplinary research centers in general and McLuhan in particular. Gordon was slated to retire soon, however, and did not actively oppose the center as long as it offered no graduate courses for academic credit. When Gordon retired the following year, Sirluck approached the executive committee and the council of the School of Graduate Studies to obtain their approval for an accredited course. The issue was not trivial. Not only was McLuhan eager to teach an interdisciplinary graduate course for credit, but the center had to offer at least one credit course before it could use the name of the university on its letterhead. McLuhan would not have been satisfied without that name.

In his efforts on behalf of McLuhan, Sirluck found himself up against the entrenched opposition of many academics to the concept of interdisciplinary centers. These professors felt that graduate courses for credit should be given by individual departments. But that opposition was only part of Sirluck's difficulty: many faculty members were opposed to McLuhan himself. "I remember people saying to me, 'With all your emphasis on academic quality, how can you sponsor that charlatan?'" Sirluck recalls.

> I would say to them, "There might be an ingredient of charlatanism in McLuhan, but he has great gifts and we have means of controlling any threat to the university's reputation." After all, if any of the departments that agreed to [give credit for] courses offered by the center... subsequently changed their minds and refused to give them further credit, I would have reported it back to the council, and the question of the center could then have been debated again in council. So the School of Graduate Studies had control. Members of the council knew damned well that if the center didn't work out I would have terminated it.[52]

With such assurances from Sirluck, the School of Graduate Studies reluctantly approved the center as an institution worthy of offering graduate courses for credit. In the 1967 calendar of the School of Graduate Studies, a description of the center finally appeared. Under "Courses of Instruction," the description included the following:

C&T 1000Y/1001F&S
Media and Society/A course considering media as man-made environments. These environments act both as services and disservices,
shaping the awareness of users. These active environments have the
inclusive character of mythic forms and perform as hidden *grounds* of
all activities. The course trains perception of the nature and effects of
these ever-changing structures.

For years that description stood as a summary of the only course offered by the
center. Any other, more elaborate summary would have been beside the point:
Media and Society, with the four sentences describing it, was a formal title for
the fluid, idiosyncratic probes that McLuhan conducted, year after year, in the
presence of the graduate students who enrolled in the course.

Since the center offered only one course, the administration of the university and the School of Graduate Studies had little to fuss over, a fortunate cir-
cumstance for all concerned. Its budget, once set, remained as a stable and
minuscule part of the university's budget.[53]

The basic understanding among deans, department heads, and everybody
else in the university was that the center was a comfortable nest, a letterhead,
and a couple of employees. As such, it need not be disturbed. A later dean, A.E.
Safarian, even exempted it from the five-year review by outside examiners that
he imposed on the other interdisciplinary centers. "I decided not even to try
with him," Safarian recalls. "And I don't remember anybody questioning the
fact that I hadn't. Everybody understood that so long as McLuhan was able to
carry on with the Centre, he had the job. And so I left him to run that thing just
the way he wanted."[54]

As soon as he was set up in his new quarters, McLuhan began developing an idea
that he had hit upon while working for the NAEB. He wanted to devise a test to
measure what he called the "sensory typology" — that is, the sensory balances
or preferences — of entire populations. When applied to an individual, the test
would determine his preferences for the use of his own senses. But the test could
also be applied to whole cultures — enabling one to predict with some precision
the effect of, say, transistor radios or television on a society that had never
known either. Educators, politicians, and others would have the most formidable tool yet known for shaping policy and altering human lives.

McLuhan relied on Dan Cappon, the psychiatrist in his scientific coterie at
the University of Toronto and a man with a long-term interest in perception, to
devise the test. Cappon came up with a series of tests to determine how quickly
people recognized something, such as a rectangle, through the various sense

modalities. For the visual test, subjects were shown a series of slides with swarms of dots on them that gradually formed a rectangle. For the tactile test, subjects felt a series of plastic sheets with raised dots in the same configurations as the slides. The auditory test presented a series of tapes in which the word "rectangle" was spoken with increasing loudness. The idea was that a person recognizing "rectangle" much faster through slides than through the raised dots or the tapes had a decided visual preference. McLuhan also envisioned the possibility of direct testing of preferences: people would fondle fabrics of different textures and colors, look at pictures, listen to tape recordings, and record directly their degree of pleasure or displeasure from their experiences.

McLuhan was excited by the prospect of Cappon's tests. If they came into wide use, they would obviously lend a good deal of substance to his media theories. They would also be of great use to advertisers, who could pitch their products in the appropriate sense modalities to the appropriate audiences. They would be useful to educators as well: aptitude tests could take on a whole new meaning. Science-oriented students could be tested to see if they had different sensory preferences than art-oriented students. Math whizzes could be tested and compared with literary types.

Once sensory deficiencies were determined by these tests, educators could figure out the proper sensory environment to teach math to the literature lover and vice versa. And if, as the high-energy McLuhan suspected, high energy itself was the result of a rich interplay of the senses, educators could probably teach that energy as well. IQs in general could be raised, since low IQs were probably the result of sensory inhibitions. McLuhan began to see again the vision he had glimpsed while working for Harry Skornia — a whole society of young brains learning Chinese and algebra virtually overnight.

McLuhan assumed that the test population would be in Toronto, but he changed his mind after taking an Aegean cruise in May 1963. The cruise was organized and sponsored by C.A. Doxiadis, a wealthy Greek architect who had established the Athens Technological Institute and conceived the idea of a yacht cruise on which all the guests would be intellectual luminaries. These guests could visit the islands of the Aegean, consume some excellent food and wine, and at the same time discuss the fate of the world. McLuhan and his wife enjoyed themselves hugely. McLuhan, as usual, dominated conversations, even in the company of people like Margaret Mead, Buckminster Fuller, and Sigfried Giedion, understandably irritating a fair number of his fellow great minds. McLuhan took note of the irritation but attributed it merely to the invasion of their print-era minds by his electronic-era ideas.[55]

While on the cruise, McLuhan learned that television was shortly to be introduced to Greece. Here was an ideal opportunity to test the sensory preferences of an entire people before and after television. Such opportunities would inevitably disappear from the globe within a few years. On his return, McLuhan solicited granting agencies and various schools and universities for funds to send a team to Greece with kits full of fabrics and tape recordings as soon as possible. The Athens Technological Institute, which now included as a senior scholar Jacqueline Tyrwhitt, McLuhan's colleague from the 1953 Ford Foundation seminar, expressed a willingness to participate, but McLuhan was not able to interest other institutions or drum up enough money to undertake the testing.

In the meantime, he had enough to do at home. Cappon estimated that the center needed $10,000 to launch his test properly. McLuhan put up the periscope again. Doggedly he wrote dozens of foundations asking for money, explaining the crucial importance of the sensory typology tests as a means of furthering the highest possibilites of humanity. In breathless and assured tones, McLuhan contacted U.S. network agencies, Canadian politicians, university officials, and local advertising agencies, all to no avail.[56]

In January 1965 McLuhan received money for the tests from the IBM office in Toronto, through the efforts of an IBM executive and personal friend, Mac Hillock. Hillock was one of those robust businessmen — handsome, square-jawed, wavy-haired, with a hearty male voice tinged with healthy aggression — for whom McLuhan had considerable respect. Hillock may have been the only man McLuhan ever suspected of alienating the affections of his wife. The suspicion was groundless, but there was something about Hillock that brought out the insecure adolescent in McLuhan. Hillock recalls one day in 1968 when McLuhan "took hold of my hand and slammed me back against the wall. He was just feeling so rambunctious that day he had to do it. He was also trying to let me know that he was still number one around this place."[57]

Roughhousing aside, McLuhan appreciated Hillock's practical advice on managing his affairs, which by the mid-sixties were becoming fairly complicated. At times Hillock functioned as a kind of babysitter for McLuhan. He recalls McLuhan phoning him one evening to ask what he should do in order to board a flight that same evening. Hillock asked McLuhan if he had his ticket and then went through with him, over the phone, the steps involved in boarding an airplane.

McLuhan liked the fact that Hillock rarely argued with him. On one unfortunate occasion, Hillock chose to take issue with his friend's Catholicism, which deeply upset McLuhan; but most of the time Hillock listened to what McLuhan said without trying to pick apart his arguments with factual objections. That was what McLuhan's colleagues in the University of Toronto were for. Hillock's entrée into McLuhan's mental world was through his stories of life in the world of big business, a world to which McLuhan was a wistful stranger. McLuhan would mull these stories over and sometimes use them to construct a small theory.

Hillock also told McLuhan that charging corporations $100 or $200 for speaking engagements was ridiculous. (Sometimes McLuhan even left it to his secretary to ask the corporation or organization how much he should be paid for his appearances.) Hillock maintained that McLuhan should be charging at least ten times as much. High fees wouldn't dampen the demand for his appearances, Hillock argued; on the contrary, they might even attract customers. Hillock also pointed out that McLuhan would not be cheating anybody if he charged, in addition, for the extra day spent traveling to a speaking or consulting engagement.

McLuhan did his best to be businesslike about these matters, but for a man raised in Depression era Manitoba it was hard to insist on what seemed astronomical fees. By 1964, he was asking $500 a day for consultations and speeches with corporate clients. Even after that, he continued to swing between blithe disregard for economic considerations — appearing before groups virtually free of charge — and an unconvincing show of hard-nosed assertion, haggling for more money.

In lining up the IBM grant, Hillock had even arranged for IBM personnel to be used as subjects. Cappon proceeded with his tests. Two years later, in 1967, he produced his report. It revealed, among other odd bits of information, that among the diverse occupational groups at IBM, executive secretaries and systems engineers had virtually the same sensory preferences. Interesting, to be sure — but the company had no idea what to do with such revelations. Certainly the report had no effect, short term or long term, on IBM. "The only person in the company who really found the report interesting was me," Hillock recalls.

Later McLuhan did his best to interest other organizations, such as the Ontario Institute for Studies in Education (OISE), in what became known as the IBM Sensory Profile Study. The attempt proved futile. OISE in particular found the test shaky in its method and reliability. Unfortunately, Hillock had remarked to McLuhan at one point that the test Cappon developed might be worth a million dollars. McLuhan immediately had visions of that million dollars pouring into the center and financing the multitude of projects his own fertile brain was conceiving. He was infuriated when he discovered that Cappon claimed his test as his property. There followed an unfortunate dispute between the two men over its ownership, complete with the soliciting of legal opinions. Cappon ultimately retained ownership, whereupon McLuhan decided to have nothing further to do with him. The test meanwhile sank into oblivion. McLuhan, still bitter, told an acquaintance a few years later that it was "ill designed and ill executed."[58]

Although the test was perhaps the major focus of research in the early years of the center, it was not the only project McLuhan promoted in those years. Inspired by McLuhan's ruminations on the phrase "screaming headlines," Cappon and Professor E. Llewellyn-Thomas of the engineering department undertook to determine the most effective use of typeface, layout, color, and so on by newspapers and magazines. Behind this attempt was McLuhan's notion that at a certain point printed words and pictures became auditory (screaming headlines) and perhaps even tactile in their effect on the sensorium.

Cappon and Llewellyn-Thomas proposed to measure this effect by exposing volunteers to different typefaces and recording their reactions through polygraph machines, dream analysis (Cappon was a Jungian psychiatrist), and a special camera developed by Llewellyn-Thomas and a colleague named Mackworth. This camera was revolutionary in its ability both to measure a subject's eye movements *and* to refer these movements to the object being viewed. (Harley Parker and Llewellyn-Thomas had already used the camera to study the difference between artists and nonartists in their viewing of paintings.)

Another project dear to McLuhan's heart was a study of dyslexia, a reading disability that he believed was a frequent consequence of watching television. The television image required different motor responses from the eye than the printed word did, according to McLuhan. Television, in fact, exerted a severe tendency to immobilize the eye muscles of the TV child. McLuhan received backing for this notion from an Ontario optometrist, W.A. Hurst, who published a study on vision and reading achievement in the *Canadian Journal of Optometry*. From 1962 to the end of his life, McLuhan retained an interest in furthering research in this area, particularly in promoting the work of Hurst.[59]

These were the major projects of the center shortly after its establishment. There were still other schemes, such as an "encapsulating chamber," designed by an architect and intended to simulate exactly the sensory environment of another culture. The projects involving Cappon and Hurst, however, took priority. For these, McLuhan campaigned tirelessly. From Fordham University, Father John Culkin raised more than $33,000 for the center in the late sixties and early seventies. A New York stockbroker named Walter Buckner (businessmen with intellectual leanings seemed to gravitate naturally to McLuhan) raised $43,600 around the same time. Most of this money went to pay the salaries of various research associates McLuhan hired from time to time. It was never enough to adequately test or develop the projects he had his heart set on.

In 1964 McLuhan availed himself of the services of a Canadian public relations man named Robert Gray to help promote his projects. Gray wrote a press release describing the sensory typology and the "screaming headlines" projects and arranged a press conference in May 1964 at which McLuhan announced the $10,000 IBM grant. A *Toronto Daily Star* reporter obligingly wrote an article, quoting McLuhan as predicting that within five years "Madison Avenue could rule the world" and that governments could manage their economies "as easily as adjusting the thermostat in the living-room." The article said that McLuhan was scheduled to appear on the cover of *Time* magazine. (*Newsweek* ran a cover story on McLuhan in 1967, but his face never appeared on the cover of *Time*.)[60]

Gray suggested to McLuhan that he help sponsor a project for Canada's centennial year, 1967. This project turned out to be the Canadian Confederation Heritage Foundation, an organization founded for the purpose of selecting a boy or girl from each Canadian province to represent Canada abroad during 1967. McLuhan, characteristically generous, agreed to the idea with enthusiasm. He wrote to Gray that the children selected by the foundation, whoever they were, should be TV children, knowledgeable beyond their years and ready to play a part in new and absorbing communal dramas.[61]

McLuhan thereupon joined the board of directors of the foundation, which secured the services of a fund raiser. The fund raiser arranged for the printing of brochures and approached eminent Canadian businessmen and politicians for funding and support. Unfortunately, the expected money never materialized. The fund raiser, who had been well intentioned but overconfident of success, soon made himself scarce, and members of the board received letters from the firm that had printed the brochures complaining that they had never been paid. McLuhan was deeply embarrassed by the fiasco. Robert Gray drifted out of McLuhan's life, never to return.

In most areas of his life in 1964, however, McLuhan could feel justifiably optimistic. He was now making about forty out-of-town trips a year for speaking engagements. His domestic life was finally beginning to reflect his increasing prosperity. The McLuhans bought their first new car, a small Ford, and took great and innocent pleasure in it — especially the driver in the family, Corinne. (McLuhan, whose spending money was still doled out by his wife, continued to borrow dimes from friends for the odd cup of coffee, sums he carefully noted on scraps of paper and always paid back.)

The greatest event for McLuhan in 1964 was McGraw-Hill's publication of his *Understanding Media: The Extensions of Man*. In 1960, McLuhan had appeared before a seminar of publishers arranged at Columbia University by Bernard Muller-Thym, where he'd none too tactfully remarked on the obsolescence of the hardcover book, leaving most of his audience profoundly irritated. A McGraw-Hill editor in attendance, however, was sufficiently impressed with the talk to approach McLuhan about doing one of these obsolescent hardcover books for his firm, rather relishing the irony of the request.

That book was essentially a rewrite and expansion of his report for the
NAEB. McLuhan worked on it in spare moments before and after writing *The
Gutenberg Galaxy*. Corinne retyped the manuscript of the book three times. The
final product was still nothing like a conventional "serious" work of nonfiction
— advancing arguments and marching coherently to conclusions — although
the seemingly tidy arrangement of chapters might have conveyed that fleeting
illusion to a reader. The book was ultimately "packaged" in much the same way
The Gutenberg Galaxy had been. It was the collocation of years of notes jotted
down on scraps of paper and slipped into folders marked "telegraph," "radio,"
and so on. (A Toronto critic, noting a few instances of blatant repetition and
factual contradictions in the book, observed perceptively, "Sometimes it seems
that he has written the book without reading it." McLuhan, in fact, hated to
review or revise anything he had written.)[62]

In its published form, the book contained an introduction, seven chapters
on the nature of the media in general, and twenty-six chapters on specific media,
including human speech, print, clocks, money, and, of course, television. (The
number of chapters appeared to be cabalistic in design — seven introductory
chapters followed by a number of chapters equal to the number of letters in the
alphabet, an illustration of McLuhan's thesis that the phonetic alphabet was the
real source of all subsequent Western technology.)

McLuhan's basic theme was stated in the first chapter, entitled "The
Medium Is the Message," in which he repeated his earlier assertion that media
were capable of imposing their own assumptions on those who used them.[63]
Shortly after *Understanding Media* was published McLuhan hit upon a better
way of expressing the idea behind "the medium is the message." One could simply say that every new medium created its own environment, which acted on
human sensibilities in a "total and ruthless" fashion.[64] A new medium did not
just add itself to what already existed; it transformed, however imperceptibly,
whatever already existed. That was because the medium was not just the physical object but all of its appurtenances and the vortex of energy it created. The
automobile, for example, had to be perceived in the context of highways, gas
stations, neon signs designed to be seen as it passed — the almost innumerable
altered habits and situations the automobile brought with it.

This new, transformed environment had very curious properties, according to McLuhan. Like the emperor's new clothes, he said, the environment was
virtually invisible and unnoticeable — subliminal, like a person's facial expressions and posture, which may change our attitude toward that person without
our awareness. The reason this new environment was invisible, McLuhan
explained, was that "it saturates the whole field of attention."[65] Insofar as people did notice fragments of the new environment, they tended to perceive them

as disagreeable, even as corrupt — the way, for example, Romantic poets viewed steam engines or bank customers regarded electronic tellers.

In the meantime, the new environment actually made the old technologies and *their* old environments more visible, very often forcing them "into a tidy art form."[66] McLuhan's favorite example was the way in which television turned old movies into an art form by repackaging them as "The Late Show." This basic process also explained why the futurists of the Edwardian era discovered machinery as an art form just at the time the new electric technology was beginning to supersede the machines they worshiped. It explained, too, why the old textile mills of Lowell, Massachusetts, became exhibits in a national park for tourists to gaze on as if they were ruined medieval abbeys.

The one person who could see the invisible environment, according to McLuhan, was the artist. Not the artist as Bloomsbury aesthete or as a member of an artistic school, like the futurists, but the artist as the real explorer of the present, the person who, unlike most people, actually *lived* in the present and thus had a very deep awareness of what was happening in the world around him. (In this regard, McLuhan was fond of quoting his old pen pal Ezra Pound, who called artists "the antennae of the race.") James Joyce, for example, used the stream-of-consciousness technique to mime and make explicit the effects on the human psyche of the nascent electronic technologies of his era. These technologies, McLuhan said, bombarded the human psyche with huge amounts of information that had no connection and no underlying rationale, just like the stream of consciousness of Joyce's characters.

Art, then, served as a kind of advance warning system of the effects of the new media on society. This theme emerged in the chapter of *Understanding Media* entitled "Challenge and Collapse: The Nemesis of Creativity." Other broad themes that McLuhan had long been pondering were also articulated in the book. The subtitle itself, "The Extensions of Man," was the summation of his notion that every human artifact was an "outering" or extension of some human sense or portion of the body. It was a notion that had been on McLuhan's mind at least since 1947, when he had stated, in an article published that year, that all human inventions were "a reflex of our most intimate psychological experience."[67]

In the chapter entitled "The Gadget Lover: Narcissus as Narcosis," McLuhan reiterated the idea that artifacts reduced their users to the status of "servo-mechanism." (One of the most flagrant examples, in his mind, was the hated automobile: he justified his unwillingness to drive by telling friends that he refused to become a servo-mechanism of that device.) In "Media Hot and Cold," he explained the idea he had developed in the NAEB project about the two basic forms of media: "the hot form excludes, and the cool one includes." Hot media did not invite the participation of the user; cool media did. As he later told *Playboy*, books were hot, but bull sessions were cool. And in the chapter entitled "Hybrid Energy: *Les Liaisons Dangereuses*," McLuhan went further in endowing media with almost animistic qualities, claiming that they mated with each other, produced offspring, and attacked and cannibalized each other. Indeed, they had all the qualities of human beings, whose reflexes and extensions they were.

The texture of the book was much less ragged than that of *The Gutenberg Galaxy*; there were far fewer quotations breaking up the flow, such as it was, of McLuhan's prose. This was the publisher's doing, not the author's. McLuhan blamed the editors at McGraw-Hill for cramping his style; they had balked at his wholesale use of quotations. McLuhan claimed their attitude was that, unless a quotation was used to illustrate a point of view an author wanted to attack, it was unnecessary.[68] But no one ever thought of anything by himself, McLuhan argued. Why not credit this fact in one's prose? As he and Carpenter had gleefully noted a long time before, the only question in picking other people's brains was whether one pillaged from great sources or from mediocre ones. McLuhan's sources were the best.

Nonetheless, he was proud of *Understanding Media* in at least one respect. He felt it was a true depiction of the present era, unlike *The Mechanical Bride*, which in discussing industrialism had depicted an era on the way out. It had been easy to deplore that era when it was already obsolete and pushed into visibility by the new electronic environment. The successor to that book was very different. McLuhan jokingly referred to it as *The Electronic Call Girl.*

The joke was a clue that McLuhan was not as entirely enamored of the new electronic era as some critics assumed. Reviewing the book for *The New Yorker*, Harold Rosenberg remarked of the author, "In his latest mood, he regards most of what is going on today as highly desirable, all of it as meaningful."[69] Meaningful, yes; highly desirable was another question. But McLuhan had only himself to blame for misleading critics. How else could they interpret such passages as the following?

> Today computers hold out the promise of a means of instant translation of any code or language into any other code or language. The computer, in short, promises by technology a Pentecostal condition of universal understanding and unity. The next logical step would seem to be, not to translate, but to bypass languages in favour of a general cosmic consciousness which might be very like the collective unconscious dreamt of by Bergson. The condition of "weightlessness," that biologists say promises a physical immortality, may be paralleled by the condition of speechlessness that could confer a perpetuity of collective harmony and peace.[70]

Read carefully, this may not be the unqualified hosanna for the future that it at first appears. More likely, it is a rhetorical overstatement of the kind McLuhan was fond of using to get a rise out of his audience or to make sure they were awake. He did believe that the new technology might enrich human awareness and generate universal ESP; and he told an audience in New York City shortly after the publication of *Understanding Media* that there might come a day when we would all have portable computers, about the size of a hearing aid, to help us mesh our personal experience with the experience of the great wired brain of the outer world. But then he was also in the habit of defending his intellectual flank by frequently insisting that his outlining the features of the new media ought to have inspired everyone with sufficient revulsion to avoid them.

That certainly applied to the all-engulfing, tribal powers of television. Even in the pages of *Understanding Media* he slipped in a plug for Gutenberg: "To resist TV," he wrote, "one must acquire the antidote of related media like print."[71] To the editor of the *Atlantic Monthly* he protested that his investigation of the electronic media performed the same service for book lovers as a reconnaissance patrol performed for a beleaguered army.[72]

Despite Rosenberg's mistake, McLuhan was delighted by the admiring tone of his review, as well as by a similar laudatory review in the *New York Herald Tribune*. Other reviews were less enthusiastic, in publications such as the *Nation*, *Commentary*, the *New Statesman*, and numerous other heavyweight intellectual journals. But whether reviews were positive or negative, *Understanding Media* introduced McLuhan as a subject for serious debate in intellectual circles in both North America and Great Britain. He became an author that anyone with intellectual pretensions should know — with the result that his book sold nearly 100,000 copies.

A review in *Time* heralded this fact with a casually obtuse look at the book, which it characterized as "fuzzy-minded, lacking in perspective, low in definition and data, redundant and contemptuous of logical sequence." This review gave McLuhan a great deal of pain. He wrote a letter of anguished protest to Silberman and the management of the magazine, attributing the reviewer's negativity to emotional shock and psychological defensiveness.[73]

But McLuhan was going to have to learn to live with this kind of response. In a very short time he would be the object of more public abuse — and adulation — than he had ever dreamed possible.

9. "Canada's Intellectual Comet" (1964–1967)

> Now, I will hazard a guess about the future of the planet. It is not quite as harrowing as you might suppose. — Marshall McLuhan, 1966[1]

An air of excitement pervaded the art world in the early sixties. It was a great age not only for galleries and art museums trying to keep up with successive waves of pop art, op art, color field abstract, conceptual art, and so on, but for art critics, historians, and professors of art, and even journalists on the culture beat. After 1960, more and more newspapers, alerted by the eerie and very public career of Andy Warhol, decided that if they were serious about claiming the allegiance of middle-class readers they had better have at least one person on their staff who knew the difference between Jasper Johns and Robert Rauschenberg.

Into this lively scene stepped an English professor who seemed to have a radically new and profound understanding of what was going on. He especially impressed Abraham Rogatnick, a professor of architecture at the University of British Columbia. "McLuhan's philosophy seemed to be supporting the notion that art can be abstract, that the medium is a message in itself, you don't have to tell a story, that art is a story, just painting on a canvas is a story in itself," Rogatnick recalls.[2] Insofar as McLuhan bolstered the case for the abstract artists, whose day was already passing, he was hardly a herald of the avant-garde; but his theories were sufficiently supple that they could also provide useful commentary on the appearance of outsized hamburgers and Brillo boxes in chic art galleries.

Rogatnick and some of his colleagues decided to dramatize McLuhan's theories with a funhouse display of multisensual experiences, labeled "The Medium Is the Message." They took over a cement-floored armory on the university campus and rounded up about thirty slide projectors. Students operated the projectors from the rafters, from the balcony, from all over the building. As visitors wandered through a maze of huge plastic sheets hanging from the ceiling, they found themselves in the midst of a barrage of random photographic images or abstract designs projected on the floor, the ceiling, the plastic sheets, and, occasionally, themselves. The students wielded these projectors as if they were weapons pressed into the service of art. Trucks that drove into the armory for regular deliveries were suddenly transformed into luminous art forms bearing obscure but meaningful messages.

In addition to the slides, a long, meaningless movie was shown, depicting the armory empty — a Warholian exercise in boredom. Dancers pirouetted through the crowd at unexpected moments to startle the spectators. The entire human sensorium was involved in the exercise. Perfumes were spewed about for the olfactory sense. Strange noises from hidden loudspeakers and a man on a podium hammering a block of wood alerted the auditory. The best part was reserved for the tactile. The Sculptured Wall was a huge piece of fabric stretched on a frame. Dancers pressed against it and the visitors were invited to touch their squirming bodies through the cloth.

McLuhan spoke at the university just before this event in his honor took place, but he left before it actually started. Such things never really interested him. All the same, the event was a kind of signal. The McLuhan craze was about to commence — and artists would be foremost in helping to spread it. Gerd Stern, prominent in a collective of artists in New York, was one of the most ardent promoters of McLuhan in the early sixties. He spread the word through the work of his collective, which pioneered the multimedia, "total theater," psychedelic, environmental art performances (they went by a variety of names) that became a fixture in artists' lofts in the late sixties. McLuhan, on a couple of occasions, spoke at such performances. By 1968 he was being acclaimed in the pages of the *New York Times Magazine* as the "number one prophet" of this consciousness-expanding art; one Greenwich Village enthusiast, at about the same time, staged a multimedia event that was climaxed by his singing quotations from McLuhan's works.

In early 1965, McLuhan was discovered by two Californians who, as a sort of hobby, used the profits of their joint business consulting firm for what they called "genius scouting." The chief scout was a San Francisco doctor named Gerald Feigen. Feigen had read *Understanding Media* and alerted his associate, Howard Gossage, to his find. Gossage at that time headed a small advertising agency in San Francisco and was well known in the industry as an avant-garde copywriter with a genius for the fanciful. A Gossage advertising campaign for Rainier Ale offered customers sweatshirts printed with the unsmiling faces of Bach, Beethoven, and Brahms; a campaign for American Petrofina responded to rival gasoline companies' claims for their wondrous and mysterious chemical additives with claims that Petrofina gas stations filled customers' tires with pink air.

Gossage and Feigen phoned McLuhan and informed the professor that they were coming to see him. The three men met in the dining room of Toronto's Royal York Hotel, at that time the most posh of local eateries. They hit it off immediately. Gossage, a tall, white-haired Irishman with a robust laugh, was the kind of vivid personality McLuhan was always drawn to. Feigen played humble inquirer, mitigating McLuhan's tendency to engage in monologue. What did you mean in *Understanding Media*, Feigen would ask, when you talked about the great pattern of being that reveals new and opposite forms just as the earlier forms reach their peak performance? Asking McLuhan what he meant by any particular statement was always a chancy affair. Sometimes he responded with explanations that were perfectly lucid. At other times his responses were even more metaphorical and confusing. It appears that this evening he was in his lucid mode.

Feigen's appetite for puns and one-liners was equal to McLuhan's, and the two swapped jokes over the table like a couple of baseball fanatics swapping statistics. Gossage contributed lively anecdotes about the great and famous he had known — people like Bertrand Russell. The three went at it nonstop until the waiter appeared with the check. Gossage, who hated being presented with a check until he had asked for it, turned on the waiter and started to dress him down until the unfortunate man interrupted by saying, "But sir, we closed at midnight. That was two hours ago." The three men finally left.

On their return to San Francisco, Feigen and Gossage began to plot their promotion of McLuhan. Gossage, a Madison Avenue veteran, knew several editors of major New York magazines, such as *Esquire* and the *Nation*. In May 1965 he arranged for McLuhan to meet several of these eminent gentlemen at a series of cocktail parties in his suite at the Lombardy Hotel in New York. McLuhan missed the first party, having decided to wait until the next day to fly so that he could obtain a better fare. Gossage, accustomed to the expense account life, could hardly believe the story when he heard it.

Once he arrived, McLuhan appeared quite at ease at these parties and impressed the guests with his self-confident manner, if nothing else. Both *Time* and *Newsweek* subsequently offered McLuhan the use of an office in their headquarters anytime he was in need of one. (These offers were unattended by any promise of a salary or a consultant's fee.)

According to John Culkin, the Fordham Jesuit and McLuhan devotee, who was also involved in some of Feigen and Gossage's plans, the parties at the Lombardy had a modest impact on McLuhan's growing reputation. Feigen and Gossage themselves later maintained that their intervention "probably speeded up the recognition of [McLuhan's] genius by about six months."[3] In fact, the guest who proved most useful in publicizing McLuhan turned out to be Tom Wolfe, assigned by Clay Felker to profile McLuhan for *New York* magazine (which was then the Sunday magazine of the *New York Herald Tribune*). Frances FitzGerald, a journalist later well known as author of *Fire in the Lake*, had told Felker about McLuhan.

In August 1965, Wolfe followed McLuhan to San Francisco, where Gossage and Feigen held what they called a McLuhan Festival in the offices of Gossage's advertising agency, housed in a renovated firehouse. The festival, a week-long affair, consisted of meetings between McLuhan and assorted notables of the city of San Francisco, most of whom were involved in communications. Advertising executives, asides from the mayor's office, the editor of the *San Francisco Chronicle*, the editor of *Ramparts* magazine, the novelist Herbert Gold — they all trooped into Gossage's offices and listened to the oracle from Toronto. At night, Gossage and Feigen threw parties in McLuhan's honor.

One day Wolfe took McLuhan to a topless restaurant, in the company of Gossage, Feigen, and Herb Caen, the noted San Francisco columnist. At the restaurant all of the men seemed faintly embarrassed, except McLuhan. As Wolfe later wrote:

> Inside of thirty seconds McLuhan had simply absorbed the whole
> scene into... the theory. He tucked his chin down.
> "Well!" he said. "Very interesting!"
> "What's interesting, Marshall?"
> "They're wearing *us*." He said it with a slight shrug, as if nothing
> could be more obvious.
> "I don't get it, Marshall."
> "We are their clothes," he said. "We become their environment.
> We become extensions of their skin. They're wearing *us*."

After the lunch, the restaurant presented a "style show," featuring women wearing topless gowns and lingerie. In Wolfe's account, McLuhan summoned the emcee to their table after the style show to tell her he had something for her pitch.

> "What's that?" she says. Her face starts to take on that bullet-proof
> smile that waitresses and barmaids put on to cope with middle-
> aged wiseguys.
> "Well, it's this," said McLuhan — and I have to mention that top-
> less performers had recently been brought into court in San Francisco
> in a test case and had won the trial — "It's this," says McLuhan. "You
> can say, you can tell them" — and here his voice slows down as if to
> emphasize the utmost significance — "*The topless waitress is the open-
> ing wedge of the trial balloon!*"[4]

The emcee did not laugh. McLuhan's wit was often labored and, like complicated shaggy dog stories, tended sometimes to irritate rather than amuse, especially when his listeners had no idea what he was talking about. But McLuhan was a shade uncomfortable in the situation, despite his ready comments. He was not unaware that a topless bar, to a scrupulous Catholic conscience, might be considered an occasion of sin.

Feigen and Gossage, meanwhile, continued to showcase McLuhan in the spirit of an exhilarating, if expensive, hobby. Feigen recalls that the two men ended up spending "a lot more" than the $6,000 they initially invested in launching McLuhan in the spring of 1965. Feigen recalls with amusement the visits McLuhan made to San Francisco in the years after 1969, when Gossage died. McLuhan would invite Feigen out to dinner, but, as Feigen comments, "I always got stuck with the check." By that time McLuhan was confirmed in his unfortunate assumption that all people who did not teach English literature and who dallied in the world of commerce had barrels of money.[5]

In 1965, Gossage took up where Mac Hillock had left off, trying to get McLuhan to increase his fees. In one instance, chronicled by Wolfe, Gossage urged McLuhan to charge $50,000 for a lecture to a group of corporate executives. According to Wolfe, McLuhan and the representative of the corporation settled on $25,000. It is probable that the reported fee was exaggerated; the most McLuhan received for individual appearances in the years following the McLuhan Festival was $5,000 or $6,000. Very often he agreed to appear before audiences for even less than his stated minimum of $500. All of his appearances, however, were sufficient to give his income a considerable boost in 1965 — to the extent that he felt obliged to engage the services of professional accountants. (Previously the bookkeeping had been handled by Corinne.)

In addition to speaking engagements, McLuhan took on the odd consulting job for various businesses and advertising concerns. In December 1966, the advertising and communications firm of Johnson McCormack Johnson of Toronto announced that McLuhan had been bestowed the title Senior Creative Consultant and Director of the firm — which simply meant that he occasionally dropped by for an hour or so to talk about some of his latest ideas.

The business value of these chats — which McLuhan would probably have done for free, had he just wandered into a room full of businessmen or professional people — is very hard to calculate. Occasionally, McLuhan's listeners felt genuinely inspired by something he said. Gordon Thompson, for example, a senior engineer in the research department of Bell Canada, saw McLuhan frequently during the sixties. Thompson recognized all the usual points that McLuhan's critics raised: that McLuhan often talked glibly about subjects he knew very little about, that his technical appreciation of the media was very weak, and so on. "He was an artist and he embellished things a bit one way or another, whatever suited his needs," Thompson comments, "but I was

content to go along and say, 'Okay, I'll suspend judgment.'" His tolerance was
well rewarded.

> One of the things that Marshall taught me was that, as a senior scien-
> tist, my contribution should be my ability to take risks. The top man-
> agement cannot afford intellectual risks, the new scientist cannot
> because he's just graduated and he doesn't know the rules. Only an
> old bugger like myself could do it. He got me to realize that I didn't
> have to keep on proving I could continue to grind out this prosaic
> crap. It was a kind of insight into my situation that even my superiors
> didn't have. As a result, I was able to make some contributions to Bell
> that I couldn't have ever made without that kind of encouragement.[6]

For most of his business audiences, though, McLuhan remained simply a
curiosity with a certain amount of prestige value. Some members of these business audiences, unlike the polite students of the Croton-on-Hudson seminars,
did not even disguise their boredom and hostility when McLuhan's talks veered
into the incomprehensible. In late 1965, Tom Wolfe attended a regular luncheon
meeting of a New York advertising group where McLuhan was speaking. "He
was a serious-faced Lewis Carroll," Wolfe wrote after the event. "Nobody
knew what the hell he was saying. I was seated at a table with a number of
people from Time-Life, Inc. Several of them were utterly outraged by the performance. They sighed, rolled their eyeballs, and began conversing among
themselves as he spoke."[7]

Wolfe himself considerably raised the prestige value of McLuhan when his
profile of McLuhan appeared in November 1965 in *New York* magazine. Wolfe
presented a dramatic picture of this tweedy Canadian academic exerting the
unspoken and unself-conscious mastery of a seer over the world of corporate
executives. He also provided, with his usual felicity, a tag line for the entire
McLuhan phenomenon: "What If He Is Right?" The phrase stuck and would be
repeated in articles about McLuhan for years afterward. McLuhan himself was
delighted with the profile, as well he might be. He congratulated Wolfe on the
fidelity of his portrayal, comparing some of its effect on him to the mild dismay
most people experience when they hear, for the first time, the sound of their
own voice on tape. Such minor deflation of his self-image McLuhan considered
rather healthy. He asked for a dozen copies of the magazine, correctly assuming
it would give his reputation a tremendous boost.[8]

Simultaneously, *Harper's* magazine published a largely favorable article by
Richard Schickel entitled "Marshall McLuhan: Canada's Intellectual Comet." A
few months earlier, Jerome Agel published a four-page profile of McLuhan in

the September 1965 issue of *Books*, his monthly tabloid newspaper. Agel's intelligently enthusiastic article further stimulated interest in McLuhan. The *New York Times* shortly thereafter ran an article about McLuhan, describing him as "an iconoclast whose atypical theories about advertising have caused many executives in the field to take a fresh look at their approach to creativity and media selection." The article offered a few savory McLuhan quotes on the subject of advertising, including the following:

> Advertising creates environments and these are very effective as long as they are invisible. Some think that an ad is good if it is noticed. This is quite mistaken. The work of an ad is totally subconscious. As soon as you realize it is an ad, it is not serving its function.[9]

Agel himself, whose forte over the years has been collaborating with fairly cerebral authors, such as Buckminster Fuller, Carl Sagan, and Herman Kahn, to present their work in popular format, was eager to work with McLuhan. He collaborated with McLuhan on two original books, *The Medium Is the Massage* and *War and Peace in the Global Village*. He also arranged and produced one of McLuhan's more offbeat projects, an LP version of *The Medium Is the Massage*. McLuhan read portions of the book to musical accompaniment. It was the 1967 equivalent of a McLuhan video.

On the other side of the Atlantic, McLuhan gained the admiring attention of such notable intellectuals as George Steiner, fellow of Churchill College, Cambridge, and Jonathan Miller, a TV director and editor for the BBC. In April 1965, Miller wrote to McLuhan declaring himself "one of your more vehement disciples," although he was uneasy at McLuhan's use of jargon, such as the labels "hot" and "cool," and took issue with McLuhan's theory of television as a tactile experience. Miller had interviewed McLuhan in New York City for *Monitor*, a BBC television program dealing with the arts, in September 1964. McLuhan had been enchanted with the interview, afterward calling Miller "a wonderful fellow." When the interview was aired on the BBC in the spring of 1965, Miller introduced it by comparing McLuhan's explorations in human sensory experiences to Freud's explorations in sex. McLuhan approved of the comparison.[10]

The increasing exposure he was receiving from the media had little effect on the way McLuhan dressed or presented himself to interviewers and audiences. His attitude toward clothes had always been complex and remained so even when he could afford to wear whatever he desired. In the forties and early fifties he had favored suits of Harris tweed — the kind of heavy, indestructible attire that all good Torontonians wore for their durability and solid British quality. As a Cambridge man, McLuhan could face the world in nothing less.

Unfortunately, he had a tendency to spoil this effect of understated elegance by wearing a tie with fire engine red maple leaves all over it, or mismatched socks, or a fedora that was too small for his head. (At one point McLuhan, who was fond of hats, chose a fedora similar to the one his idol James Joyce had worn.)

Corinne further complicated McLuhan's appearance by insisting that he buy a toupee when his hair started to thin. The toupee kept falling off. McLuhan was simply not a man to expend a good deal of energy on getting these things right. Ties and sharply creased trousers made him feel stiff and confined. In some ways, he resented the very idea of clothes. He thought of them as uniforms, sensorily depriving, nullifying the self of the person who wore them. On one occasion, in 1974, he amused his graduate class by coming into class from a formal banquet and partially disrobing in front of them, yanking off some cheap horror of a burgundy velvet tie, shedding his jacket, leaning against the photocopy machine and removing his lace-up shoes — all the while intoning elaborate poetical curses against fancy dress, culled from the stock of verse he had memorized as a child.

When McLuhan began appearing before audiences of important people, he relied on his standby, a brown tweed suit. He had worn such a suit since his first year at Cambridge. So frequently did McLuhan appear in this outfit that friends began to speculate he had closets full of reddish-brown tweed suits. In fact, McLuhan once confessed to a reporter, "I cling to whatever I've been wearing and I wear it again and again until someone screams in protest."[11] (That someone was usually his wife.)

Occasionally he varied his wardrobe with the careful deliberation of a child play-acting. At one point he had a yen for a blazer in Scottish tartan material — it was, he told friends in his best Scottish accent, the proper attire for dealing with business matters. In 1969 he discovered a tartan material incorporating the Canadian maple leaf and had a jacket made from it. He proudly donned this garish red and black blazer when the Scottish poet Hugh MacDiarmid visited the campus — and also, of course, whenever he wished to convey the impression of Scottish canniness in affairs of money. (He heartily recommended the material to Canadian Prime Minister Pierre Trudeau.)

McLuhan's attitude toward addressing an audience was as idiosyncratic as his dress. He usually began with a joke or two. Like most speakers, McLuhan valued jokes as a way of easing into a lecture and getting a feel for the audience, but he took this ploy to extraordinary lengths. For years he collected cartoons and jokes from newspapers and magazines and put them in a special file, along with jokes he had heard and written down on scraps of paper. (Gerald Feigen was a particularly fertile source of material. In 1972 he mailed about 2,500 cross-referenced jokes to McLuhan.) In search of more and better jokes, McLuhan even wrote to the astronaut Alan Shepard to ask him to pass on any good ones making the rounds at NASA.

When Ernest Sirluck left the University of Toronto to assume the presidency of the University of Manitoba, McLuhan typed out a batch of jokes for him to use in his presidential speeches. He himself kept jokes on scraps of paper in his pockets to pull out on opportune occasions. It became a habit for McLuhan to tell five or six jokes before starting a conversation or before faculty meetings. He claimed, toward the end of his career, that he had not said a trivial thing in years — and it was true that McLuhan was very poor at small talk. The little conversational gaps that most people fill with small talk McLuhan filled with jokes, particularly in his later years, when the habit became almost compulsive. In a letter demanding money from a client who had reneged on a fee, he couldn't resist including a joke about the cross between a parrot and a tiger who said, "Polly want a cracker. AND I MEAN NOW."

He also seemed to have a particular fondness for bad puns. Throughout 1965 he liked to tell audiences the one about the first telephone pole — Alexander Graham Kowalski. In his memoirs, Peter Drucker writes, "The only time I remember him paying attention to any one *person* was when one of our children quizzed him on 'Bible jokes': 'When is baseball first mentioned in the Bible? When Rebecca goes to the well with a pitcher.' He was enchanted and kept on repeating the feeble pun for hours."[12] Part of the appeal of bad puns for McLuhan was precisely their ability to evoke groans from literary souls. There was nothing like a bad pun, after all, to tear from words the aura of respectability conferred on them by print — to destroy what I.A. Richards had called the "proper meaning superstition."

After introducing himself to an audience with bad puns and jokes, McLuhan usually proceeded to offer a volley of what many considered to be outrageous and contradictory statements. In 1958, after a talk at Hart House, the athletic and social center of the University of Toronto, a student rose up and claimed that McLuhan had contradicted himself twenty-eight times in the span of a half-hour lecture. McLuhan is reported to have replied, with his characteristic air of amused tolerance, that the observation was the result of the student's habit of thinking in a linear fashion.

McLuhan was certainly never tempted by the academic habit — or virtue — of carefully qualifying his statements. ("If I made a cautious, measured statement," he once said, "someone might mistake me for a stable character.")[13] When accused of purveying half-truths, he often defended himself with the remark, worthy of Lenin, that half a brick could break a window quite as well as a whole brick. He never intended to offer his audience eternal verities, and he was upset when he felt his remarks were taken that way. He was much happier

when people reacted to his provocations by rethinking their own eternal verities. To that end, he would do almost anything to get under their skin.

McLuhan had long since acquired intellectual backing for this approach
with his studies of rhetoric. He had also picked up notions from the French
Symbolist poets, such as Baudelaire, who believed that the poet spoke through
the reactions of his readers. From these notions McLuhan developed his idea of
"putting on" an audience. The idea was a complicated one. Basically, it meant
that McLuhan always tried to avoid a purely didactic role — the role of telling
something to others. He rarely went before any group with a prepared text.
Armed, at best, with a few headings, he "put on" the audience by appearing to
know something they did not know about the very things they were most certain they knew. (McLuhan believed that speakers who read from prepared texts
put on the texts, not the audience.) He chipped away at the identity of the audience — their assumptions, their expectations. In doing so, he consciously
assumed the oracular pose. If the members of the audience were irritated or
their sensibilities offended, all the better; chances were good that their awareness, in the end, would be heightened.

Like most oracles, McLuhan loved nothing better than being interviewed.
(He certainly preferred it to writing his thoughts down on paper. At the very
least, being interviewed was "multisensory.") If he exaggerated and simplified
his points by asserting, say, that Madison Avenue could rule the world in five
years, that was only part of the game. If the "put-on" was stinging and aggressive, that was also part of the game. "The only way you can reach people,
whether you're a preacher or a professor, is to hurt them," he told an interviewer
in 1974. "You really have to cut very close to the bone. It's like surgery."[14] At
times, McLuhan wielded his surgical tools with the relish of a bully. He told one
friend that his references to Flaubert and the Symbolist poets were always useful for putting an audience in its intellectual place.[15] (The success of this ploy
may be one reason that McLuhan, unlike many academics, never underplayed
what he knew.)

McLuhan could get away with these tricks in front of an audience because
his verbal agility was unfailing. Anyone who had grown up debating Elsie
McLuhan had nothing to fear from a roomful of angry university professors or
IBM executives. If the audience asked him sharp questions, he tossed the questions in the air like a juggler or simply ignored them and answered the questions
he would like to have been asked. After his initiation at Croton-on-Hudson, he
never fumbled or vaporized again. As Buckminster Fuller once commented, "I
have been present when hostile audiences thought they had him on the run only
to discover themselves chasing themselves up dead-end alleys, as he himself
reappeared far down another highway."[16]

McLuhan always had the last word. Usually he countered a verbal attack with a deft flip into some larger context that the attacker hadn't dreamed of. Certainly, no stubborn refusal on the part of an antagonist to concede a point ever daunted him. In an argument with W.H. Auden, for example, McLuhan asked the distinguished poet about the relation of English language and prosody to jazz and rock 'n' roll music. Auden said he knew nothing about it. McLuhan insisted that Auden must have heard jazz and rock music, since it was impossible for anyone to avoid it. Auden replied with equal vehemence that he deliberately and successfully avoided hearing any such thing. Unable to override Auden on this point, McLuhan, without missing a beat, directed the poet's attention to a certain fact that Auden would doubtless have noticed had he allowed himself to notice rock music at all — namely, that rock could be sung only in English. No other language on earth, McLuhan said, could carry rock rhythms. Auden did not bother to argue this dubious assertion, especially since he had just admitted to total ignorance of rock music. McLuhan had drawn a circle around him; if Auden had tried to draw a larger circle around McLuhan's, he would have responded with a still larger circle, and so on. He was unbeatable at that game.

McLuhan also developed very early a simple and effective strategy to deal with hecklers: he invited the heckler to come forward and address the audience. He used a variation on this strategy in his Monday night seminars at the Centre for Culture and Technology, which were more or less open to anyone who dropped in. If someone interrupted McLuhan's remarks to denounce them, McLuhan would appear to take the interruption very seriously and then, as if that person were an authority, ask him a question on an entirely unrelated topic. With unfailing regularity, the critic would be wholly nonplussed. McLuhan's tactic not only took the wind out of the heckler's sails, it also emphasized that everyone has vast areas of ignorance and that fruitful discussions often begin when people recognize the frontiers of their ignorance rather than harping on what they know.

By "putting on" audiences and trying to outmaneuver opponents in debate, McLuhan projected the protean face of an actor, capable of assuming different expressions without being committed to any one of them. Aside from tucking in his chin before speaking, he avoided any kind of pose or set of mannerisms that would leave an audience with a well-defined impression of him. Leslie Fiedler comments, "The first several times I met him, I would not recognize him. He had a funny way of assuming roles and faces, and it was rather disconcerting. The features were the same, but this was not exactly the face which had looked at me on previous occasions." Richard Kostelanetz, a sympathetic McLuhan critic, put it a different way: "He has a face of such negligible individual character that it is difficult to surmise what his personality might be or even remember exactly what he looks like; different pictures of him rarely seem to capture the same man."[17]

Most photos of the middle-aged McLuhan show a man strikingly detached from whatever surrounds him. He was very conscious of cameras aimed at him and regarded them with the wariness of an animal sniffing bait in a trap. "I find cameras very annoying," he told one reporter accompanied by a photographer. "They intrude, like the telephone. They eat you alive, like a piranha, in little bites."[18] His response was sometimes to mug deliberately in front of cameras.

He similarly distrusted tape recordings. When he was being recorded he tended to put on different tones of voice with the same facility he put on facial expressions, as if he were protecting himself, like a savage, against having his essential self reproduced in some way that would destroy his soul. He did not like seeing photographs or videotapes of himself any more than he enjoyed rereading his own books. All these things, he said, were like seeing one's life replayed on the Day of Judgment.[19]

The moment's performance was what McLuhan relished. At his best — perhaps when he was not so conscious of "putting on" his audience — these performances could make a roomful of people come alive. Arthur Porter recalls McLuhan's address to a University of Toronto audience composed mostly of Harvard students on a weekend cultural exchange in the spring of 1966. The students, who had not yet heard of McLuhan, were completely perplexed when he began his remarks with a few Polish jokes and then a series of comments about some newspaper clippings he had in his pocket. None of his comments seemed to have any connection or point to them. "This went on for at least four minutes," Porter remembers.

> And then he struck. I can't recall how he struck, but somehow his voice became more charged and he seemed, suddenly, with his one-liners and aphorisms, to be opening a new world for these students. It was absolutely electric. His listeners were practically bouncing off their seats. From then on he had them in the palms of his hands.[20]

It was not surprising, in a way, that McLuhan came to feel that such performances were wasted on his colleagues. By the mid-sixties he had grown rather contemptuous of their failure to appreciate the realities he pointed out to them. He did not deliberately set out to startle his colleagues with his originality, he protested. Why should he be blamed because they had never noticed the effects of the media before? McLuhan sometimes seemed almost pleased by the scorn of his colleagues, taking it as proof that he still possessed intellectual vitality.[21] On one occasion, he told an interviewer, "Some of my fellow academics are very hostile, but I sympathize with them. They've been asleep for 500 years and they don't like anybody who comes along and stirs them up."[22]

When cartoons about McLuhan began appearing in *The New Yorker* in 1966 — the final confirmation of his celebrity — he wrote a member of the magazine's staff that his appearance in "the benign and urbane *New Yorker* pages" might reassure his colleagues that he was "not a totally destructive force." (This tongue-in-cheek remark had a hidden bite to it. McLuhan's attitude toward *The New Yorker* had not really changed since he had lambasted the magazine, along with *Time* and *Life*, in the days of his Leavis-style moralism.)[23]

Mixed resentment and amusement at his colleagues' attitudes were very much a part of McLuhan's mood in January 1965 when he visited his friend John Wain in Oxford. Wain recalls the two of them running into the noted literary critic Christopher Ricks while strolling the streets of Oxford. The three retired to a pub, where Ricks, according to Wain, "put forward a few objections and qualifications" to McLuhan's theories. "He did so good-humouredly," Wain wrote,

> but Marshall's answers were much more brusque. "I don't think I agree with you about simultaneity being the effect of the two-dimensional and homogeneity being the effect of the three-dimensional." Or whatever it was. "That's because you've never considered these matters," replied Marshall off-handedly. Period. He was not going to say anything else. Another point from Ricks, and another very similar answer. We drank up our beer and left.[24]

Wain neglects to mention that in the month previous to this encounter, the *New Statesman* had published a review of *Understanding Media* by Ricks in which the Fellow of Worcester College savaged McLuhan's book with that high superciliousness only critics aged and mellowed in Oxbridge common rooms can attain. Few authors manage to exercise Christian forgiveness when critics draw blood. McLuhan was certainly not among them. Privately he characterized Ricks as one of the "pedant and aesthete" types who haunted English universities.[25]

Speaking invitations continued to pile up through 1966. In mid-May, with Muller-Thym and Peter Drucker, McLuhan was a speaker at the five-day Laurentian Conference held at the Chantecler Hotel in the resort country north of Montreal. Organized by Stewart Thompson Ltd., a Montreal management consulting firm, it was the most prestigious management conference held in Canada that year and perhaps the high-water mark for such affairs north of the forty-ninth parallel. These conferences were still a novelty in Canada in 1966; corporate executives were just beginning to discover "personal growth" as a career necessity and were telling one another that they had to be much smarter than their counterparts had been in the fifties. Great change of a pressing and mysterious sort was in the offing — in classrooms, in churches, in the tastes of the public — and someone had to help executives increase awareness and foster their "growth." By 1966 some large corporations were even forcing their executives to spend a week in T-groups (where the marketing director would end up venting his rage at the company controller, and the assistant vice-president would leave the room in tears.)*

The forty-odd senior executives who had paid $450 each to attend the Laurentian Conference had a far more cerebral experience. McLuhan was the first to speak, followed by Muller-Thym and then Drucker. It was a descending order of the abstruse. (One participant described McLuhan as being "on cloud 99," Muller-Thym as "on cloud 9," and Drucker as almost "down to earth.") McLuhan began his talk by saying, "Greetings from the DEW-line." The Distant Early Warning line — a fixture of the Canadian-American defense system in the north of Canada — was then becoming a persistent McLuhan analogy for the activity of the artist and the man with ESP in clearly perceiving the present. His audience, of course, was mystified. "DEW-line?" whispered one vice-president. "I thought this guy was an English professor from the University of Toronto."

McLuhan then announced, "No longer will I say that the medium is the message. I've changed my thinking." He paused, with practiced timing, while his listeners waited, their pens poised above their notebooks. "From now on, I believe that the medium is the *massage*." The pens remained poised, and the faces of the audience assumed a vacant look, tinged with annoyance. McLuhan then let loose with a barrage of jokes and aphorisms, not excluding his Alexander Graham Kowalski joke. "We don't know who discovered the water, but we're pretty sure it wasn't a fish. ... A newspaper without a deadline is a surrealist poem. ... Kids don't have goals anymore; they want to know what's going on. ... In this all-at-once world, everybody's involved in everything. Kids make movies. Soon they'll be making books. ... Our generation is observers. The young today are participators. Your young workers aren't satisfied with slots. They want to participate in depth in the operation. ... Why can't you get ten million people on prime-time TV to work on cancer research? Some kid is sure to say: 'Hey, why don't you do it this way?' and discover the cure."

---

* McLuhan had nothing but contempt for encounter groups and their like, which proliferated in the sixties. He associated them with phenomena described in Wyndham Lewis's *The Diabolical Principal and The Dithyrambic Spectator*, in which Lewis attacked the displacement of art by ecstatic and communal rights. He also dismissed the work of psychologist R.D. Laing, which enjoyed a vogue in the sixties, as a remorseless indoctrination in the joys of tribalism. (Letter to Mary Jane Schoultz, April 6, 1970.)

The last suggestion, of course, had first been hatched by McLuhan in the days of Idea Consultants. No one had listened to it in 1955. But McLuhan's speculations and the appetite of businessmen for clues to the nature of change had now, at last, come together. "A few years ago, he'd have been kicked out bodily," one executive at the conference remarked. "But business has changed. You've got younger, more aware men at the helm." He then added, echoing Tom Wolfe, "I don't pretend to understand him, but I want him around just in case he's right."[26]

To nearly all of his corporate audiences McLuhan offered one important suggestion. If they wanted to avoid confusion, he said, they should ask themselves what business they were really in. It was not an obvious question. Bell Telephone, for example, was unaware that it was not in the business of telephones. IBM did not know, for a long time, that it was not in the business of business machines. Such companies, McLuhan told them, had to look at what changes they actually made in the human environment. Only after looking at such changes could IBM, for example, begin to perceive that it was really in the business of information. Only then could it begin to come up with adequate strategies for operating in the present. Like many McLuhan techniques, this one had little immediate practical application for corporate executives, but it was useful in shaking up their attitudes and broadening their appreciation of reality.

Corporate groups continued to seek out McLuhan. In November 1966 McLuhan gave two lectures for the Container Corporation in San Francisco at $2,500 apiece. In these lectures he barely concealed his contempt for the products of his host. Its plastics and nonreturnable containers had contributed a great deal to the littering of the landscape, which thoroughly repulsed him. Once in an interview he had referred to the "massive garbage apocalypse" created by the Container Corporation.[27] On this occasion he told its executives, in his most inscrutable and oracular fashion, to get into another line of business.

He also tried to enlighten reporters at a press conference in the city's St. Francis Hotel, where he announced, "People make a great mistake trying to read me as if I were saying something. I poke these sentences around to probe and feel my way around in our kind of world." He also ventured a prediction. "Color TV will mean more involvement," he told the reporters.

> You know, we see color with the cone of our eye, black and white with the edges, and color is more in demand in a primitive society. So are spiced dishes. I predict a return of hot sauces to American cuisine. With color TV the entire sensory life will take on a whole new set of dimensions.[28]

(McLuhan believed that the cone of the eye — the photoreceptor in the retina

— was much more "tactile" than the edges of the retina. Hence color was more "involving" than black and white.)

That same year McLuhan also addressed the American Marketing Association, the American Association of Advertising Agencies, and the Public Relations Society of New York. He gave a talk in Washington, D.C., before approximately twenty assistant secretaries in the Johnson administration, under the auspices of the United States Civil Service Commission. He gave numberless interviews to magazine and newspaper reporters, appeared on television and radio talk shows, spoke at various colleges, and, as part of his continuing campaign to annoy the literati, addressed the annual meeting of PEN (the international association of writers and editors) in New York City. One French delegate spoke for many in the audience when he told a reporter that, after hearing McLuhan's remarks, "I was amazed, baffled and rather sickened."[29]

The following year McLuhan began speaking to an entirely new kind of audience: Canadian politicians. Several prominent members of Canada's Liberal party, then in power under Prime Minister Lester Pearson and anticipating another federal election in one or two years, began a series of informal talks with McLuhan in the summer of 1967. The first talk occurred in the penthouse apartment of a Toronto lawyer, Gordon Dryden. "I wasn't married at the time and there was nobody living with me to interrupt us and tell us to shut up and go home or go to bed," Dryden recalls. "This was just as well, because McLuhan could stay up and talk all night."[30]

It was primarily for the joy of talking all night that McLuhan began these consultations. McLuhan himself had no partisan affiliations; indeed, three years earlier he had given an address at a conference sponsored by the other major Canadian party, the Progressive Conservative. Certainly, McLuhan was allergic to Marxism, while at the same time retaining a general antipathy toward mega-corporations. His own work he considered a satirical examination of the monstrous, swollen institutions beloved, it seems, by both big business and socialism.[31]

Politics itself, as a process, he tended to view the way Poe's sailor viewed the maelstrom. Nonetheless, McLuhan would have been happy to work with virtually any government. He felt that his insights were true and important (despite what he occasionally said about the tentative nature of his "probes") and should therefore be brought to the attention of the powerful. If the powerful misused them, the responsibility for that would lie on their souls, not on McLuhan's.

The talks with Dryden and his Liberal colleagues — Boyd Upper, Robert Stanbury, and a handful of others — were reasonably lively. McLuhan provoked an argument from Dryden when he proclaimed, with his typical assurance, that political parties were a thing of the past. McLuhan also railed against the deterioration of Toronto's comfortable cityscape. He was particularly incensed by the building of a huge hotel overshadowing the public square in front of Toronto's new city hall. On the whole, McLuhan's conversations with these men were not very different from his talks to businessmen, academics, and students. "Most of the session would start with a general discussion and then veer off into a McLuhan monologue which would go on as long as he wanted to talk and we wanted to listen," Dryden recalls. "We would start off raising questions and batting them around and arguing them and then he would take off."[32]

One of these political sessions was interrupted by an event that was altogether unexpected to McLuhan's host but that McLuhan himself had come, from experience, to know and dread. In the course of McLuhan's monologue his listeners, sitting in a hotel room, noticed that he occasionally hesitated, a blank look in his eye, as if his train of thought had been broken by some fleeting memory or association. Then he would shake his head in an effort to recapture his train of thought and proceed with the monologue. This happened two or three times in a half hour. McLuhan excused himself to go to the bathroom, and a moment later the men heard a dull, unsettling noise from inside. When they opened the door McLuhan was lying on the floor. He had lost consciousness momentarily and collapsed. He seemed to recover immediately, however, and they took him home, where Corinne received him with concern but with no evident surprise at what had happened. (In retrospect, Upper, a medical doctor, wondered if McLuhan had been suffering from a petit mal epileptic seizure.)

In fact, McLuhan had been suffering from blackouts since the late 1950s. His coloring in those years had taken on its characteristic pale, tense, apoplectic hue. As the sixties wore on, his coloring became worse and the blackouts became more severe and more frequent. An episode in 1966 was typical.

He had been invited to appear before a class in bibliography taught by a colleague at St. Michael's, Sister Geraldine. When McLuhan, sitting at a desk, was asked a question by a student and he did not respond, Sister Geraldine turned around to see him with his head rolled back, one arm raised stiffly in front of his head as if to ward off a blow, and one of his legs stuck out at an awkward angle. Sister Geraldine stood closely in front of him, to shield him from the curious gaze of the students, and as she stood there she heard him sigh almost imperceptibly. Then a silence. She thought for a moment that he was dead. Then a light reappeared in his eyes, and the stiff arm and leg relaxed.

He looked at a stack of papers piled on the desk in front of him. "What are we going to do with these papers?" he asked, with the confusion of an extremely fatigued child. Sister Geraldine, trying to get him out of the classroom, said, "Didn't you tell me you had an appointment for 2:30?" He looked at her, made an effort at thought, and replied, very slowly, "Yes, I guess so." Sister Geraldine helped him out of the classroom and down a flight of stairs, but he refused to let her go with him back to his office.

McLuhan did his best to ignore these blackouts. If anything, he became even more active in 1966 and early 1967 and placed more demands on his mind and body. He was determined to capitalize on his sudden vogue before it disappeared. To that end, he worked simultaneously on at least six books. McLuhan considered this to be a perfectly sensible procedure, feeling that the insights derived from working on one subject stimulated related insights into other subjects.[33]

One such project was a book commissioned in the early sixties by Clay Felker, then a consulting editor with Viking Press. Another book was commissioned by Harcourt, Brace and World on the subject of business management. McLuhan was supposed to collaborate on this book with Ralph Baldwin, who had acted throughout 1966 as McLuhan's unpaid agent in New York City.[34]

A third book was to be a collaboration between McLuhan and William Jovanovich, president of Harcourt, Brace and World, on the future of publishing. (Jovanovich had suggested the project to McLuhan in the fall of 1965.) A fourth book was *Culture Is Our Business*, a study of advertisements on the order of *The Mechanical Bride*, but with a series of one-liners instead of essays accompanying each advertisement and with a completely altered tone and approach. McLuhan had been working on this book since the late fifties. A fifth project was a book, later published under the title *Through the Vanishing Point*, that McLuhan had been working on for several years in collaboration with Harley Parker. This book consisted of a series of reproductions of paintings — also accompanied by one-liners and aphorisms — in which McLuhan and Parker attempted to illustrate the nature of acoustic versus visual space.

In addition to books, McLuhan poured out a series of articles. No longer were these limited to scholarly journals such as *Renascence* and the *Sewanee Review*. In 1966 and 1967 McLuhan produced articles, or articles in the form of interviews, for the periodicals *TV Guide*, *McCall's*, *Family Circle*, *Glamour*, *Look*, *Vogue*, and *Mademoiselle*. (In the two years after that, he would add *Playboy*, the *Saturday Evening Post*, and *Harper's Bazaar* to the list of improbable publications carrying his byline.)

The quality of the writing, in this torrent of prose, did not boost McLuhan's reputation, even among those sympathetic to his ideas and willing to overlook the absence of conventional style. Overproduction was particularly unfortunate for McLuhan, who did not enjoy writing pieces of any length and considered that writing articles interfered with his thinking, since most of them retrod old ground. To facilitate the process, he dictated his work to his secretary, Margaret Stewart, or even to his wife. Often he did so while reclining on the green mattress of his chaise longue or on the floor. (He claimed that lying on the floor helped his back.) On one of the first occasions he dictated to Stewart, he startled the woman by disappearing behind his desk and speaking from somewhere near the region of the wastepaper basket.

From 1966 he could have used at least half a dozen secretaries, as he worked simultaneously on books, articles, and correspondence, suddenly switching from one to the next as his mood dictated. Margaret Stewart, who revered McLuhan and protected his time and energy as much as she could — often shielding him from bad news or problems in connection with the center — was invaluable. It was she who corrected the grammar and untangled some of the rougher syntactical knots in McLuhan's dictated sentences. Since the sentences were obscure enough — like many of McLuhan's statements in the classroom, they assumed that everyone had been present at the birth of McLuhan's ideas and had assisted with the midwifery — any help Stewart could give McLuhan's prose was welcome. Her help could extend only so far, however. McLuhan was being entirely candid when he told a friend that he was not very proud of his books.[35]

The selling of McLuhan was most clearly demonstrated in two articles he wrote with *Look* editor George Leonard for that magazine in 1967. The first one, on the future of education, was an interesting specimen of McLuhan's prose simplified, as it were, for the reader by Leonard. The article itself was full of the breathless predictions of sweeping change characteristic of the sixties — "By the time this year's babies have become 1989's graduates (if college 'graduation' then exists), schooling as we now know it may be only a memory" — and a kind of progressive, New Age optimism that was far more characteristic of Leonard's mentality than McLuhan's.[36]

The second article, on the future of sex, contained some rather good predictions. The authors clearly saw that the image of the he-man and the John Wayne type of masculinity would continue to decline and that gender roles would blur and meld, with men having to make the more wrenching adjustments. They saw that the pill would have a devastating effect on romantic love and sexual reticence. They even predicted marriage contracts.

The article also contained some rather bad predictions. McLuhan's hopefulness that the Dagwoodian male would finally disappear prompted a prediction that homosexuality, which McLuhan had viewed in the forties as an inevitable by-product of Dagwood and his aggressive mate, would also fade. Clearly McLuhan did not relish, or predict, any feminist revolution that would result in "a brittle feminist dame" — as opposed to a "womanly woman" — becoming the president of the corporation or the secretary of state. Like the education article, "The Future of Sex" was full of a dreamy, progressive optimism: "It is possible that the family of the future may find its stability in constant change, in the encouragement of what is unique in each of its members; that marriage, freed from the compulsions and restrictions surrounding high-intensity SEX, can become far more *sensual*, that is to say, more integral."[37] Eight years after its publication, in response to a request to reprint the article, McLuhan asked that his byline be removed since he said he had no part in writing it.[38] This was only a slight exaggeration. In the mid-sixties McLuhan was entering a period of his life when he was willing to have others turn out material, more or less related to his thought, under his name, particularly if the money was good.

McLuhan was extremely poor at establishing priorities. In many ways, he seemed to lose control of his schedule and was constantly engaging in projects that even he felt, after a while, were a waste of time. At one point, he made efforts to cut down on distractions by setting aside mornings at the center for writing, kicking out all the hangers-on and visitors walking in off the street so he could get some work done. But McLuhan could never discipline himself to any remotely tight schedule. For one thing, he hated to wear a wrist watch and it was years before he broke down and put a clock in his office. His sense of time was decidedly pre-Gutenberg. Corinne frequently had to act as his schedule patrol, coming into his office with her lilting "Where's my Marshall?" and hauling him off to some appointment.

Throughout 1966, as the great North American publicity machine continued to promote his reputation, the demands on his time increased tremendously. In the space of a year — from February 1966 to March 1967 — lengthy profiles and treatments of McLuhan appeared in *Newsweek*, *Fortune*, *Life*, the *Saturday Review*, *Esquire*, and the *New York Times Magazine*. It was becoming impossible for a literate person to avoid hearing or reading his name. The great majority of intellectuals and academics, it is safe to estimate, remained hostile to McLuhan, despite the flurry of publicity and his intense following among a minority of their class, not to mention his following among intellectually inclined journalists, advertisers, businessmen, entertainers, and cultural revolutionaries like Abbie Hoffman and John Lennon. For every Susan Sontag who proclaimed her admiration of McLuhan, there seemed to be a dozen Dwight MacDonalds and Christopher Rickses straining to achieve the definitive put-down of the man.

"An obscure professor of English from the Canadian provinces has succeeded in perpetrating a hoax so gigantic that it shows every sign of becoming an international intellectual scandal," wrote the editor of the *Journal of Existentialism*. In doing so, he spoke for legions of McLuhan detractors in the universities and in the cultural journals.[39] The cry of scandal usually boiled down to three points. The first was that his prose style was deplorable. The second was that he was ignorant and totally fanciful, as proved by his cavalier treatments of facts. (It didn't help when McLuhan defended himself by saying of his works, "If a few details here and there are wacky, it doesn't matter a hoot.")[40] The third was that he was complacent about the phenomena he described and indifferent to matters of social justice.

McLuhan's response to the last criticism was simple. He maintained that if people allowed their senses to be walked all over by private manipulators of the media, they really had no rights left worth talking about. He did not, however, attempt any convincing counterattack against his critics or detailed rebuttal of their arguments. Such an attempt would have been foreign to his entire approach. Instead, he did his best to ignore the criticism. He knew that if he took it seriously he would end up paralyzed by it. Moreover, he felt, with considerable justification, that a great deal of the criticism was intensely personal. Something about McLuhan drew forth astonishing venom from that class of individual that T.S. Eliot described as "the mild-mannered man safely entrenched behind his typewriter."

In the midst of this vortex of talk and writing, McLuhan was from time to time reminded that he was head of a family. As the pressures of McLuhan's celebrity increased, they threatened to turn his private life into a shambles.

The two eldest daughters, the twins, twenty-one years old, had left home to pursue their own careers. Teri was studying sociology as an undergraduate at the University of Ottawa and was working for the Company of Young Canadians, a government-sponsored volunteer organization devoted to social work projects. Mary was secretary to the dean of the Golden Gate University Law School in San Francisco and was apparently living a rich version of the California life, with summer vacations at Lake Tahoe, winter treks to Squaw Valley, and New Year's Eve parties with Frank Sinatra.

The two youngest daughters, Stephanie and Elizabeth, as well as Eric and the youngest child, Michael, remained at Wells Hill Road, ensuring that the household remained reasonably lively.

Occasionally, McLuhan made his presence felt in a more or less benign fashion. When the daughters brought home their dates, for example, McLuhan would sometimes summon them to the living room, where he lay stretched out on the floor in front of the fireplace, so he could talk to them. This friendly invitation usually succeeded in making the young men extremely nervous.

In 1965 Eric became a paid assistant to his father. He had spent three years, from 1961 to 1964, in the U.S. Air Force and had returned to study at the University of Toronto. Always the most rebellious of the McLuhan children, Eric began a remarkable transformation at this time that led to his becoming, on and off for the rest of McLuhan's life, his father's faithful confederate. His early passions had been model airplanes and electronics. Now he discovered a passion for his father's literary patron saint, James Joyce. He immersed himself in the depths of *Finnegans Wake*, to the point where his pleased father began telling people, in the late sixties, that his son was the world's best interpreter of *Finnegans Wake* and possibly the best Joyce critic there was.[41]

In 1965, with some advice from his father, Eric began writing a book on *Finnegans Wake* that, not surprisingly, showed this arcane masterpiece to be a complete guide to the media, just as McLuhan Senior had insisted all along. With his father's help, Eric "deciphered" the key to *Finnegans Wake*: the ten "thunders" that announce the beginning of new cycles of human history. Joyce had predicted that no one would decipher the key for three hundred years, McLuhan informed one correspondent; he and his son had just proved the great man wrong. To another correspondent he predicted that Eric's book would have great impact on Joyce studies: it was an incomparable contribution to the field.[42]

The manuscript was submitted to Viking Press. Alan Williams, then an editor at Viking, recalls that the manuscript received a reasonably warm response, including a favorable report from one reader, Joseph Campbell, the great scholar of mythology and co-author of McLuhan's old standby, *A Skeleton Key to Finnegans Wake*. Evidently, the response was not quite warm enough for Viking to publish it. In December 1967 Eric signed a contract to publish the work with McGraw-Hill, and three and a half years later McLuhan was assuring correspondents that the book would be coming out at some point. (In fact, the book was not published until the spring of 1997, under the title *The Role of Thunder in Finnegans Wake* by the University of Toronto Press.)

There was no doubt that McLuhan enjoyed Eric's company and found his conversation extremely gratifying. When Eric returned to Toronto from the air force, McLuhan discovered that the young man was interested in the same things he was; seven years later he insisted to a correspondent that Eric had become a world-class authority on media.[43] With his son, McLuhan felt he could share the ideas and insights closest to his heart.

As his father's assistant, Eric also performed other functions. He was a chauffeur and companion-at-arms in settings outside the home or the center, where McLuhan in later years often felt the need of somebody who could offer him unqualified support. When McLuhan was addressing a roomful of people, he would sometimes look at Eric after speaking for a few minutes, and Eric would nod in agreement.

Although he was profoundly appreciative of the support Eric gave him, McLuhan could be rude and impatient toward his son in public. (This was especially true in the last decade of McLuhan's life.) "At the Monday night seminars, McLuhan used to treat Eric like a rag doll," one participant who attended those seminars in the late seventies recalls. "He would verbally abuse him in front of all his friends, and Eric would take it."⁴⁴ McLuhan, alas, was not kind to those who decided to become, in the language of *Understanding Media*, extensions of himself.

In 1965, however, when Eric first signed on, McLuhan was grateful for his help. The following two years were among the busiest in McLuhan's entire career, culminating in the publication of books he'd been steadily working on since the publication of *Understanding Media*. The first, *The Medium Is the Massage*, was published in March 1967. This book had in fact been composed by Jerome Agel, who had written a profile of McLuhan in 1965, and Quentin Fiore, a first-class book designer. The two selected or commissioned photographs to accompany excerpts they culled and reshaped from various writings and statements of McLuhan's. McLuhan himself contributed the punning title and approved the text and layouts. Agel and Fiore evidently did their work well: McLuhan changed only one word. Their mix of text and visuals was indeed a virtuoso feat. They used arresting photographs and artwork and performed interesting experiments with type, laying it upside down, on the slant, or in mirror image, switching its size from page to page, varying between regular and boldface, and so on.

Agel referred to the result as a "cubist" production. McLuhan recognized that it was an effective sales brochure for his ideas. Seventeen publishers turned down the proposal before Bantam Books agreed to take on the project, as a book by Marshall McLuhan and Quentin Fiore, "co-ordinated by Jerome Agel." Anyone could read it in an hour. As with the articles McLuhan produced for *Look*, *The Medium Is the Massage* was McLuhan Made Easy. Perhaps for that reason the book sold very well — the only one of his books, McLuhan complained, that did so. The first printing of the Bantam paperback version (which was released a week or so before Random House's oversized hardcover edition) was 35,000 copies. It was quickly followed by two more runs of 35,000 each. Eventually it sold nearly a million copies worldwide.

Almost simultaneous with the publication of the book was the release of the CBS Records version of *The Medium Is the Massage* and a one-hour NBC television documentary on McLuhan, which aired on March 19, 1967. The NBC film featured clips of McLuhan delivering one-liners, thrown into a stew of pop art, animated visuals, newspaper headlines, and other images and edited with the fast-cut technique coming into vogue in both movies and television. (Beatles' movies had led the way.) McLuhan detested the film, produced by Ernest Pintoff, calling it "grotesque trash."⁴⁵ He had reason to be concerned. It was a sign that he was now being processed by the great North American publicity machine, just like any other sensation of the decade, from Haight-Ashbury to Mao's *Little Red Book*. Even Marshall McLuhan could be turned into Instant Trivia.

10. New York City (1967–1968)

> Never could notoriety exist as it does now, in any former age of the world; now
> that the news of the hour from all parts of the world, private news as well as
> public, is brought day by day to every individual...by processes so uniform, so
> unvarying, so spontaneous, that they almost bear the semblance of a natural law.
> — John Henry Newman[1]

Since at least the beginning of 1966 John Culkin, S.J., director of the Center for
Communications at Fordham University, had been plotting to lure McLuhan
to New York City. Ralph Baldwin, an associate of Culkin's, wrote to the
McLuhans in June of that year, explaining that Culkin wanted to establish a
"research council," loosely associated with Fordham University, to employ
McLuhan in a way best calculated to exercise his "genius," while keeping distractions to a minimum.[2] Culkin's research council was to be a think tank
devoted to research in the media and communications in general — a high-
powered version of McLuhan's center in Toronto. Its chief virtue, in theory,
was that it could offer McLuhan a post connected to Fordham University without delivering him into the hands of his former employers the Jesuits. In this
post, McLuhan would be surrounded with research assistants, left free to pursue
whatever ideas interested him, and weaned from his modus operandi of appearing for two or three hours before virtually any group that asked him to speak.

   The device that eventually brought McLuhan to New York, however, was
not this research council (which had yet to be funded) but the promise of an
Albert Schweitzer Chair in the Humanities at Fordham University. In 1964 the
New York State legislature, to attract top scholars to the state, had set up five
Einstein chairs in the sciences and five Schweitzer chairs in the humanities. A
grant of $100,000 — for personnel, research facilities, and so on — was
attached to each chair. In January 1967, the New York State Board of Regents
approved McLuhan's nomination to a Schweitzer chair at Fordham University
for one year.

   Despite the lure of this generous grant, Culkin had a great deal of trouble
persuading McLuhan to accept the chair. McLuhan was quite content in his leafy

Wells Hill Road neighborhood, walking down to his office on St. Joseph Street and his faithful secretary every day, stretching out on his chaise longue, and talking to students or dictating an article. He was comfortable there. If he wanted to look up a back issue in his *Scrutiny* collection, for example, he knew exactly where to find it.

Ralph Baldwin had the answer for Culkin. He knew that, in the fuss made over McLuhan by corporate executives, public relations directors, journalists, and so on, Corinne was not always included. McLuhan himself, in his preoccupation with his work, did not always make sure that she got a break from the house and the children. Baldwin recalls, "I said to Culkin, 'It's very simple. If you ask Marshall, you'll never get to first base with him. All you have to do is to discreetly ask Corinne. I know she'll be enamored of the idea of coming to New York. She'll persuade him.' And she did."[3]

Culkin had many concrete inducements to offer. Elizabeth McLuhan, who was beginning her first year of university, was offered a scholarship at the new women's college at Fordham. Most important, the McLuhans were provided with a beautiful house in another leafy neighborhood, in Bronxville, not too far from Fordham and reasonably sheltered from the clamor of the city. Finally, there was the matter of the $100,000 grant. The money was sufficient for McLuhan to hire his old friends Ted Carpenter and Harley Parker and his son Eric as members of a research team that would teach a course entitled Understanding Media and would undertake various communications projects.

On September 1, 1967, the McLuhans moved into the house at 1015 Kimball Avenue, next door to Jack Paar. ("Jack Paar," coincidentally, were the first words in the introduction to the paperback edition of *Understanding Media*.) A few days later, the attorney general of New York State blocked the payment of the grant to Fordham, on the grounds that it violated the state's constitutional ban on aid to sectarian schools. Fordham officials expressed suitable outrage at the decision, but to no avail: the university was obliged to make up for the loss of state money out of its own funds. McLuhan himself ended up with a $40,000 salary that year, the rest of his grant going for salaries for his assistants, including Eric.

The brief controversy had little effect on him personally, but it was a portent of an unsettling year in the making. The figure of $100,000 in particular, mentioned in the lead sentence of innumerable news stories about McLuhan and Fordham, was as much a curse as a benefit. To McLuhan's colleagues, toiling away in English departments at $14,000 a year or less, it was evidence that he had been drawn into the crasser zones of celebrity. Those who hated him were confirmed in their belief that he had sold out, abandoning the effort to explicate Francis Bacon in favor of tickling the fancy of advertising men. In New York

City, even outside academia, McLuhan was increasingly a target of criticism. The favor of the gods seemed to be deserting him.

On September 18, McLuhan gave his first lecture at Fordham, to 178 undergraduates who had registered for Understanding Media. A *New York Times* reporter in attendance reported afterward that one student was "in a mild state of shock. 'The poor kid,' another student said. 'He tried to take notes like it was a normal lecture.' Dr. McLuhan termed the students 'a rather alert sort of group — I thought they were sort of turned on.'"[4]

Later in the semester, another reporter visited McLuhan's classroom and remarked that McLuhan's "topical arguments seem to evoke only a depressing chorus of stares, stirrings and coughs.... As for his coughing classmates, [Anthony] Perrotto retorts 'a lot of them feel 141 is a waste. They think the course is disorganized.'"[5] Still, McLuhan attracted his admirers and, as always throughout his career, they came from the most unlikely quarters. Philip Romano, then a sergeant with the NYPD, claims that he and his fellow police officers who were taking undergraduate courses at Fordham University loved McLuhan. "When McLuhan spoke, we understood his language — we knew about clues, we knew about probes, we knew about pattern recognition," Romano recalls. "He was an inspiring teacher, no doubt about it."

Of course, Romano and his colleagues were more than ordinarily alert to developments in the neo-primitive electronic world they patrolled. They were just beginning to learn, for example, that demonstrations were strictly a media phenomenon, aimed at the six o'clock news. And they knew about the interaction of the senses. "In the car, we were listening to the [police] radio, and looking outside the car, processing information from many different points of view, and using our sense of smell — because a lot of buildings were going up in smoke at that time," Romano remembers. "McLuhan talked about acoustic space — we knew about that. If you didn't decipher those radio runs, you were in trouble."[6]

Others, however, were much less receptive to McLuhan. Reporters enjoyed quoting McLuhan when he commented about his theories, "I don't pretend to understand them — after all, my stuff is very difficult."[7] It was not often that giant intellects made such good copy in such self-damaging fashion.

Those Fordham undergraduates who felt McLuhan's course was "disorganized" were only reliving the experience of generations of St. Michael's students. In 1967, however, the situation was considerably worsened by McLuhan's deteriorating health. The turbulence inside his skull, and the anxiety connected with it, drove him to uncharacteristic displays of petulance. "He would lash out at the students," Harley Parker recalls. "He would say things like, 'Ask me a stupid question, you'll get a stupid answer.'"[8]

Nor did McLuhan ingratiate himself with other institutions in New York. At a conference entitled "Media and the Museum" at the Museum of the City of New York on October 9, he succeeded in alienating his entire audience. Never particularly interested in the contents of museums, McLuhan was far more concerned with the effects of museums on their visitors. Why, he asked himself, did museums make people tired? Because, he answered, there was a constant clash between the acoustic space of the cultures often featured in museums and the visual spaces of the exhibits, mounted by literate nineteenth-century-oriented museum designers.

At the Museum of the City of New York, McLuhan attempted to convey this insight to his audience, which included the director of New York's Metropolitan Museum of Art, Thomas Hoving. "Museums are a magnificent illustration of the Western bias toward the world of artifacts," he told that audience. "The Westerner, a visual man, thinks of art and language as representational. But we are the only civilization of which that is true."[9]

McLuhan cited the example of a museum draining the excitement from an old fire engine by enclosing it in a glass case. He followed this remark with reminiscences about the pleasures of growing up near a firehall on Gertrude Avenue in Winnipeg and then startled his audience by pronouncing that horse-drawn and motorized fire equipment caused urban congestion and slums. "I'm sorry," one member of the audience interrupted, "but I must have misunderstood you. I thought you said that the advancement of the fire engine produced slums, not the reverse. Surely I'm mistaken." McLuhan gave the questioner a look of vague annoyance. "Definitely the fire engine caused the crowding and congestion, and definitely not the opposite," he insisted. Hoving went away from the talk shaking his head. "He seldom allows reason or common sense to get in the way of his unquestionable brilliance" was his verdict on the performance.[10]

McLuhan believed that advertisers were among the few people who read him seriously. They had a vested interest in learning about the effects of the media, about why the soft drink can, for instance, had to hiss when it was opened and the liquid had to go glug, glug when it was poured. McLuhan was not unwilling to help advertisers, just as he was not unwilling to help politicians, consoling himself with the thought that they were doubtless too incompetent to take full advantage of the insights he bestowed on them. McLuhan's ambiguous respect for and fascination with their black arts continued throughout this period of his life. Advertisements, he maintained, were the single most rewarding source of information about society. More than that, they were "the greatest art form in human history."[11]

The latter statement infuriated the literati. Unfortunately, McLuhan did not always clarify what he meant by "greatest." In his own mind, he made a sharp distinction between the art he truly cherished — the art of Joyce and Eliot, which served to sharpen human perception and reveal the essential features of the contemporary environment — and the art of popular entertainment and advertising — which, like the art of preliterate societies, tended simply to intensify the hypnotic effects of that environment. (It was not just a question of highbrow versus lowbrow. It was also the difference between "Li'l Abner," which McLuhan felt was genuine satire, and "Blondie," which was an unthinking reflection of the cultural patterns it portrayed.) Advertising was "great" in its resourcefulness in inducing certain states of mind, but in no other respect.

Fewer people in the business were more resourceful and more aware of what they were about than Tony Schwartz, a well-known advertising wizard specializing in audio effects. Schwartz had read *Understanding Media* shortly after its publication. "I was playing the same ball game — commercials — as other people were playing," Schwartz recalls, "but I was playing in a different ball field. They were playing in a print-oriented ball field and I was playing in an auditory-structured ball field. After reading McLuhan's book, I understood that. It was almost as if it had been a rainy, cloudy day and my discovering this in his book just made it the clearest blue-sky day you could ever have."[12]

McLuhan asked Schwartz to appear with him at a couple of his lectures in New York, including one at the Museum of Modern Art. Schwartz presented his material at these lectures with a good deal more lucidity than McLuhan. "People would come up to me afterward and speak and not go to him," Schwartz recalls, "and he sensed that and stopped asking me to speak with him anywhere." Nonetheless, McLuhan continued to drop in at Schwartz's studio and participate in his classes in "auditory perception," in which students performed experiments such as finding various ways to tape interviews.

Schwartz himself taped innumerable hours of McLuhan "rapping" — with Ted Carpenter, with John Culkin, with Harley Parker, with Eric — about innumerable topics. By the time McLuhan left New York City, Schwartz had a wall full of McLuhan cassettes. Those tapes probably represented the most significant effort put out by the McLuhan "research team" assembled at Fordham.

The most public effort was a spread in *Harper's Bazaar* featuring McLuhan-esque reflections on fashion and beauty. Although blazoned with McLuhan's name, the spread was almost entirely the work of Ted Carpenter. It featured much agreeable nonsense of the type McLuhan was often accused of purveying:

> We are no longer interested in the surface appearance of things, least
> of all in the surface appearance of people. Wrinkles are ignored; the
> old, the blemished, are accepted for themselves. We are concerned
> with depth; we want to know the inner person.

Hopi farmers don't grow beans; they relate to beans so as to relate to
the bean-ness within each seed — and thus food comes into being.[13]

Accompanying the test were numerous photographs. A picture of solemn
East African warriors gripping their spears was contrasted with a shot of a
Twiggy-like model puffing on a cigarillo. The girl with the eyeliner and the go-
go boots was, the photos proclaimed, as wild and earthy as the queen of an
African village. It was a wonderful sixties product, suffused with the spirit of
those years. It was also one of the most blatant examples of McLuhan Product
— shrewdly counterfeited McLuhanisms used as a kind of soundtrack for an
exercise in the far-out.

The extreme of McLuhan's commercialization was yet to be attained, however. The apogee was the brain child of Eugene Schwartz, an entrepreneur who
in the fifties had pioneered the idea of selling books — mostly of the self-help
variety, in the fields of health and finance — through the mail. Like everyone else in New York publishing circles, Schwartz had become alerted to the
McLuhan phenomenon sometime in the mid-sixties. He conceived the idea of
publishing McLuhan, as Schwartz puts it, "in a medium that could be delivered
faster than a book but had more inherent depth than television" — that is,
through periodic newsletters. He phoned McLuhan in Toronto with the idea
and then flew up to meet the professor in early 1967.

At lunch, the two got on well. At one point, however, McLuhan broke off
in the middle of a sentence, hand frozen in air, as if warding off a blow to his
head. Everything stopped. Startled, Schwartz leaned forward to see if McLuhan
was making an intense effort to concentrate or was having difficulty swallowing
or something of that sort — but all he could see was that McLuhan was completely immobile. The light had gone out of his eyes. A minute passed, perhaps
two. Schwartz grew increasingly uncomfortable. He was about to motion to a
waiter when, again without warning, life returned to the body. The arm relaxed
and the rest of the sentence came out from McLuhan's lips. The lunch proceeded as if nothing had happened.

This unsettling episode aside, Schwartz was eager to proceed with his idea
for a newsletter, and, shortly after McLuhan's arrival at Fordham that summer,
the two signed a contract. The venture was under way by the summer of 1968,
the first issue appearing in July of that year. Schwartz, like Ralph Baldwin, had
early realized that the key to McLuhan's approval or disapproval of any given
project was his wife, Corinne. Treat the woman nicely, appeal to her, and one
was halfway home. Corinne therefore found herself riding in limousines in
Manhattan, courtesy of Eugene Schwartz, during her stay in the city. Schwartz
also offered Eric, whose employment was always a pressing concern to his

parents, a well-paid job ($15,000 a year, according to Schwartz) and an office on the top floor of 200 Madison Avenue, as editor of the "DEW-LINE" newsletter.

The newsletter was offered to the public at $50 for a year's subscription; more than 4,000 people eventually signed up. Schwartz considered this figure to be "relatively low." He assured McLuhan that this was only the beginning and that circulation would climb as the newsletter took off. The readers were primarily top-flight executives in advertising, in firms like IBM. Schwartz even made sure a copy was sent to the White House (to an obscure Nixon aide named Fred Panzer).[14]

In the next two years Schwartz would propose a number of other projects for his client. One was a scheme in which McLuhan would nominate those books he considered of great importance either to his own intellectual development or to an understanding of the contemporary world. Schwartz would then obtain the rights for those books and would market them under the title "Marshall McLuhan Book Shelf." Another scheme was the "Marshall McLuhan Show," in which McLuhan would chat on nationwide television with people like Buckminster Fuller or with his fans in the entertainment world such as Dan Rowan or the Smothers Brothers.

Another scheme involved making educational films using television ads and old film clips — a sort of movie version of *Culture Is Our Business* (a book of commentaries on ads that McLuhan would publish in 1970, more than a decade after starting it). Schwartz also proposed the establishment of "sensory retraining centers" for the McLuhan-approved rehabilitation of individuals with hopeless sensory biases.[15]

None of these projects came to anything, although not for want of trying on Schwartz's part. He was eager to be a McLuhan impresario, a role that had been assumed in the past, with varying degrees of selflessness and intelligence, by people such as Felix Giovanelli, Robert Gray, Howard Gossage, and Ralph Baldwin. It was not an easy part to play. If the person who assumed the part did so for anything more than the disinterested pleasure of promoting McLuhan, the relationship ended unhappily. Ultimately, this would be Schwartz's fate. Throughout the year McLuhan spent in New York, however, he rather appreciated Schwartz's enthusiasm.

In the meantime, other luminaries from the media world sought out McLuhan in New York. Stanley Kubrick arranged a private screening of his new film *2001: A Space Odyssey* for the McLuhans. Unfortunately, McLuhan detested science fiction: it might be ingenious, he felt, but it lacked art's rich perceptiveness. His daughter Teri kept him from walking out ten minutes after the movie started, but the film did not alter his opinion. Midway through, the Strauss waltz on the soundtrack was punctuated by the sounds of snoring.

(McLuhan had a similar reaction to many of the avant-garde, multiscreen works he found himself subjected to in New York.)

McLuhan also met NBC vice-president Paul Klein, who was a fervent admirer of his work. Considered NBC's intellectual, Klein at one point had purchased twenty copies of *Understanding Media* and had sent them to his fellow NBC executives, along with a strongly worded memo urging them to read the book. For Klein, McLuhan's major insight concerning television was that a dramatic show "did not need a linear exposition. Viewers did not even need to see the beginning of a show — they could still figure out what it was all about. In fact, they liked it better if they had to figure it out for themselves." Some producers, according to Klein, got the message.

> Norman Felton was the most important producer of his age. He used to make his shows — *The Man from U.N.C.L.E.*, for example — the way he thought McLuhan would envision them. The rule was, the more you made it so that viewers had a challenge figuring out what was going on, the more you didn't tell them, the more they wanted to watch what was going on.

When some of Klein's colleagues at NBC asked the guru of the electronic age why a particular person was successful on television, McLuhan would readily answer in terms of the individual's hotness or coolness. (McLuhan's powers of divination in this respect were not infallible. When CBS removed Walter Cronkite from the anchor booth at the 1964 political conventions, for example, McLuhan had quickly come up with a theory about why this man was unsuited to the television medium.) Klein's colleagues were unimpressed. "They all ridiculed McLuhan," Klein recalls, "I think partly because of his personality and partly because of what seemed to be his outrageous statements."[16]

As the fall of 1967 wore on, McLuhan's blackouts grew more severe. A few times Culkin had to step in and take over for a few minutes when McLuhan blacked out in front of the two hundred students in his classroom. Even his fabulous energy began to wane. By the middle of the day he was obviously tired. But McLuhan refused to see a doctor. Although his condition had been diagnosed at one point in Toronto as epilepsy, he had made no effort subsequently to obtain a different or more accurate diagnosis. In late October, however, Ted Carpenter explained to a doctor how McLuhan seized up when he lost consciousness, demonstrating the arm bent and held rigidly in the air in front of his head. "The first thing I would check for," the doctor said, "is a brain tumor."

A few days later Fr. Culkin sat down with McLuhan and Corinne and told them about the doctor's suspicion. McLuhan finally agreed to enter Columbia Presbyterian Medical Center for tests. The X-rays showed a meningioma, a tumor of alarming size — "as big as a tennis ball," according to Carpenter — under his brain. Fortunately, it was benign and operable, the doctor explained, and they had discovered it before it caused irreversible damage. McLuhan, however, heard only the word "operation" and, in a tormented gesture, shoved the doctor aside. When surgery was scheduled for him, he checked himself out of the hospital and went back to his classes at Fordham.

Corinne and Culkin urged McLuhan to go back to the hospital. The doctors had laid it on the line, they told him: blindness and perhaps insanity lay in store for him within four months if he did not have the tumor removed. Finally McLuhan agreed, pleading to be spared any details of the operation, anything that would enable him to visualize the workings of the surgeon's knife. On November 18, he reentered Columbia Presbyterian. The surgery was scheduled for November 25. Corinne lived in fear that her husband would walk out again before the operation. McLuhan asked Culkin if his surgeon was a drinker. Although Culkin assured him that the sixty-two-year-old Dr. Lester Mount — perhaps the most distinguished brain surgeon in the world at that time — was a model of sobriety, McLuhan was still terrified.

The operation started at 11:30 on a Saturday morning. After five hours, the first team of assistants and nurses was replaced by a fresh team. Dr. Mount toiled on. Five hours later a third team took its turn with Dr. Mount. The surgeon maintained his unhurried pace. He had, first of all, virtually to lift the brain of his patient to get at the tumor, in the process exposing some of the surface cells to oxygenation and therefore a degree of damage. He was extraordinarily cautious. He would make a cut and then take fifteen or twenty minutes to decide where exactly to make the next one. Finally, at five o'clock on Sunday morning, Dr. Mount finished his work and declared the operation a success. It had been the longest neurosurgical operation in the history of American medicine.

When McLuhan awoke a little less than an hour after the operation was finished, he looked at the clock on the recovery room wall and thought that it was 6:00 p.m. on the same day he had been wheeled into the operating room. The surgeon came in and asked him how he was feeling. Still groggy from the anesthesia and perhaps groping toward a pun floating far off in the haze, McLuhan replied that it depended on how one defined "feeling." The remark went the rounds of the hospital. Almost all patients after an operation of this kind spent their first conscious moments hallucinating. McLuhan's relative presence of mind under the circumstances endowed him with almost heroic stature in the eyes of the neurosurgical personnel at Columbia Presbyterian.

McLuhan's gallantry was reflected in his comments to worried friends that the operation had taken a load off his mind. The reality, however, was far from humorous. As the last numbing effects of the anesthesia wore off, McLuhan was immersed in a bath of unimaginable pain. He told friends that he would never have had the operation if he had known what the pain would be like — blindness and insanity notwithstanding. Shortly after he left the hospital, McLuhan composed, for his own interest, a note about his experience recovering from the operation. The note began with the phrase "no atheists in recovery rooms." The experience of pain, McLuhan wrote — especially the huge and lasting pain made possible, paradoxically, by anesthetics and the cruel versatility of modern surgery — created in the human creature an indelible sense of his closeness to, and dependency on, God. The pain revealed to him the emptiness of the world constructed by human beings and their media of communication. The toiling and the lust for such human constructs, perceived by the soul close to God, was as absurd and repulsive as a "bouquet of roses...in a death camp."[17] The message of McLuhan's operation was, in short, the hidden message that underlay all his work on media: put not thy trust in the worldly extensions of man.

On December 1, McLuhan noted in his diary that life was "bearable" for the first time since the surgery. His three-week recuperative period at the hospital was further brightened on December 9, when Eric signed his contract for publication of his book on Joyce (the book that would appear thirty years later). On December 12, McLuhan left the hospital. He had been scheduled to leave the following day but had flatly refused to begin his postoperation life on the thirteenth day of the month.

Although life was now bearable, the effects of the operation would linger for the rest of McLuhan's life. In the months immediately following, it was dramatically obvious to his associates that McLuhan had changed. For one thing, his senses, always acute, became hypersensitive. After the operation, he seemed to be one exposed and quivering nerve. Entering the kitchen of his own home was like stepping into a chemical factory, he told friends. Noise particularly tormented him. When planes flew over his house in Bronxville — unfortunately it was under a flight path to one of the New York airports — McLuhan literally screamed in agony.

Another effect was the disturbance of his memory. McLuhan had always possessed what people often refer to as a "photographic" memory. Along with his energy, it had been his most obvious professional asset. After his operation, however, he noticed that memories from the year or so preceding the operation were particularly fuzzy. It took a long time for them to come into focus. Moreover, he discovered that "several years of reading got rubbed out," as he put it to one correspondent, and he was obliged to reread many books.[18] A year or two after the operation, he found it helpful to purchase stacks of those student study guides that give the plot and theme of works like *Moby-Dick* and *Paradise Lost*.

Other memories were actually stimulated by the surgeon's knife. Soon after the operation, for example, McLuhan discovered a passion for the songs of the Scottish tenor Harry Lauder, who had been popular in McLuhan's youth, and expressed this passion with his powerful, but unmelodious, voice. Some childhood memories seemed closer to him. The memories were quirky, however. At times he remembered faces but could not conjure up names, and at other times he read names of people he realized were long-time friends and could not conjure up their faces.

On the whole, McLuhan recovered with a rapidity that amazed his doctors. This was due, more than anything else, to his strength of character. Even so, his comeback was by any measure long and slow and partial. For more than two years after the operation, he took sedatives and depressants such as Librium and Valium to curb the high-strung responses from his central nervous system. He retained his acute sensitivity to noise. The man who once was able to fall asleep at any time was now wakened by the slightest sound.

He also never regained the degree of emotional and intellectual resilience he had possessed before the operation. His friend John Wain described him as "nervous, fragile, tense" the year after his operation. To some extent, he remained that way for the rest of his life. Marcel Kinsbourne, a neurologist who knew McLuhan in the last year of his active career, recalls that he looked and acted older than his sixty-eight years. "He was sort of querulous and irritable in his later years," Kinsbourne recalls. "He didn't come across as being particularly mentally alert or flexible." It was after his operation, too, that McLuhan increasingly dealt irrational, and uncharacteristic, abuse to students and colleagues.[19]

It was no wonder, then, that the new year of 1968 was in some respects hellish for McLuhan. He was obliged to navigate a world that had suddenly become, even in a physical sense, threatening to him. Occasionally, he lost his way. Shortly after his operation he went to visit his friend Claude Bissell, then teaching at Harvard and living in Cambridge, Massachusetts. McLuhan stepped out one day to buy some roses for Mrs. Bissell and did not return. Hours later, Bissell found him wandering around the streets of Cambridge, trying to find his way back to the house.

As these incidents multiplied, McLuhan became seriously upset at any suggestion that there was a problem. But he was also panicked — it seemed as if each wrong turn was a symptom that he had lost a certain control over his life. When the McLuhans were invited to dinner at the house of his secretary on

Long Island, Corinne took care to drive there alone beforehand to make absolutely sure that she did not get lost on the night of the dinner. She knew that if she did, her husband in the seat beside her would react like a man threatened with shipwreck.

The worst incident of this kind occurred when McLuhan, Harley Parker, and Ted Carpenter found themselves in a cab on the expressway after the driver had taken a wrong turn. As they sped along with the traffic and the meter clicked away, McLuhan, according to Carpenter, went "berserk." He began to shout for the driver to let them out, until even this hardened New York cabbie became frightened and tried to pacify McLuhan by turning off the meter. The cabbie's promise that there would be no charges whatsoever had no effect. McLuhan kept demanding to be let out, as if he were being kidnapped. As soon as the cabbie managed to ease out of the traffic and bring the car to a stop, McLuhan leaped out. When Carpenter started to pay, he shouted, "Don't you give him any money." Carpenter had to slip the shaken cab driver some money when McLuhan wasn't looking.[20]

McLuhan's recovery was not helped by other irritations in his life. His son Michael dropped out of Grade 10 in Roosevelt High School in March of 1968 to crash with friends in an un-rented apartment on the east side. He did not return to his family until a year later, after panhandling, experimenting with drugs, and otherwise fully immersing himself in the joys of the counterculture. McLuhan seems to have felt that lax post-Vatican II theology — including much that was now being taught by hip Jesuits at Fordham — somehow played a part in all this. When McLuhan returned to St. Michael's College in the fall of 1968 he transferred his anger at the Jesuits to some of his clerical colleagues. One unfortunate Basilian got a punch in the arm, by no means playful, from McLuhan one day when the subject of the Fordham Jesuits came up.

The chief target of McLuhan's wrath, however, was the theologian Gregory Baum, Canada's most prominent Catholic "progressive." At lunch in the faculty cafeteria at St. Michael's one day in late 1968, McLuhan made a few disparaging remarks about the Fordham Jesuits, obviously directed at Baum. "I said, 'Marshall, what did the Jesuits teach your boy?'" Baum recalls. "He turned to me and he said, 'They taught him that you can be saved outside the Roman Catholic Church.'"[21]

Among his friends, McLuhan referred to Baum as a KGB agent. His friends were never sure just how seriously he meant the remark. What was beyond doubt was McLuhan's outrage at Baum's highly politicized, left-leaning theology. When Baum endorsed the New Democratic party, Canada's "social democratic" party, despite its advocacy of legalized abortions, McLuhan, who would become an active member of the pro-life movement in Canada in the seventies, was confirmed in his view that Baum was beyond the pale. As far as he was concerned, Baum's talk about "solidarity with the oppressed" was nothing more than an appeal to what McLuhan called "unemployed emotions" eager to be enlisted in some self-righteous crusade. This eagerness was antithetical to the traditional wisdom of the urbane and self-possessed Catholic.[22]

In 1968 many Catholics of progressive bent tended to associate McLuhan with another innovative theologian, the French Jesuit Pierre Teilhard de Chardin. Chardin maintained that organic evolution was converging toward what in traditional Christian theology is termed the parousia — the second coming of Christ. This idea seemed to fit very well with McLuhan's more cosmic pronouncements on the effect of electronic technology. In fact, McLuhan did credit Teilhard with realizing that this technology was an "outering" of the human nervous system and therefore a new stage in human development.

At times McLuhan made statements that seemed to indicate a belief that this new stage was, in fact, a Teilhardian move toward the cosmic Christ. The Mystical Body of Christ was, he told the interviewer Gerald E. Stearn, "technologically a fact under electronic conditions." In his 1969 *Playboy* interview, McLuhan amplified this statement by remarking,

> The computer thus holds out the promise of a technologically engen-
> dered state of universal understanding and unity, a state of absorption
> in the Logos that could knit mankind into one family and create a
> perpetuity of harmony and peace. . . . Psychic communal integration,
> made possible at last by the electronic media, could create the univer-
> sality of consciousness foreseen by Dante when he predicted that men
> would continue as no more than broken fragments until they were
> unified into an inclusive consciousness. In a Christian sense, this is
> merely a new interpretation of the mystical body of Christ; and
> Christ, after all, is the ultimate extension of man.[23]

This echoes the paragraph from *Understanding Media* about the powers of the computer that disgusted Dwight MacDonald and other critics. In the *Playboy* interview, the theme has even more of a Christian interpretation; a follower of Teilhard de Chardin might be forgiven for thinking that McLuhan was giving new substance to the good Jesuit's planetary visions. It is hard to know just how committed McLuhan was to such statements — he reserved the right to disown these "probes" at any time, and it is even possible that they were basically put-ons, teasing exaggerations, from the beginning. Indeed, the entire *Playboy* interview has the flavor of an extended put-on, as when he put on the mask of the academic futurist to assure *Playboy* readers that, "projecting current trends, the love machine would appear a natural development in the near future — not just the computerized datefinder, but a machine whereby ultimate orgasm is achieved by direct mechanical stimulation of the pleasure circuits of the brain."[24]

In any case, it is certain that, even at the time he talked about computerized absorption into the Logos, McLuhan was condemning those Catholics who did, in fact, mistake the new electronic environment for the Mystical Body of Christ. He warned more than one coreligionist (including his old mentor Jacques Maritain, who most assuredly needed no such warning) that "the Prince of this World is a very great electric engineer." As for Teilhard de Chardin, McLuhan increasingly came to view him as "science fiction" — that is, as a futurist who was devoid of genuine perception. With the fading of the sixties, he came less and less to share anything remotely resembling the optimism of Teilhard de Chardin.[25]

Of course, McLuhan also recognized that much of the traditional theology of the Church, based on Christianized reworkings of Plato and Aristotle, was derived from the Greco-Roman world now finally dissolving in the electronic world. Such theology was obviously in danger of becoming useless. But that was no excuse, according to McLuhan, for the Protestantization of the Catholic Church that occurred after Vatican II. This process — marked by the abandonment of Latin in the Mass, the de-emphasis of such Catholic doctrines as purgatory, the invocation of the saints, and so on — thoroughly disgusted him.

The devastation of the Mass wrought by post–Vatican II changes was particularly painful. He did not like the fact that the priest now looked at the congregation, "with his face hanging out," as McLuhan put it, and a microphone around his neck. That microphone, according to McLuhan, automatically eliminated Latin from the Mass, since the Latin was not really intended to be the focus for participants at the Mass but rather a kind of audio backdrop. The mumbled Latin was almost a subliminal element, playing off against the vernacular in such a way as to provide a rich and almost magical multilayered expressiveness. In addition, this backdrop freed participants to meditate or pray during the Mass. Not so the vernacular, amplified through the wretched microphone — a "hot" medium killing off the "cool" medium of the nearly inaudible Latin. The new Mass, it is true, did afford McLuhan the chance to belt out old, familiar Protestant hymns (he wondered why no one else in the pew joined him), but even that did not compensate for the loss of Latin or the new, folksy attitude of the priest trying to "communicate" with the congregation. The Mass, he noted in May 1972, was getting "longer, limper, lumpier."[26]

If McLuhan was unhappy about the assault on the Church by theological revolutionaries, he was not particularly pleased about the use of his work by cultural revolutionaries such as Abbie Hoffman, who in 1968 was saying, "The Left is too much into Marx, not enough into McLuhan." When Hoffman published his *Revolution for the Hell of It* in 1968, McLuhan regarded it simply as a manifesto for the new tribalism. What was absurd, according to McLuhan, was that Hoffman seemed to think it meritorious to embrace this tribalism, when such embrace was almost as automatic, in the new electronic environment, as taking off one's sweater in a warm room.

Around this time, McLuhan met another noteworthy counterculture figure, Timothy Leary, who later remarked that there was no need to turn McLuhan on to LSD since the professor got high, as he put it, on the yoga of his art form — talk. "He talks in circles, and spirals, and flower forms and mandala forms," Leary commented.[27]

Unlike Leary and Hoffman, McLuhan had no desire to overthrow the United States government or even to protest the war in Vietnam. As far as the war was concerned, McLuhan, in keeping with his avoidance of moral stances, never said a word about whether the United States was right to wage it. He knew that such pronouncements robbed one of much power over audiences — the power of putting them on. McLuhan had learned from Joyce that oracles don't take sides. While in New York, McLuhan characteristically saved his moral indignation for issues like local pollution. He came up with an idea for an antilitterbug slogan and passed it on to New York Mayor John Lindsay in June 1968. The city could distribute a poster with a picture of a littered area. Under this picture could be the caption "But you should see the inside of their houses!" (Lindsay wrote back politely declining the suggestion. "My initial reaction," he said, "is that the phrase may be misinterpreted, particularly in minority neighborhoods.")[28]

As the 1968 U.S. presidential election neared, McLuhan was pressed for comment on the various candidates. He responded by writing an article for the *Saturday Evening Post* entitled "All of the Candidates Are Asleep." Nixon, Humphrey, and Wallace, according to McLuhan, were too "hot" — too intent on pushing their points of view, too single-minded, too sharply defined in their personalities — to do well in this age of tribalism and television. By November, McLuhan predicted, the electorate would be ready for a "non-professional" politician who would seem to have come out of nowhere.[29] McLuhan was, perhaps, eight years too soon in that prediction. By the late seventies, the American public was entirely receptive to little-known politicians, such as Jimmy Carter, who gloried in their remoteness from the old centers of power.

McLuhan's comments on individual politicians were nearly always astute. "Poor old Richard Nixon hasn't got a chance," he predicted three years later.

> He has no mask. He just has his bare face hanging out, just private
> Dickie Nixon. The public are dying to see themselves there. They

don't want to see Nixon; they want to put him on. If he would even
grow sideburns, if he would do anything, they would be transformed
in their attitude towards Nixon.[30]

Pierre Trudeau, on the other hand, surging to power in Canada in 1968 on a
wave of mass enthusiasm, was a politician who might have been invented to
illustrate McLuhan's theories on successful television presences. Cool, insou-
ciant, equipped with his famous shrug to dismiss shrill criticism and bitter
personal attack, Trudeau possessed a "corporate tribal mask" for a face and
personality. "Nobody can penetrate it," McLuhan wrote, admiringly. "He has
no personal point of view on anything." Perhaps more pertinent than his comments on these individual politicians, however, was his final comment on politics in the electronic age: "The politician will be only too happy to abdicate in
favor of his image, because the image will be so much more powerful than he
could ever be." McLuhan did not quite live to see the Age of Reagan and the full
vindication of that statement.[31]

     In 1968 an apocalyptic mood — the product of burning cities, radicalized
youth, and a hopeless war in Vietnam — pervaded American politics. McLuhan
was not immune to that mood, any more than he had been immune to the invigorating optimism of the early sixties. Despite his previous notions that television actually cooled people down, McLuhan began to fear that television
heightened racial tension in America, since the television image of the black
person was immensely superior to the television image of the white — the
black face was more "iconic." He recommended at least a temporary ban on
all television broadcasts to save the millions of lives that would be lost in
the racial wars sure to come within a few years. Shortly afterward, in 1970,
when North American campuses were in turmoil over Nixon's invasion of
Cambodia and the Kent State shootings, McLuhan predicted that the generation of students who had been raised from infancy on television would arrive
on campus in two or three years and complete the job of tearing down all the
campus bricks and mortar, so profound would be their alienation from pre-
television institutions.[32]

     It was in this atmosphere of violence and apocalyptic upheaval that *War and
Peace in the Global Village*, the second book McLuhan undertook with the
graphics designer Quentin Fiore and book coordinator Jerome Agel, appeared
in September 1968. The book had an unusual genesis. In late 1967, Agel and
Fiore had collaborated to produce a manuscript entitled *Keep in Touch* —
another presentation of McLuhan's thoughts, this time on the subject of
automation. (As conceived by Agel, half the book would have been printed
upside down, in such a way that it would have been impossible for a reader to

determine which end was up.) The book was to have appeared under the McLuhan / Fiore co-byline, "coordinated by Jerome Agel."

McLuhan approved the manuscript and Bantam Books scheduled its production. After McLuhan's operation, however, he asked that the book he scrapped and offered in its stead *War and Peace in the Global Village*, which he then wrote himself. Fiore and Agel designed and illustrated this work, and Bantam Books published it. Unlike *The Medium Is the Massage*, it did not entertain the reader with far-out experiments in typography or layout. As the title implies, it offered an explanation for some of the more pressing problems of the contemporary world, though it did not refer in great detail to the war in Vietnam. Instead, the book presented broad approaches to the themes of violence and identity.

Chief among these was the notion, as McLuhan put it, that "when our identity is in danger, we feel certain that we have a mandate for war. The old image must be recovered at any cost."[33] Because the new electronic technology was rapidly eroding cherished American images of selfhood — Wally and the Beaver had become long-haired dropouts, for example — it was not surprising that U.S. military adventures resulted. (The same process could be used to explain revolution and war in Third World countries.)

There was also, according to McLuhan, very little difference between "war as education and education as war." The Vietnam War was a kind of all-out educational effort, an attempt to westernize an Eastern culture. At the same time, the education of the young in the West was similarly rife with aggression in its attempt to impose on students "the patterns we find convenient to ourselves and consistent with the available technologies."[34]

McLuhan frequently employed in this book a metaphor he had devised to express what Wyndham Lewis had once taught him: "The present cannot be revealed to people until it has become yesterday."[35] McLuhan termed this the "rearview mirror phenomenon." People went through life looking into the rearview mirror — seeing the present in terms of the past — instead of paying attention to the reality confronting them. (Years later someone pointed out to McLuhan that drivers can see, in the rearview mirror, a car about to overtake them. This fact rather altered the terms of the metaphor, which he cheerfully accepted. He rarely had any problem with criticism that was not personal and that actually pointed out some aspect of reality he had overlooked.)

Like most of McLuhan's books, *War and Peace in the Global Village* is full of fascinating and perceptive comments ("Bad news concerns few, but good news can upset a whole culture") as well as his famous dubious details. McLuhan articulated the notion, for example, that the economic slump of 1929 was caused by the rise of a generation of Americans, influenced by jazz and radio, who rejected the goals and objectives of their visually oriented elders. This was certainly stretching things; but McLuhan was almost preoccupied with the notion, partly for personal reasons. He saw his youngest child, Michael, burning incense before a statue of Buddha and reading Hermann Hesse. None of his older siblings had ever displayed such behavior. McLuhan attributed it to the fact that Michael had been much more influenced by television than they had been. When Michael and his peers reached maturity and were ready to enter the job market, McLuhan predicted, their lack of goals and need for "total involvement" would make the 1929 slump look mild. Throughout his stay in New York, McLuhan predicted a terrible economic depression in approximately five years, when Michael would be twenty years old.[36]

McLuhan published another book in 1968, *Through the Vanishing Point: Space in Poetry and Painting*, which he and Harley Parker had been working on together since the fifties, when the project had been inspired by their long talks about art and the human sensorium. The focus of the book was the spaces created by each of the senses — visual space, acoustic space, and so on. The difference between these spaces and the implication of this difference for human perception and understanding were demonstrated by the poems and reproductions of paintings included throughout. Although it was a substantial work, *Through the Vanishing Point* was, curiously, ignored by the press. To McLuhan's puzzlement and disappointment, it received virtually no reviews.

Father Culkin had toyed hopefully with the idea that McLuhan might stay on at Fordham after his first year. McLuhan's health problems removed that possibility, however. By the spring of 1968, he was anxious to return to Toronto and to resume something resembling his comfortable routine. An excess of patriotic feeling for Canada played no part in this decision.

McLuhan's feelings for his native land, in fact, were always highly mixed. Something of their early flavor is conveyed in a comment he made after returning from Cambridge in 1936: "I am going to tear the hide right off Canada some day and rub salt into it."[37] The stodginess and simple-minded smugness of his native land drove him to distraction. Moreover, he had very little respect for its official literary culture. He told interviewers in the seventies that there was no "*serious* writing going on in Canada today — by *anybody*" and called the country's more publicized writers "a fifth rate bunch of people who are very smug, very happy about themselves."[38] In an interview he did with *Mademoiselle* while he was living in New York, he was even more scathing. Speaking of the hypothetical "serious Canadian," McLuhan proclaimed, "Nothing comes out of Canada that he would dream of taking seriously." English Canada in particular was "the most apathetic and unenthusiastic territory in all creation.... The Canadian is mildewed with caution!"[39]

Nonetheless, this unpromising land had its advantages. "The Canadians are a lazy bunch — really lazy. The idea of exerting themselves, in school or elsewhere, that's not for them," he explained. "No, there is a huge inertia in Canada. I like this because it enables me to sort of move ahead of the rest of them." Canada also provided a "quiet, restful place in which to luxuriate without contemporary pressures."[40] Some of this was pure put-on for the benefit of both his Canadian and American audiences. Some of it was deeply felt.

It was no wonder, in any case, that McLuhan was eager to return to Toronto after his year in New York. He needed a rest. His friends and colleagues could have predicted that nothing of the sort awaited him.

11. Unsold Books (1968–1972)

To have ideas is paradise, to work them out is hell. — Maurice Maeterlinck

McLuhan returned home in 1968 to complete chaos. For one thing, the family chose this time to move from their Wells Hill Road home to a new house in the Wychwood Park area of Toronto. It was poor timing but a desirable move, befitting McLuhan's new status as a worldwide celebrity. Wychwood Park is an oddity in Toronto, a wooded twenty-two-acre enclave of Edwardian mansions in the middle of a crowded neighborhood in the heart of the city. On a hill overlooking a pond stands number 3 Wychwood Park, the house McLuhan moved to in the fall of 1968. He would live there for the rest of his life.

McLuhan grew to love this house. Like the house on Wells Hill Road, it was within walking distance of the University of Toronto campus. The house itself, described by John Wain as "baronial," was full of mellow oak paneling and high ceilings: Elsie McLuhan would have adored it. Her son enjoyed showing it off to visitors with a simple-hearted pride. Even the address was pleasing to him, with his mystical reverence for the number 3.

The chief attraction was the park itself, for it reminded him of Gertrude Avenue in Winnipeg, where he and his friends had once played in the streets uninterrupted by traffic. Of course, this was a Gertrude Avenue with no chilling reminders of poverty and empty coal cellars. The sight of the pond below soothed him as well. He called it "Walden Three," successor to the great good places celebrated by Thoreau and B.F. Skinner.[1] The park, he later commented, is "built on a circle around a pond, which turns it into a theatre. Living on a street is not a community. In a circle round a pond you have a community. It's the first time I've lived in one."[2]

So attached to his home did McLuhan become that he felt unhappy to leave it even for relatively short trips, especially in the summer, when the park was at its most bucolic. On pleasant nights it became a ritual for McLuhan and Corinne to take a stroll around this little urban retreat before dinner.

The disruption of the move to Wychwood Park was compounded by the move of the Centre for Culture and Technology from the office on St. Joseph Street to an old coach house tucked away at the end of a lane off that same

street. The move was arranged by Arthur Porter, who had acted as director
of the center during the year McLuhan spent at Fordham. In McLuhan's
absence, Porter had decided to take the opportunity to beef up this mini-
institution that had fallen to his trust. McLuhan tended to dismiss the physical
and administrative paraphernalia of institutions, calling them "hardware" as
opposed to the "software" of information processing. Porter, somewhat more
earthbound, realized that such paraphernalia lent institutions a certain indispensable aura of reality — and there would never be a better time than 1968,
when the university was relatively flush, to make a bid for some of this expensive paraphernalia.

He put in a request to the university committee in charge of accommodations and facilities for the coach house, which had a seminar room capable of
holding three dozen or so participants plus separate offices for McLuhan, his
secretary, and one or two assistants (for whom Porter requested salaries). It was
not a lot to ask, and Porter's requests were granted. The coach house, hidden
away in the midst of several other buildings, was a shabby, two-story building
only slightly more impressive than the office at 96 St. Joseph Street and smaller
than most people's summer cottages. But it did contain more space.

Very quickly McLuhan felt as comfortable in the new office as he did in
his old brown tweed suit. He filled it with his cherished artifacts, including
his Cambridge oar and a large, semiabstract, incompetent mural done by a
friend that captured, it was said, the effects of television on the modern world.
He filled the bathroom with clippings of cartoons. He grew to love the building even more than he loved his manor at Wychwood Park. It was totally his.
Even the fact that it was on the fringes of the campus endeared it to him. Had
he not always insisted that the "fringe" was the best place for a true explorer
to be?

In the meantime, however, the moves of his office and his house imposed
an intolerable strain on a man barely recovered from harrowing surgery.
Hauling McLuhan's books to the coach house was a major ordeal; they ended up
in piles stacked everywhere in the building, including the bathroom. In the confusion, McLuhan's beloved collection of *Scrutiny* was lost or stolen. All the
while, he was under siege by mail and telephone. Every five minutes the phone
rang with someone asking to speak to him. Often it was an anxious voice at the
other end of the line describing an invention or a new scheme to remake
Western civilization, sure to interest Professor McLuhan who, as the world
knew, had a head full of utopian visions.

Under the circumstances McLuhan was sometimes rude, sometimes for-
bearing. With serious callers or with acquaintances — and no one had a wider
field of acquaintances than Marshall McLuhan — he listened to their requests

with an impatient ear, as he put it, for the caller's "main verb." The main verb was usually "speak to my association/college/staff." McLuhan's brother, Maurice, who had been hired as a research assistant upon McLuhan's return to Toronto, urged him to accept only "significant" invitations. (Maurice was quickly dubbed Sheriff McLuhan by students.) The effort was futile. McLuhan seemed to be governed almost by whim in this regard. The money offered and the size or importance of the group seemed to have little to do with his decisions to appear or not.

In the fall of 1968, however, he seemed to be more than usually erratic and uncontrolled in his responses. At one point on his return he told an acquaintance he would like to speak to educators about some of his pressing concerns in their field. The man duly organized a speaking date with officials in the Ontario Ministry of Education and other worthies. Just before this date the North Koreans captured the American spy ship *Pueblo*. McLuhan was fascinated by the incident and its revelation of the vulnerability of American hardware to the most "backward" countries. In his address to the educators he spoke of nothing but the *Pueblo*. Part of his audience was bothered that McLuhan said not one word on the topic of education; others rather enjoyed his comments. All of them were puzzled.

It would have been perfectly reasonable for McLuhan to tell everyone he was taking six months off to lie in bed until the pain in his head subsided and he felt human again — reasonable, but not in character. He still had a few dozen things he wanted to do, and they could not be put off. Yet it took him a long time to shake off the disorientation and loss of memory he felt after the operation. He worried friends by occasionally walking straight into walls or forgetting the way to his own house. He was perpetually mislaying things about the office and generally acting the part of absent-minded professor, which had never before been quite in character.

Other professors and university administrators did not rush to assist their famous but beleaguered colleague. Was this not the man, after all, who had gotten $100,000 from the state of New York? In their eyes, McLuhan had received quite enough special treatment already, especially since the University of Toronto had agreed to appoint two English professors from the University of Alberta, Sheila and Wilfred Watson, for the academic year 1968–69 as assistants to him. If Sheila Watson found herself, at the beginning of that year, spending part of her time obtaining ashtrays and doormats for the coach house and the other part shielding McLuhan as best she could from the crowd of people who wanted something from him, that was no concern of theirs.[3]

McLuhan had other helpers as well. In addition to Maurice McLuhan, Harley Parker was hired as a research assistant. They were supposed to help

McLuhan in his probes by becoming thoroughly conversant with his approaches. Their role, however, was never a happy one, and they tended to be written off as lackeys. Other assistants would trot around with McLuhan, notebooks in hand, to take down anything of significance the great man said. Parker himself became known as the person who handled the overhead slide projector at McLuhan's talks. During the fifties and early sixties he and McLuhan had formed a fast friendship, lighthearted in tone and based on the genuine exchange of ideas and humor. By 1968, however, the lightheartedness had vanished. After Parker was hired as a research assistant (he held the position until 1975), McLuhan tended to be brusque and irritable with his old friend. Sometimes the irritation was understandable — Parker occasionally showed up at the center over-fortified with alcohol — and sometimes McLuhan's snappishness was gratuitous. Once McLuhan threatened to punch Parker out for telling a caller that the professor was "resting" at the time. (McLuhan hated any suggestion that he was not exceptionally vigorous.)

Maurice McLuhan's experience at the center was almost as unhappy, although it was, mercifully, much briefer (he lasted only three years). Besides assisting at probes, Maurice made himself useful by doing secretarial chores, answering the phone, and replying to some of McLuhan's correspondence. The man who had been the perpetual prankster of Gertrude Avenue still could not resist a little fun. Since his voice was strikingly similar to his brother's, he had some opportunity to confuse callers on the phone. When the mood seized him, he would, as he put it, "turn ribald" in conversations with people who assumed they were speaking to Marshall McLuhan.

Maurice quit when he realized, as he put it, that "nothing grows under an oak tree." A few years later his older brother referred to him as an "anti-force" at the center.[4] The resentments entertained by both brothers over the period of Maurice's employment were understandable. It took an extraordinary individual to remain in proximity to McLuhan for any length of time and not feel dominated, intellectually and emotionally. McLuhan seemed oddly untroubled by the thought that he was, intentionally or not, turning out intellectual clones of himself. In fact, he always tried to solve the problem of his numerous speaking invitations by asking groups to accept some of his assistants as substitutes for himself. McLuhan reasoned that his assistants were so thoroughly versed in his thought that audiences would virtually get the same product. These assistants represented the deterioration of the notion of a "fine circle of intellects" that McLuhan had cherished as an undergraduate. He did not even seem to realize that he was insulting his assistants when he wistfully told a reporter for the *New York Times* in 1970 that it would be "fun to have a little company" in his work as a "metaphysician of the media."[5]

But late in 1968 McLuhan was in dire need of as much assistance as he could get. Several chores were extremely pressing. One of them was the production of "The Marshall McLuhan DEW-LINE Newsletter" with Eugene Schwartz. The "DEW-LINE" had made its appearance at what was probably the peak of McLuhan's celebrity status. There was no doubt, in the summer and fall of 1968, that it had a certain marketability, and initially at least it was received with genuine curiosity by many powerful and influential people. The first issue, in July 1968, had consisted of several 8-1/2-by-11 sheets enclosed in a vinyl cover with a plastic spine, like a high school term paper. Other issues, partly as a measure to prevent the newsletter from being photocopied for nonsubscribers, came complete with posters, vinyl recordings, and slides of advertisements (accompanied by a printed booklet analyzing the ads).

One issue provided a copy of *War and Peace in the Global Village*, with a pamphlet explaining McLuhan's prose style in that book. This style, the unsigned author (probably either Schwartz or Eric McLuhan) maintained, consisted of "the sudden disclosure of previously undetected similarities and relationships, by eliminating transitions and forcing thoughts into abrupt interface with each other." Where any other writer would take all his sentences dealing with subject Y and lump them together in a section devoted to subject Y, McLuhan scattered his subject Y sentences among subjects, A, B, C, and so on. Examples from *War and Peace in the Global Village* were adduced.

The most novel feature "DEW-LINE" offered its readers (for an extra five dollars) was a pack of playing cards with a text, like the message on a Chinese fortune cookie, on each card. The text, written by McLuhan, was intended to prompt spur-of-the-moment thinking. Among these messages were some groaners ("Fulton's steamboat anticipated the mini-skirt: we don't have to wait for the wind anymore" on the three of spades) and some interesting aphorisms ("Propaganda is any culture in action [Jacques Ellul]" on the five of spades). The player was supposed to think of a personal or business problem, shuffle the deck, select a card, and apply its message to the problem. "If you get your breakthrough in thirty seconds or less," the instructions accompanying the pack read, "you are a top Dew-Liner!"

The set of newsletters produced over the period from 1968 to 1970, put together in New York by Eric McLuhan and Eugene Schwartz from copy mailed by McLuhan, contained much incisive commentary. ("Instead of having a line or a policy to follow, the political candidate must become an image capable of including all the hopes and wishes of the electorate.") The trouble was that this commentary was hardly different from the remarks McLuhan had been giving away for years, gratis, to interviewers. Moreover, the commentary was sweeping and nonspecific and rather contradicted the logic of a newsletter, which is to present topical information — the inside dope.

Nonetheless, Schwartz maintains that the newsletter was a success and believes it could have gone on for two or three more years, had McLuhan been willing. "We could have had a renewal rate of seventy to eighty percent," he believes. Nor did the words of the newsletter fall entirely on stony ground. McLuhan later complained that Alvin Toffler scored big by recycling the contents of the newsletters for books like *Future Shock*. He himself managed to get a $25,000 advance from Doubleday for a book (never written) based on the "DEW-LINE" newsletters. On the whole, however, the episode was severely damaging to McLuhan's reputation. Advertisements for it, like the one in the *New York Times* that began "This is an invitation to join a select group of business, academic and government leaders who are about to receive what must be the most startling newsletter ever printed," did nothing but intensify suspicions that McLuhan was a charlatan and a man out to exploit his reputation as a media wizard for every penny he could get.

The suspicions increased with the announcement of the three-day McLuhan Emergency Strategy Seminar on Grand Bahama Island, starting January 1, 1970. Organized by Schwartz and "strictly limited to 500 top executives," as one letter to "DEW-LINE" subscribers put it, the seminar was described as an occasion to "explore the most frustrating breakdowns in your organization, your market and your environment — and restructure them into the kind of breakthrough you may have been waiting for for years." Price: $500 per customer. Speakers were McLuhan, Buckminster Fuller, Bernard Muller-Thym, Harley Parker, and Barrington Nevitt.

The seminars turned out to be enjoyable for most of the participants. Fuller, garrulous as usual, talked for hours one evening, embarrassing some of the participants but inspiring others. "Bucky Fuller was in his prime," one of the latter group recalls. "He just took off and he brought the audience with him. It was incredible to see a room full of people who didn't know the first thing about physics talking about angular momentum and procession and things like that. It was a very uplifting experience."[6] McLuhan entertained the guests by singing "Yes, We Have No Bananas," a reference to the miserably cold and wet weather on the island during their stay.

The Emergency Strategy Seminar was the last venture McLuhan ever undertook with Schwartz. McLuhan was annoyed that he received no payment for his participation. Schwartz argued that the money from the seminar should go to offset costs incurred by "DEW-LINE" and other activities of Schwartz's Human Development Corporation, which he had formed, basically, to market

McLuhan. The argument confirmed McLuhan's suspicion that Schwartz was taking undue advantage of him. McLuhan wrote to Schwartz invoking a clause of their "letter of understanding," stating that if he did not receive a minimum of $10,000 in the first year and a minimum of $20,000 for every year thereafter as his share of the profits, he could cancel the agreement. He had not received this money, and he was canceling. (Schwartz invited McLuhan to examine the books himself: he would see that there were no profits.) Thereafter, McLuhan forbade any mention of Schwartz's name in his house. He was not only angry at Schwartz, he was also disappointed that the newsletter had not been more popular. It might have been a way, after all, to exert some influence on the thinking of important people.

The newsletter was not the only urgent assignment that awaited McLuhan on his return from New York. He was under great pressure to finish the book he had agreed to do, in collaboration with Wilfred Watson, for Viking Press back in 1963. Although Clay Felker had commissioned the work, another Viking editor, Alan Williams, later assumed responsibility for the project. Williams recalls that McLuhan would pop into his office unannounced from time to time and "look out the window and make generalizations about women's skirts in relation to high-rise buildings — out of context, you know. Everything was context to him, and nothing was."[7]

Mostly Williams wrote anxious letters to McLuhan in Toronto, inquiring about the progress of the book. McLuhan and Watson, who lived in Edmonton, Alberta, where he was a professor of English at the University of Alberta, had been working on it for five years, exchanging ideas by mail. In 1965, McLuhan told Williams he anticipated finishing the book by February 1966. In April 1966, Williams wrote McLuhan that "it would be very nice if we could have a line from you as to when you and Mr. Watson hope to complete the book." A month later he was offering to visit McLuhan in Toronto if it would help matters. A year later he was still writing to McLuhan in hopes of eliciting a firm date for delivery of the manuscript.

When McLuhan returned to Toronto in the fall of 1968, his top priority was to give Viking its book. It was more than just a question of meeting an obligation — McLuhan had high hopes that this work would stand as a major presentation of his thought and help restore his intellectual reputation, now weakened by nonbooks like *The Medium Is the Massage*. It would be his first solid work since *Understanding Media*. Indeed, it would do for literature what that book had done for technology: it would reinstate him as a man who appreciated print and still made use of his library card. It would also, coincidentally, be a slap at his old enemy Northrop Frye, the man whose criticism trafficked in literary "archetypes." McLuhan's book would inform Frye, and the world, just what an archetype was.

The idea for the book was based on McLuhan's discovery of the process whereby the ordinary technology and activity of one era become the artistic and archetypal forms of the following era. Such ordinary technology and activity, characteristic of the contemporary environment, McLuhan called "cliché" — that is, forms of expression that are pervasive in a culture and are used with little or no examination of their meaning. Archetypal forms, on the other hand, compel attention and seem to be charged with significance.

The process was obviously multifaceted and needed much clarification if it was to be convincingly presented as a basic dynamic of human culture. There was hope in the fact that McLuhan's collaborator was a Canadian poet of great wit, erudition, and virtuosity. When the two men got together in New York and then in Toronto, where Watson spent the academic year 1968–69, they began a series of "dialogues" — considering, say, the example of the flag and the flag-pole, asking themselves which was the cliché and which the archetype. In the process they seemed to generate fascinating ideas.

When they sat down to dictate the book to Margaret Stewart, McLuhan's secretary, however, they immediately ran into trouble. McLuhan did most of the dictating and ignored almost entirely every idea that had developed in the dia-logues with Watson, reverting to his original thoughts on the subject. Watson was not sure whether McLuhan actually forgot what they had talked about or whether he was simply ignoring their joint conversations. Given McLuhan's state of mind at the time — his loss of memory and the tormented state of his nervous system — either possibility was likely.

As the year wore on, McLuhan seemed less and less tolerant of Watson's participation. Margaret Stewart, trying to take down the comments of the two men, was utterly bewildered. McLuhan would produce a quote out of a file and ask Stewart to put it down for inclusion in the book. Then Watson would say something that might or might not be related to the quote. Then McLuhan would make a comment completely unrelated to both the quote and what Watson had just said. The "dialogue" had gradually become two monologues. It was out of Stewart's notes of this contrapuntal pair of monologues that the first draft of the book was finally produced. Eventually Eric was given the tran-scribed notes and asked to "revise" them.

In this manner, *From Cliché to Archetype* found its way to Alan Williams's desk in the spring of 1969 and then to the printer. McLuhan still had high hopes for it, and in the summer of that year he wrote Frank Kermode that the book might be "a bit of a blockbuster." Not that he made special efforts to promote the work once it was done. Robert Silver, then director of sales at Viking, recalls that McLuhan was so supercilious in his talks with him that "he made me not give a damn whether I sold the book or not." McLuhan, in turn — and perhaps cognizant of Silver's attitude — nursed the suspicion that Viking was going to sabotage his book by ignoring it.[8]

Even if Viking had promoted the work extravagantly, it is doubtful it would have made much of an impression. McLuhan's method of "forcing thoughts into abrupt interface with each other" had simply gone out of control. There are hints of brilliance in the text, but for the most part it is unreadable. A few years later, McLuhan called it "stupid" and "pointless" in his diary and blamed Watson for derailing it.[9]

From Cliché to Archetype appeared in the fall of 1970, the same year as *Culture Is Our Business*, which McLuhan had been working on for ten years and had finally finished in July 1967. The latter book, a collection of reproductions of advertisements, annotated largely by McLuhan's one-liners, evinced no pride of authorship in McLuhan either. The book's comments were the fruits of much thought and observation, but on the page the wit seemed labored and the conclusions merely puzzling rather than stimulating. Many of the comments were all too evidently the pronouncements of a man who was talking to himself. They certainly did not make the old-fashioned essays in *The Mechanical Bride* suffer by comparison. Moreover, McLuhan claimed that the book was marred because its publisher, McGraw-Hill, had not sent him proofs before publication and that the text was full of errors. In any case, *Culture Is Our Business* ended his relationship with McGraw-Hill; he felt the publisher had unconscionably neglected him for years.[10] McLuhan had a tendency, not unknown among writers, to leap to the conclusion that his publisher was a malign force.

These two books, along with another brief collection of his more recent insights, published in 1969 by Harcourt, Brace and World and entitled *Counterblast* (in honor of his manifesto of the 1950s), brought to an end the spurt of books McLuhan published between 1967 and 1970. With the appearance of each book, McLuhan's prestige among intellectuals sank a little further. With the possible exception of *Through the Vanishing Point*, the books seemed contrived and — surprisingly for a writer so interested in classical rhetoric — produced with little thought of the impact they might have on their readers.

By 1970, commentary on McLuhan had finally lost its breathless journalistic tone. No longer did magazines feel compelled to do a piece on him. He had already been "done" and there seemed nothing more to say. Indeed, comment on McLuhan now took on a somewhat cynical cast. In January 1969, the *New York Review of Books*, which had never been a fan of McLuhan, carried a review of a number of books by and about him. The editors no doubt hoped the review would be the final word on the McLuhan episode, which the reviewer characterized as — what else? — another symptom of the intellectual disorder of our times. The September 26, 1970, issue of *The New Yorker* ran a cartoon of a young woman saying to a man as they left a cocktail party, "Ashley, are you sure it's not too soon to go around parties saying, 'What ever happened to Marshall McLuhan'?"

McLuhan continued his work in Toronto, meanwhile, much as he had done before and during the explosion of interest in his theories. He taught classes, he ate his lunches in the faculty room at St. Michael's College, and he continued to hold weekly seminars as he had done since 1953. After 1968 the Monday night seminars in the coach house became a fixture of McLuhan's life. Even the most trivial details of these seminars were stamped with the impress of McLuhan's personality and reflected his obsessions. He determined, for instance, that the fluorescent lights in the seminar room impeded dialogue, so he turned them off, stood a floor lamp on a table, minus the shade, and declared the lighting much "warmer."

Invariably, he had one or two people on hand as a kind of support system. Not that he particularly needed one: as the years went by and the merely curious ceased coming, the people on hand tended to be mostly hard-core devotees. As for the occasional interlopers looking to challenge McLuhan — students and scholars from the nearby Ontario Institute for Studies in Education, in particular, often came gunning for him — he was more than capable of handling them on his own.

Nevertheless, McLuhan seemed to need a faithful sidekick. Usually it was Eric or his friend Barrington Nevitt. Driving to the seminar, McLuhan would ask Nevitt what topics they might discuss that evening and would sometimes ask Nevitt to deliver a half-hour introduction to whatever topic they hit upon. McLuhan usually interrupted this introduction after the first five minutes, launching into a long disquisition of his own. Then Nevitt would function as a prompter, saying at fitting moments, "Look, Marshall, this might be a time to talk about acoustic space" or some other standard theme.

Eric and Nevitt occasionally came in for some abuse from McLuhan who every once in a while became fed up with anyone who seemed to be parroting him or trying to anticipate what he might say on a subject. Recapping McLuhan in his presence was always a risky affair. If he was indulgent enough to abide the attempt, he might still react angrily to an old catch phrase improperly repeated. (One never, for example, referred to the five *parts* of rhetoric; it was always the five *divisions* of rhetoric.) It was not only Nevitt and Eric who were targets of his annoyance in this respect. "There seemed to be a coterie of individuals at the seminar who almost masochistically thrived on that treatment," one participant recalls.[11]

Occasionally McLuhan brought guest speakers to the seminar — colleagues at the University of Toronto or intellectual celebrities in town, such as Malcolm Muggeridge. The invitations were always casual and sometimes slipped McLuhan's mind. It was not unusual for Nevitt and McLuhan to thrash out what they wanted to talk about en route to the seminar and discover waiting for them when they got there a guest of honor whom McLuhan had forgotten. Often, McLuhan closed his eyes while the guest speaker talked, as if in rapt contemplation of his words, and then, after five minutes, interrupted with a question or a comment. The guest was fortunate if he had opportunity to say much more after that.

In the casual atmosphere of the seminar, McLuhan relaxed and indulged in boy-from-the-prairie mannerisms that would have made Elsie McLuhan wince. Always fond of cigars, he made a small drama of lighting them. The drama sometimes lasted as long as forty-five minutes: thirty minutes for McLuhan to find some matches, talking non-stop all the while, and another fifteen actually to light the cigar between sentences. He put the ashes in the cuffs of his trousers. He liked to unbutton the top three buttons of his shirt and scratch his chest; he said it was "good for the nerves" and helped maintain energy. When he was about to make some particularly pertinent observation, he sat with his knees far apart and grabbed his suspenders, like a regular gent sitting around the hot stove.

His most notable signature was his tendency, as the years went by, to repeat his ideas and "percepts" over and over, as if they were irritants lodged in his nervous system that had to be worked loose. Often this was the reason for his infamous 2:00 a.m. phone calls to friends and acquaintances, in which he would go over something he had talked about countless times before — hoping, perhaps, that a slight nuance might emerge, unrecognized in all his previous articulations of the notion.

Even his reading took place, as it were, through the filter of his preoccupations. The indexes he made at the backs of books were virtually in code: "C/A" (cliché/archetype), "SI/SC" (sensory input/sensory closure). These mental nodes became his all-purpose tools for grappling with unfamiliar intellectual phenomena. With a few phrases from this tool kit he could put even Kant or Hegel in his place. But then the man who scorned "concepts" was really as easy with them as a mechanic is with wrenches and screwdrivers. Anything new and intellectually heavyweight — the works of Derrida, for example — McLuhan would quickly reduce to five or six basic concepts, which could then be dismissed as, say, the responses of somebody unknowingly caught in the tug of war between visual and acoustic space.

Few things annoyed McLuhan more, in fact, than grand philosophical propositions of any kind. He valued James Joyce far more for the puns in his work than for any profound themes that could be extracted from it. "He had a wonderful way of putting you down when you came to him hot with an idea and you were going to present it to the master and make him respond," one associate recalls.[12] By contrast, he lit up when you brought him any sort of odd believe-it-or-not fact that promised to stimulate "perception."

One of the high points of the seminar in later years was Canadian Prime Minister Pierre Trudeau's visit in November 1977. McLuhan had met Trudeau on two or three occasions after his election in 1968, and Trudeau's famous charisma had worked its effect on McLuhan. He wrote the prime minister, after dining with him in 1970, that the experience had been "euphoric." Thereafter, in his letters to Trudeau, McLuhan praised the man in terms bordering on the grossest flattery. No British prime minister, much less any of Trudeau's Canadian predecessors, could compare with him, according to McLuhan. Reviewing Trudeau's 1968 book *Federalism and the French Canadians* in the *New York Times*, McLuhan had placed it in the same league as the works of Edmund Burke.[13]

He once compared Trudeau's facial expressions to those of the archetypal North American Indian. Afterward, he worried that the great man might take this amiss and wrote to reassure him that the public would never hear anything disparaging about Trudeau from Marshall McLuhan.[14] He was as good as his word.

In 1968, McLuhan saw himself in the role of adviser to this luminous politician. As a student of the Renaissance, McLuhan was aware of the hazards of the courtier's life. On prominent display at the center was the text of Erasmus's advice to a courtier friend: "Observe the prince's likes and dislikes. Smile when he speaks, and if you can say nothing, look admiringly." Trudeau was a philosopher king to tempt any intellectual to attend court. While not quite of the stature indicated by McLuhan's praises, he was literate and intelligent. And he did genuinely appreciate people who had made a name for themselves in their chosen fields.

Yet Trudeau was a consummate dialectician, fond of displays of logic and possessing the kind of mentality McLuhan had always found antipathetic in scientists and academics. McLuhan was never quite convinced that Trudeau actually understood anything he said. After their first meeting in 1968, McLuhan told a confidant that "there was no meeting of the minds. He didn't have the least comprehension of what I was saying."[15] Nonetheless, the two maintained a steady correspondence during the seventies.

McLuhan loved to write letters to eminent people he had met — Hubert Humphrey, Henry Ford II, King Carl Gustav of Sweden — and to some, such as Jimmy Carter, whom he had never met at all. (His most unusual correspondent was Eppie Lederer, a.k.a. Ann Landers, who gushed in her reply to McLuhan, "I'm zonked by the way you put words together.")[16] Usually he would receive a polite note from the eminent personage he had contacted, not encouraging of further correspondence. Trudeau, however, always responded as though he had carefully read what McLuhan wrote to him. His correspondence with Trudeau was the closest McLuhan ever came to an ongoing relationship with someone in power.

McLuhan was careful not to lobby the prime minister on specific issues; instead, he tried to give Trudeau advice on the use of television, much of which was very shrewd. He usefully passed on his own techniques of inviting hecklers to come up to the podium and address the audience. Shortly after the 1968 election when "student power" still seemed a threat to peace, order, and good government, McLuhan suggested that Trudeau chat on television with student leaders in groups of four or five in a casual, intimate setting, without the use of scripts. Trudeau never followed this suggestion, but McLuhan was right to point out that such a format was far more suited to the television medium than "debates" in which participants stood behind podiums in an auditorium. Television cameras loved small studios, McLuhan insisted, not large auditoriums.[17]

Only a few times was McLuhan tempted to bend the ear of the prime minister with directly political appeals: once when he was upset with the numbers of "colored" immigrants coming into Canada under liberalized immigration policies (he felt this immigration should be frozen until the country had some policy in place to help native-born Canadians deal with it), and a second time to suggest a "national humor program" to combat the grievances of French Canadian separatists. McLuhan postulated that all humor is based on grievance and that ethnic jokes actually help to diffuse ethnic tensions by dissipating the grievances behind those tensions. Under a national humor program, writers would deliberately churn out such jokes for national consumption.[18]

Aside from writing books and articles and holding his weekly seminar, McLuhan continued to meet his regular university teaching obligations. After 1968 he began to feel some anxiety over his position in the graduate school. No longer was he the sole authority on the moderns; other professors from other colleges wanted to teach graduate courses in Pound and Joyce. McLuhan began to fear that he might lose his old courses; he suspected, with justification, that other professors in the graduate school were increasingly discouraging students from enrolling in his classes. He also felt keenly that his virtual exclusion from Ph.D. examinations, an exclusion that grew more pronounced after 1968, was hurting his prestige with students.

In their defense, McLuhan's colleagues could point to episodes such as one involving a graduate student named Roger Lewis in 1969, in which McLuhan had seemed vindictive and unreasonable in refusing to give a mark to the student. Somehow Lewis had struck a nerve in McLuhan, who condemned Lewis's work in his course for showing "no competence in relating media to society" — that is, Lewis had said or written something that jarred with a McLuhan Theme.[19] The episode ended amiably, after much distress on Lewis's part, but it confirmed the old suspicion among many of McLuhan's colleagues that the man was erratic and unsound. It was certainly true, in any case, that a number of McLuhan's students in later years found him less than helpful. Occasionally he could surprise a student by spending hours on a problem that had piqued his curiosity; most of the time, however, he tended to neglect his students, both graduate and undergraduate.

In the last decade of McLuhan's teachings, the undergraduate students tended to be more puzzled than stimulated by his classroom performances. The exams and essays he required of them were alternately strict and indulgent. Ted McGee, one of his teaching assistants in the mid-seventies, recalls McLuhan giving an open-book, fill-in-the-blank test ("Complete the line 'Let us go then, you and I . . .'"), the type of exercise imposed on high school students but unheard of at the University of Toronto. When students did write essays, McLuhan could be finicky. He failed students for using the aphoristic style he himself extolled in the classroom and insisted not only on the traditional style but on the traditional framework of the undergraduate essay, the diligent comparing and contrasting of this theme and that in Yeats and Eliot and Dylan Thomas.

McLuhan occasionally invited his undergraduate classes to a party at his home, where the students drank punch awkwardly in the presence of this now venerable and slightly intimidating figure. He still longed for "community," but he found St. Michael's College, full of younger professors he barely knew and, like the other colleges, swollen with increased enrollment, less and less a congenial environment.

A mood of disgruntled isolation grew on McLuhan as the seventies wore on. Nonetheless, he retained his innocent and generous desire to share in the life of the college. He was the only professor to show up at the student's pub at the college during Lent, where he overcame his scruples about drinking beer during the penitential season. "He would like to have been included more as one of the boys," an associate in his later years recalls. "He often had this wistful look when he heard of a group going to a pub without him, for example. He would like to have taken part in group shenanigans in a way that he had never been able to."[20]

McLuhan's isolation on campus was increased by the changes that took place there in the late sixties and early seventies. The university expanded dramatically with the arrival of the baby boom generation; the English department grew to four times the size of the department McLuhan had known in the forties and fifties. Much as he had suffered at the hands of that older department, he was not happy to see it swamped by new arrivals.

Moreover, the basic nature of the university was moving away from McLuhan's idea of what a university should be. He envisioned the ideal university as a cluster of small teams of researchers working in a milieu of students eager for intellectual dialogue — an ideal that he had first formulated when he found himself with virtually nobody to talk to at the University of Manitoba.

The University of Toronto had never exactly embodied this ideal, but its old structure of semiautonomous colleges, based on the Oxbridge model, had at least encouraged some diversity and fostered some sense of community among professors and students. In the seventies, however, the colleges were under severe financial pressure to surrender much of their autonomy to the university. That pressure culminated in the University of Toronto's 1974 "Memorandum of Understanding," according to which the subjects taught by the colleges — basically the humanities — were taken over by university-wide departments, themselves firmly under the control of the central administration of the university. With this step the distinctive differences between the colleges were virtually wiped out.

McLuhan bitterly resented this change, viewing it as a miserable sellout on the part of the colleges and a triumph for the bureaucrats and centralizers within the university. It finished the process of converting the University of Toronto into what McLuhan termed "Ontario State University" — that is, a sprawling, anonymous megaversity in which the most vital and complex component seemed to be the administration building. (The symbol of this new megaversity was a towering concrete library building constructed during the seventies. Of faintly Babylonian grandeur, the library was dubbed "Fort Book" by the students and heartily detested by McLuhan. In his view, it was nothing but "hardware" on a monumental scale.) But then McLuhan detested anything suggestive of a giant institution. So darkly did he view the workings of these institutions that he termed even a comparatively trivial act, like the changing of the first three digits in university telephone numbers in 1976, "a blind act of psychic violence."[21]

If the university around him was becoming unrecognizable, McLuhan at least had his coach house and his home in Wychwood Park as inviolable retreats. Within these sanctuaries he conducted his life pretty much as he wished. At home, he was usually up by four or five in the morning, sometimes waking after a dream shaking with laughter.[22] McLuhan then read his Bible, with the aid of lexicons, in Spanish, Italian, Greek, German, or French. (He had formerly read it in Latin as well but had become so used to the Latin that he found it insufficiently challenging.)

At breakfast Corinne read bits of the Toronto *Globe and Mail* to him. Although she retained almost as much vitality as her husband, her hearing had begun to fade by 1970 and she was confronting other serious health problems, such as deterioration of the cartilage between some of her vertebrae. The loss of hearing was most disturbing to her; she felt extremely self-conscious about it. In the hope of restoring her hearing the McLuhans enlisted the aid of a "spirit healer" in Great Britain by the name of George Chapman; they also tried acupuncture and transcendental meditation.[23] Corinne's hearing, unfortunately, never improved.

His son Michael, who had acquired the nickname "Acid Mike," remained alienated in a manner typical, McLuhan felt, of a generation that had no purpose in life. In April 1969 McLuhan wrote to Ted Carpenter complaining that Michael now had shoulder-length hair and constantly played the electric guitar — with an amplifier. In the fall of that year he was arrested for selling hashish in Yorkville, then Toronto's hippie district. The following summer he was put on trial and received a jail sentence of two months. McLuhan assumed that the incident was a setup aimed at embarrassing him and assured his colleagues that his boy was innocent and had been framed.[24]

The episode did nothing to lighten McLuhan's distress at the turn Western civilization was taking. His dire predictions of race war and epidemics of mass murder frightened even his daughter Elizabeth, who turned to Sheila Watson, her father's former Ph.D. student and now colleague and family friend, for reassurance that he was not serious. Never was the discrepancy between McLuhan's personal feelings and his public pronouncements so great as at this time. In marked contrast to his private gloom were lyrical statements like the ones in his *Playboy* interview in March 1969. "I feel that we're standing on the threshold of a liberating and exhilarating world," he told the interviewer,

> in which the human can become truly one family and man's consciousness can be freed from the shackles of mechanical culture and enabled to roam the cosmos. I have a deep and abiding belief in man's potential to grow and learn, to plumb the depths of his own being and to learn the secret songs that orchestrate the universe.... There is a long road ahead, and the stars are only way stations, but we have begun the journey.

The conclusion of that journey, McLuhan warned, would bring an end to the era's indulgence in drugs, sexual promiscuity, birth control, and abortion:

In the transition to a retribalized society, there is inevitably a great
explosion of sexual energy and freedom; but when that society is fully
realized, moral values will be extremely tight. In an integrated tribal
society, the young will have free rein to experiment, but marriage and
the family will become inviolate institutions, and infidelity and divorce
will constitute serious violations of the social bond.[25]

This prediction seems to have been more wish than genuine "percept" on
McLuhan's part. His personal attitude toward sex remained that of the young
man who had been appalled at the jazz dancing of his contemporaries in
Winnipeg. The frantic insistence of modern society that sex was "healthy" and
"beautiful" left McLuhan cold. A frank admission that sex was basically evil, he
thought, would give it more dignity and interest than this kind of vacuous stamp
of approval. As for feminism, that other product of social change in the late sixties and early seventies, McLuhan did his best to explain it — and thereby dismiss it. "Women's Lib is just a return to a more normal and traditional form of
human relationships," he told one interviewer in 1974. In a tribal society it was
entirely appropriate for matriarchs to make all the real decisions, as in the case
of the Iroquois.[26]

Resigned as he was to the turmoil of a dissolving society, McLuhan still
played an active role in political issues close to his personal life. He continued
to campaign against litter and pollution and tried to interest Ralph Nader in
the idea of a direct tax that might be imposed on individuals for the sake of
clean air.[27] He publicly opposed increased congestion in the heart of the city,
whether in the form of new expressways or high-rise apartment buildings,
which he particularly despised. He was outraged when four such buildings were
erected near Wychwood Park, threatening, he thought, the very existence of
his neighborhood.

In these and other civic concerns, McLuhan found an ally in Jane Jacobs, a
leading North American authority on cities and a pioneer of the notion of pre-
serving neighborhoods. He consulted her in 1970 when he collaborated with a
local filmmaker, David MacKay, to produce a twelve-minute film called "A
Burning Would." (The title is from a Joyce pun in *Finnegans Wake*: "A burning
would has come to dance inane.") The film, financed by a friend of McLuhan's,
was intended to convey a brief impression of the nature of the city. It turned out
to be largely a plea to stop the construction of the Spadina Expressway, then
threatening the city core.

McLuhan's most unusual exercise in civic responsibility was his joining
a three-man commission to investigate the death of six ducklings and four
adult ducks on Ward's Island, a small island just outside Toronto harbor. The

commission had been formed in 1969 at the behest of a student group called Pollution Probe and a citizens' group called Group Action to Stop Pollution (GASP). An unofficial hearing was held at Toronto's city hall in the presence of about seventy students, residents of Ward's Island, university professors, journalists, and the idly curious. Various expert witnesses testified on a wide range of environmental issues not necessarily connected with ducks. Some of this testimony verged on the ridiculous. The chairman of the University of Toronto zoology department, for instance, revealed that there was enough DDT in human beings to render them inedible. McLuhan enjoyed himself thoroughly. (The commission eventually determined that the ducks had perished from Diazinon, a pesticide used at Ward's Island.)

By the time McLuhan served on this commission, he was begininng to resign himself to the grim role of the seer who is sometimes derided, sometimes petted, but never heeded. He admitted to one correspondent that his success in alerting people to their environment was almost nil.[28] A few years later, paraphrasing William Empson, he complained. "The only form of response that people in general *ever* make to anything is a numb, somnambulist response. They do whatever happens to come into their heads — like tulips. A tulip does just whatever comes into its head, that's all."[29]

In 1970, McLuhan's campaign to raise awareness about the media received a boost from an unexpected quarter. In a paper delivered to the American Association for Public Opinion Research, a General Electric research scientist named Herbert Krugman cited his experiments validating McLuhan's claim that the experience of watching television was something new and fundamentally different from any other form of human communication — especially that form known as print.[30]

Krugman's experiments were indisputably the most serious attempt yet made to test McLuhan's theories. Krugman had become interested in television after a stint with an advertising agency well known for its work in TV. The agency's ads were based on the assumption that viewers of television advertisements carry away small bits of impressions, solidified by repetition and having nothing to do with "thought." Krugman tested viewer reaction by measuring brain waves produced in response to print and to television. Brain waves were a far more sophisticated indicator of response than pupil dilation, heartbeat, or respiration. Delta waves, for example, signified relaxation and drowsiness, while beta waves signified alertness and arousal.

Krugman attached a tiny electrode to the back of the head of a twenty-two-year-old secretary and recorded her brain waves as she read advertisements in a magazine and then as she watched a series of commercials on television. The

TV commercials might feature a baseball pitcher throwing fastballs at what looked like a sheet of glass, followed perhaps by a gentle, almost dreamy shampoo ad. The difference in the secretary's response to the two media was dramatic. The print ads induced fast waves; the television ads, the blazing fastball as much as the lyrical shampoo ad, induced slow waves. "It appears that this subject's mode of response to television is very different from her response to print," Krugman wrote in his report. "That is, the basic electrical response of the brain is more to the media than to content differences with the TV commercials or to what, in pre-McLuhan days, would ordinarily have been called the commercial message."[31]

He characterized the difference in response as "active" with regard to print and "passive" with regard to television. After reading his results to the American Association for Public Opinion Research, Krugman told a reporter, "I'm on a trip I didn't plan. I never set out to confirm McLuhan's hypothesis — I just kept falling over him."[32]

McLuhan, of course, was delighted with Krugman's report and kept up a correspondence with him for several years afterward. He was especially grateful to Krugman for relieving him of the burden of proof for his theories. Not that McLuhan ever thought of doing a lot of laboratory work himself; but now he could proceed with his other projects, blessed, as it were, by an authentic member of the white-coated fraternity.

And McLuhan had, as usual, a great many projects ahead of him. He was interested in doing more short films and videotapes with David MacKay for use on educational television, on topics like how to read a newspaper, how to see TV, how to understand painting. (He wrote mystery writer Ross MacDonald, a fellow Canadian and personal acquaintance, asking if he would like to do a ten-minute script on the detective story as vision of the world.)[33]

On a more commercial note, he conceived an idea for a television program that would feature interviews with people in desperate need of money, for business or personal reasons. The interviews would be climaxed by the presentation of checks for the requested sum to the interviewees. (McLuhan insisted that this show would have to be telecast live.) The title of this interesting program, which never succeeded in tickling the fancy of a producer, was "Up Against the Wall."

Another commercial project McLuhan tried to promote was a deodorant called PROHTEX. This scheme was somewhat closer to McLuhan's heart than the TV shows; since the early fifties he had railed against the North American cult of the squeaky clean, odorless body. He himself resented the pressure to scrub one's flesh daily in a shower, and in *The Mechanical Bride* he pointed out that body odor had been considered an aphrodisiac and a bond between human beings until the Puritans, with their hatred of the body and their mechanistic view of life, went to work on the culture.

The idea of PROHTEX, which McLuhan developed with his nephew Ross Hall, a biochemist, was originally to remove the odor of urine from underclothes. Hall found a chemical formula that would absorb such odors while not damaging skin tissue, as other deodorants did, by stopping perspiration or inhibiting odor-producing bacteria. "Legitimate body odor," in McLuhan's words — an important olfactory means of communication — would be enhanced, not destroyed, by PROHTEX.[34] McLuhan and Hall had patented the formula and registered the trade name PROHTEX. They attempted to interest various corporations in the product, but McLuhan's hopes for it were disappointed.

In the midst of all these schemes, he was still pursuing various book ideas. One project he was very keen on throughout 1971 and for a number of years afterward was a series of books that would outline the theories of communication held by the great figures of Western civilization, from Plato to Pound. By "theory of communication" McLuhan meant the *effect* a person was aiming to produce in his audience and in his culture. Thus a book in this series on, say, Isaac Newton would concentrate not on Newton's laws but on the effect Newton sought from his work — that is, proving the existence of God through mathematics.

McLuhan was struck by what he considered the absence of any overall theory of communication among philosophers or historians of culture and technology, and he hoped this series might be a contribution toward the elucidation of such a theory. He interested a publisher friend, William Jovanovich of Harcourt Brace Jovanovich, and then began casting about for co-workers. He asked his former student Walter Ong if he would like to share the editorship of the series and tried to recruit sympathetic graduate students to write the books. Like many of McLuhan's other good ideas, this one was abandoned before he really began to pull it off.

All of these projects in 1970 and 1971 were overshadowed by the deterioration of his health. High blood pressure and what looked suspiciously like minor heart attacks plagued McLuhan sufficiently that he was once again hauled into the hospital for tests. The results indicated a serious problem with his internal carotid arteries, the major suppliers of blood to the hemispheres of the brain. They had narrowed dangerously, and surgery seemed to be called for.

The thought of another skull operation was more than McLuhan could bear. In desperation, he appealed for medical advice to Hans Selye, the great endocrinologist and authority on stress whom McLuhan had known since Selye had published in *Explorations*. Selye, however, was unable to recommend another course of action, and surgery seemed to be the only option. McLuhan found himself in the hospital in May 1971, this time in London, Ontario, under the care of a trusted neurologist named H.J.M. Barnett.

Before scheduling surgery, Barnett performed another angiography (a test in which dye is injected into the arteries to trace the flow of blood) and reported astonishing news. McLuhan's internal carotid arteries were still blocked, but his external carotid artery — the artery that supplies blood to the face, scalp, and jaw — had formed huge connecting channels through the base of his skull and inside the skull. Such channels are normal in cats but are almost never seen in humans to the extent revealed in McLuhan's angiography. Barnett had never known anything like it and could only compare McLuhan's exterior carotid arteries to those of a tiger. They were supplying enough blood to the hemispheres of his brain to make an operation unnecessary.

When McLuhan heard the news of his reprieve, he was overcome with joy. He considered it to be literally a miracle, an answer to his prayers. In jubilation he sent his friend Sheila Watson a photocopy of a diagram showing the carotid arteries in a tiger and penned on it, in imitation of William Blake, "What immortal hand or eye / did bring this novel artery!"

Despite his spectacular arteries, McLuhan never fully regained his health. This may explain in part why he no longer responded to critics — now becoming more and more contemptuous — with the lighthearted certainties of his earlier years. The certainties he did expound began to sound increasingly like the obsessions of a crank. More and more he felt that he was denied not only critical sympathy but even a reasonable acknowledgment of the scope and intention of his work. By 1971 he was reduced to complaining to his old friend John Bassett that book reviewers in Bassett's newspaper, the *Toronto Telegram*, were using its pages for broadsides against him.[35]

The specific occasion for this protest was the appearance in the *Telegram* of reviews of two books, *The Medium Is the Rear View Mirror* by Donald Theall and *McLuhan* by Jonathan Miller. Both were attempts to more or less cut McLuhan's reputation down to size, and both had been written by men who had once seemed to be in thrall to McLuhan's vision. Theall, a former student and colleague of McLuhan's at St. Michael's, had sharply broken with the master in the mid-fifties, and the two men feuded for years afterward.

The title of Theall's book, published by McGill–Queen's University Press in Canada, referred to Theall's belief that McLuhan had examined media without using the rich resources of contemporary psychology and sociology. In McLuhan's thought, therefore, media became an autonomous force in human society that, properly understood, reinforced a conservative view of change. The book was obviously based on a thorough knowledge of McLuhan's intellectual tastes and background. Its prose, however, was not of a kind to ensure much readership outside small academic circles within Canada. In this respect, Miller's book was far more damaging. While not as knowledgeable as Theall's, it was written in the lively prose of a clever man who was more than capable of scoring points off McLuhan. On the other hand, there were depths in McLuhan that Miller breezily ignored.

Miller had been commissioned to do the book as part of the Modern Masters series, edited by Frank Kermode. Pleased that his old disciple had been chosen for the task, McLuhan wrote Kermode that he was "enormously flattered" that Miller considered his work worthy of attention.[36] The result was therefore all the more unpleasant. McLuhan realized that Miller's attack was plausible and effective, although to friends he dismissed it as the reaction of a man repulsed by McLuhan's religion. Miller, he claimed, was committed to the view that McLuhan was an agent of Rome. McLuhan returned the compliment, suggesting to one friend that Miller was an agent of Moscow and that the tone and approach of the book had been dictated by Miller's superiors.[37]

McLuhan's friend Barry Nevitt urged him to reply when Miller partially replayed his criticism in an article in the July 15, 1971, issue of the BBC publication the *Listener*. McLuhan had long believed that enemies were to be cherished because they functioned as superb PR agents, indirectly moting one's work far more effectively than friends did. And he usually realized that replying to specific attacks, on the same principle, was giving your enemy free advertising. This time, however, McLuhan felt the need to strike back. He and Nevitt sent off a letter to the *Listener*, which Miller simply dismissed in reply.

In 1972, the McLuhans participated in another Aegean cruise, joining the likes of Margaret Mead and Arnold Toynbee, whom McLuhan afterward described as "really quite a dandy old sport."[38] Despite an attack of bronchitis, McLuhan enjoyed the chance to socialize. Once again, however, he felt unable to get his message through to the participants and compared the cruise unfavorably to the one he had enjoyed almost ten years earlier, when the tone had been decidedly more intellectual.[39]

The intervening nine years had seen great changes in his fortunes. On the first cruise McLuhan had been a rising star, just beginning to impress the world with a cultural theory of great originality. Now he was beginning to feel obsolete, as if people were no longer interested in hearing from him or about him. Although he was still being sought out by people from all over the world — many of whom were important figures in academia, mass media, or big business — interest in McLuhan had diminished in North America, and he was forced to confront the fact that he was becoming something of an acquired taste for a nearly invisible intellectual minority.

It was about this time that McLuhan contemplated producing a new version of *Understanding Media* using the approach he had long since abandoned — the approach of a man of letters whose sensibilities had been outraged by the effects of the media. He thought such a crowd-pleasing approach might be just the thing for reviving interest in his ideas on the media.[40] Although McLuhan made the revision of *Understanding Media* a top priority throughout the seventies, he never did complete it. As usual, he was keeping too many other balls in the air. The one that occupied most of his time from 1968 until 1971 was a book he had originally undertaken to write with Ralph Baldwin on corporate management. After his relationship with Baldwin had gone sour, McLuhan turned to Barry Nevitt as a collaborator.

Nevitt was extremely well qualified to coauthor such a work. An electrical engineer, he had worked worldwide for the telecommunications industry, private corporations, and public utilities. Since 1965 he had become a familiar figure at McLuhan's center, one of the men whom McLuhan most trusted to speak for him and alongside him.

The book, which McLuhan eventually entitled *Take Today*, was mainly concerned with the changes in organizations wrought by electronic media. McLuhan and Nevitt outlined three major shifts. The old hardware — the buildings, bureaucratic structures, filing cabinets, and so on — was being over-ridden by software — the design and programming of microchips and electric circuitry that created the new information world. People within organizations were shifting from specialized jobs to multifaceted role playing. And all of this was accompanied by a shift from centralized structures to decentralized forms of organization. As a corollary, McLuhan and Nevitt maintained that top exec-utives were increasingly losing touch with their organizations and becoming more and more isolated from the real action of the corporations and institutions over which they presided.

As soon as McLuhan returned from New York in 1968, he and Nevitt met several times a week, often over lunch, where McLuhan, a soup fanatic, would concoct a repast out of two different varieties of Campbell's soup. Then they sat down to talk. For two or three years they continued these dialogues, taking notes all the while, until Corinne, in her capacity as manager of her husband's affairs, insisted they actually begin writing the book. According to Nevitt, "Corinne said, 'You boys have had enough good time, now it's time to get to work. I'll type out the first chapter for you to make sure you get started.'"

It was not that simple. In their years of conversations, McLuhan and Nevitt had accumulated several hundred pages of notes and a stack of essays they had written together. Whenever McLuhan and Nevitt decided to write on a particu-lar topic, they would select headings from their notes and McLuhan would usu-ally begin writing, with Nevitt functioning, as he did in the seminars, as a

prompter ("Wouldn't this be a good time to write about electric speedup reversing the forms of human organizations?"). McLuhan usually got tired of holding the pen in about half an hour and would then lie down and think out loud
while Nevitt continued to prompt and suggest and write, until McLuhan would
finally say, "You finish it off, Barry, and give it to Marg tomorrow." Margaret
Stewart, McLuhan's secretary, would then type out the handwritten pages.

The prospect of somehow converting these essays and notes into something
resembling a book was overwhelming. Nevitt comments, "Marshall said, 'Look,
we should just give it to the editor. What are editors for? That's their job. Let
them worry about it, Barry, we've got all the material there.' I said, 'Marshall,
there's no editor on God's green earth that's going to be able to make head or
tail of what we've written.'" It fell to Nevitt to try to arrange the material into
a book.

In April 1971, a 600-page manuscript arrived on the desk of William
Jovanovich at Harcourt Brace Jovanovich. Jovanovich looked it over, sighed,
and flew up to Toronto with another editor, Ethel Cunningham, to talk to the
authors about reducing the manuscript to publishable size. McLuhan and Nevitt
agreed to cut between 150 and 200 pages.

Back in New York, Cunningham went through the manuscript and began
to suggest cuts to Nevitt, who had told her, with considerable justification, that
McLuhan was not to be bothered because he had no patience for this sort of
thing. Nevitt himself rejected most of her cuts, however, often suggesting a
cut somewhere else and then adding something even longer at another point in
the manuscript. While the two of them struggled and got nowhere, McLuhan
began to complain that he wasn't being shown anything and had no idea what
was happening to the book. Unfortunately, Nevitt was reluctant not only to
do major surgery on the book on his own but also to force McLuhan to make
cuts. Although he had been as accommodating a collaborator as McLuhan could
have wished for, their working relationship showed signs of strain. Privately,
McLuhan grumbled that Nevitt was "allergic to anecdotes" and, in his editing
of their work, was determined "to putty over every nook and cranny."[41]

Eventually the two managed to make the required cuts. *Take Today* went to
the printer in November 1971 and was published the following year. The book
was intended to sum up all of McLuhan's insights about patterns of human
organization in the twentieth century. He hoped it might function in much the
same way as the encyclopedic wisdom of the ancient Sophists and rhetoricians,
swaying the minds of "governors and executives."[42]

Even an exceptionally intelligent governor or executive, however, could
have made little of this book. The extreme condensation of the thinking process represented in its prose, as well as the abundance of private jokes and

references, make it very hard to read. "Writing, naturally, has its limitations, and you can't get the effect of simultaneity except in the form of a joke or an aphorism, and we were inclined perhaps to overdo that when we were writing together because it gave us so much fun," Nevitt remarks. "*Take Today* was perhaps written more for our own entertainment than for what any other audience might share."[43]

It did not take McLuhan long to realize that he had turned out another unreadable book. He advised people to "dip into" the book, to "sample" its contents, rather than to read it straight through. (It was good advice. The genuine insights of the book seem to catch the attention of the casual browser; the reader trying doggedly to follow the argument of the book from cover to cover simply gets lost.) To one friend he put the blame for the opacity of the final product, once again, on his collaborator.[44] And once again the criticism was unfair, for Nevitt had been put into an impossible position to begin with. McLuhan was now simply paying the price for refusing to tackle the problem of how best to translate his thinking to the medium of the printed word. It was a problem for which he never really found a solution.

Take Today sold little more than 4,000 copies and went virtually unreviewed in major North American periodicals and newspapers. (The book did enjoy an odd sort of revival when it was partially serialized in 1975 in the pages of a trade journal called *Modern Office Procedures*.) Its failure signaled the end of an era — when corporate executives no longer felt that they might be missing something very important if they did not hear what this professor of English from the University of Toronto, Marshall McLuhan, had to say.

12. The Sage of Wychwood Park (1972-1979)

> Cognitive license, like most things, is habit forming. Anyone who has been in
> unchallenged possession of it for too long is liable to end up using it with less and
> less restraint, and in the end perhaps with no restraint at all.
> — Ernest Gellner, *The Psychoanalytic Movement*[1]

The day after the American presidential election in November 1972, McLuhan took note in his diary of the "Bloody Nixon victory."[2] A month earlier, in the Canadian general election, his exemplary television-age statesman Pierre Trudeau had lost his majority in the House of Commons and barely hung on to power with a minority government. The fortunes of these two politicans would shortly be reversed: Nixon would be destroyed by Watergate, Trudeau granted another majority government after an election in 1974. Nonetheless, the elections in the fall of 1972 seemed to flatten, once and for all, the exhilaration of the sixties. To be sure, profound social change begun in the sixties continued in the seventies, primarily in the sexual sphere; but the apocalypse hovering around the corner, the long-haired, Day-Glo, ultimate revolution spearheaded by youth, was evidently postponed forever.

McLuhan still kept his ear cocked for signs of it, but this time the apocalypse he anticipated was an extremely unpleasant one. Despite the back-to-business air on campus — which spectacularly falsified his predictions of increasing student violence — he still believed, as he told one audience in 1976, that our society stood on the verge of "another binge of slaughter," which would overshadow even Dachau and Buchenwald.[3] To the end of his life he kept waiting for epidemics of mass murder to break out.

He was also anticipating apocalypse in the more literal sense of the word — that is, the revelation of the second coming of Christ. When asked if he was optimistic or pessimistic, he inevitably replied that he was neither, he was apocalyptic. He meant it. As early as 1968 he had advised a priest who was about to leave the priesthood not to quit, on the grounds that the world was going to end very soon. Along with many Protestant fundamentalists, he regarded Israel's role in the Middle East — primarily the recapture of Jerusalem in 1967 — as an unmistakable portent of the last days. The Yom Kippur War in 1973 he noted in his diary as "obviously apocalyptic."[4]

This hankering for apocalypse and its inseparable counterpart, a gloomy view of present reality, were heightened by a development in his thinking about the new acoustic world, in which he at last perceived the full nature of what he now called "discarnate man." Discarnate man, according to McLuhan, was electronic man, the human being used to talking to other humans hundreds of miles away on the telephone, used to having people invade his living room and his nervous system via the television set. Discarnate man had absorbed the fact that he could be present, minus his body, in many different places simultaneously, through electronics. His self was no longer his physical body so much as it was an image or a pattern of information, inhabiting a world of other images and other patterns of information.

The effect of this reality was to give discarnate man an overwhelming affinity for "a world between fantasy and dream" and a "typically hypnotic state," in which he was totally involved in the play of images and information, like a small child fascinated by a kaleidoscope. Psychically, discarnate man suffered a breakdown between his consciousness and his unconscious. (According to McLuhan, the unconscious was simply that state of mind in which perceptions and thoughts coexisted without any connections whatsoever. Television, in particular, might have been designed expressly for enlarging the domain of this unconscious.) Under these circumstances, the self, or identity, of discarnate man was virtually swamped by the barrage of images and information in a phantom electronic world.[5]

This destruction of private, personal identity was the unexpected — and toxic — side-effect of the integrated sensuous life McLuhan had happily proclaimed in the early sixties. Now he saw several unpleasant consequences. The children who experienced this destruction were incapable of civilized pursuits. "I myself think they are sinking into a kind of world where satisfactions are pathetically crude and feeble, compared to the ones we took for granted thirty years ago," he told one interviewer in the seventies. "Their kicks are on a seven- or eight-year-old level."[6]

These TV children seemed aimless, undisciplined, and illiterate. Even worse, the identityless inhabitants of the acoustic world reacted to their state by acts of violence, physical and psychological. Violence, McLuhan pointed out, was the unfailing remedy for those deprived of their identities; it was one method, often futile but always available, of grasping for the meaningful. No wonder McLuhan kept waiting for the blood to flow.[7]

In 1977 McLuhan further elaborated his perceptions of discarnate man by bringing in the concept of natural law. Natural law, according to Catholic theology, is that expression of divine law which we apprehend by virtue of our being rational creatures. Natural law tells us to seek health and avoid injury, to cherish the bonds of family and social life, and so on. It virtually upholds the entire edifice of Catholic morality. In the case of discarnate man, McLuhan asserted, natural law was almost wholly obliterated.

Without it, McLuhan wondered whether there was still any basis for morality. For the rest of his life he was unable, it seemed, to decide whether the disappearance of natural law meant that discarnate man would somehow tap directly into supernatural law or rather give his allegiance to terrible spiritual substitutes for divine law — to the occult, for example, or to the modern superstate. Only one thing was certain: this discarnate activity meant we were in for a great religious age. It might well be a destructive or diabolically religious age, but religious it would be.[8]

There was yet another twist to the phenomenon of discarnate man, as McLuhan saw it. In an age when people were translated into images and information, the chief human activity became surveillance and espionage. Everything from spy satellites to Nielsen ratings to marketing surveys to credit bureau investigations was part of this intelligence-gathering, man-hunting syndrome. So pervasive was the syndrome that discarnate man worried whether he existed as nothing more than an entry in a data bank somewhere.[9]

According to his mood, McLuhan viewed this situation as a comedy or as a tragedy. An associate, Scott Taylor, remembers that McLuhan almost wept on one occasion in 1977, from merely contemplating the future of a humanity absorbed into a surveillance network of satellites and computers. That same year, McLuhan's apocalyptic frame of mind made him receptive to *Spear of Destiny*, a book that claimed Hitler was a devotee of the occult. McLuhan was aware that the book might be one of those pseudoscholarly productions like *Chariots of the Gods*. Nonetheless, he was fascinated. Something in it resonated with his darker contemplations of the world, feeding his old suspicion of gnostic, Rosicrucian-like mysteries.[10]

Regardless of mood, McLuhan's driving pace never let up. It was as if he could not live without half a dozen projects — usually what he referred to as "wee books" — on the go. In 1973 he devoted himself primarily to two such projects. One was a book that he referred to as a "twentieth-century Baedeker," a kind of tour guide to the major scientific and artistic achievements of the century. In a "structural analysis" of these achievements, the structure of artistic breakthroughs would be closely related to the structure of the scientific breakthroughs.[11]

In a sense, Wyndham Lewis had first shown McLuhan the possibilities of such an enterprise with *Time and Western Man*, in which he pointed out the affinities of Einstein's theory of relativity with the work of writers such as Joyce. McLuhan's prime examples were Max Planck's quantum theory of physics in 1900 and Freud's *Interpretation of Dreams*, published the same year. According to McLuhan, the former demonstrated basic discontinuities in the properties of matter and the latter demonstrated the discontinuity of conscious and unconscious life. Both were obviously related to the work of his own all-time favorite painters, the cubists, who broke up the continuities of visual perspective.

McLuhan signed a contract for his "Baedeker" in 1971 with Doubleday, for what he called a "handsome advance." He intended to do a series of thousand-word essays on the various achievements he inventoried. These quickly transformed, in his mind, into one-hundred-word "groupings," highly aphoristic, that emphasized two basic themes. The first reflected his long-held belief that there was no real split between artistic and scientific enterprises, particularly in the twentieth century. The second theme underlined his belief that artists, far from merely being influenced by scientific discoveries, tended actually to anticipate these discoveries, by several decades, in the patterns and approaches of their own work.[12]

In 1973, after the publication of *Take Today*, McLuhan started work on this new book in earnest, spurred on by his reading of Thomas Kuhn's noted book *The Structure of Scientific Revolutions*. Confident that he already knew quite enough about the artistic breakthroughs of the century, he set out to bone up on the scientific breakthroughs, rummaging through copies of *Scientific American* going back to 1900, quizzing his scientific friends, like Arthur Porter and Hans Selye, and urging his graduate students to help with the task. Harley Parker did research on the book for a while before he left the center in 1975, but he was convinced that it would never be published. The whole scheme of it required the kind of dogged, systematic research and organization of thought that was utterly foreign to McLuhan, especially at this point in his career.

In 1974 McLuhan applied to the Canada Council, the country's chief granting body for work in the humanities, for a $50,000 grant to complete the work. When his application was turned down, he professed not to be surprised by the rejection. He knew very well that the academics chosen to evaluate his proposal would have had a field day dismissing his sweeping assertions and his highly improper "methodology." (His articles, which lacked the scholarly standards required by most university journals — the rigorously documented and footnoted conclusions and so on — manifested the same refusal to play by academic rules.) Even if he had gotten the grant, however, it is unlikely he would have been able to finish the book. A few years after signing the contract, McLuhan told his editors at Doubleday that he was "putting aside" the book for the time being. In fact, he knew then that he would never complete it.[13]

In its place, McLuhan offered a willing Doubleday another book that had been much on his mind in 1973: the formulation of what he called "the laws of the media." The term had its origins in the contention held by the philosopher Karl Popper that every truly scientific theory necessarily had to be couched in a form capable of being disproved. McLuhan, who had long been somewhat jealous of the prestige of science, seized on the notion. He determined to cast his insights into a form that would rise to the challenge of the falsifiable hypotheses supposedly characteristic of the scientific world. He would formulate scientific laws about the recurring dynamics of the principal human artifacts.[14]

McLuhan first articulated these laws while working with Nevitt on *Take Today*, and they were originally intended to be part of that amorphous work. The first two laws were fairly obvious. Any major artifact enhanced or accelerated a certain process or thing. At the same time, the second law ran, it tended to render obsolescent another process or thing. Money, for example, enhanced the process of trading goods. At the same time, it rendered the process of barter obsolete. These were fairly obvious dynamics, but the two laws that followed gave McLuhan's laws of the media their real originality.

The third law grew out of McLuhan's work in *From Cliché to Archetype*, which attempted to articulate the process whereby those clichés retrieved from the scrap heap of the past became archetypes of the present. The law stated that any major artifact retrieved some process or thing that had once been obsolete. Money, for example, retrieved the potlatch — that is, revived the spirit of conspicuous consumption. McLuhan's fourth law grew out of a principle he had noted as early as his articles in *Explorations*: that of the great pattern of being, as he put it in *Understanding Media*, "that reveals new and opposite forms just as the earlier forms reach their peak performance."[15] The law stated that any major artifact, when pushed to the limits of its potential, flipped into something entirely new. Money, for example, flipped into credit — thus, MasterCard.

Briefly stated, then, McLuhan's laws of the media, when applied to any major artifact, posed four questions: what did it enhance, what did it make obsolescent, what did it retrieve, and what did it reverse or flip into? McLuhan at one point stated flatly that he considered his discovery of these four questions — or "tetrad" — to be his greatest achievement.[16] The tetrad was not terribly hard to comprehend, and yet it seemed to resonate with limitless and fruitful intellectual applications.

It also had a satisfying numerological shape. Its four-part structure seemed to him superior to Hegel's triad of thesis, antithesis, synthesis. Indeed, McLuhan considered that Hegel's triad was merely a truncated version of his tetrad, obtained by eliminating his third law, the law of retrieval. The triad was a tool for visual man, concerned more with forming conclusions than with reaching understanding. By contrast, the four parts of the tetrad constituted a kind of total perception of things. They were not, like the three parts of the triad, sequential in time but simultaneous. They were not logically connected but related like the parts of a metaphor, in a living, vibrant ratio. (In fact, McLuhan had always maintained that every metaphor has a four-part structure: A is to B as C is to D. Four-part structures of this kind seemed to be inherent in the human psyche and also inherent in all of the things that humans made — most especially in language.)[17]

Once McLuhan devised the tetrad, he began to apply it to endless phenomena, as if the tetrad were another all-purpose intellectual tool kit. It became something of a parlor game in McLuhan's company to come up with tetrads for things, from the wheel to the novel; visitors were always in danger of being asked to play. In a 1973 letter McLuhan informed a friend that he had just come up with sixty-three examples of the four laws of the media.[18]

He revealed his findings to the world in January 1975, in an article entitled "McLuhan's Laws of the Media," published in the journal *Technology and Culture*. He claimed that the laws were arrived at through the scientific method, since they were the fruit of induction, and issued a general challenge to readers to disprove them, if they could. He was, he said, inviting this attempt as a substitute for personal abuse (the usual form that criticism of McLuhan took). He added that the attempt might also lead to some profitable new insights.

No doubt McLuhan felt that he had seized the high ground, and he waited hopefully for battalions of scientifically minded critics either to attempt to storm his positions or else to acknowledge, in effect, that he was not the intellectual buffoon he had been made out to be. He was desperate, in short, to be taken seriously. Unfortunately, the sole response was a long, rambling letter from one William Henry Venable, described as an "engineer and practicing attorney" from Pittsburgh, Pennsylvania.

In a way, it was absurd for McLuhan to think that his laws of the media were any more "inductive" or "scientific" than any of his other pronouncements. They could not be disproved, only endlessly argued. There were too few factual constraints on this tetrad for it to resemble, however faintly, a true scientific hypothesis. And yet, no more than the rest of McLuhan's work, it could not be dismissed as simply fanciful. There was too much intellectual life compressed within it.

Ernest McCulloch, director of the University of Toronto's Institute of Medical Science at the time, was fascinated by the tetrad. He thought the concept was brilliant, and when he left the university to become one of Canada's foremost cancer researchers, he claimed he used the tetrad as a method of analysis in the formulation of some of his novel views of leukemia. Later, he invited McLuhan to give a seminar on the tetrad at the hospital where he worked. McLuhan jubilantly wrote to a friend — with characteristic overstatement — that McCulloch's response indicated the "tetrad approach works 100% in the case of cancer." This proved, according to McLuhan, that disease was a human artifact, just as much as houses or airplanes.[19]

Reconciled to the loss of the twentieth-century Baedeker, Doubleday agreed, in 1974, to publish instead McLuhan's book on the laws of the media. Doubleday Canada's editor in chief, Betty Corson, met with McLuhan several times to nudge the manuscript along. "It was not in publishable shape at the time that I saw it," she recalls. "It was very obscure and in a state of flux. Every time he got something finished, he decided that there was something more to be done to it. And the arrangement was changing constantly. He had written a very long preface to the book which was totally obscure. I asked him if he could clarify and shorten it, and then it turned out to be one page, which, of course, was too short."[20]

As in the case of *Take Today*, McLuhan tried to satisfy his publisher without any real idea of how he might be able to do so. "He was very obliging, but we didn't seem to get anywhere," Corson recalls. The manuscript — also like *Take Today* — consisted of a small core of "percepts" utterly resistant to any shape or structure. (McLuhan also had the odd notion of adding illustrations by Canadian sculptor Sorel Etrog to the book. Whatever Etrog came up with, the illustrations would certainly have been another interesting puzzle for readers to meditate on.)

Despite her best efforts Corson did not succeed in extracting a preface that was a happy medium between the first long, confusing essay and the even more confusing one-page summation that followed. The work never really got off the ground, despite Eric McLuhan's promises to bring in revisions to the manuscript. After a series of delays it became obvious that this book, too, would not materialize. (In 1988 the University of Toronto Press finally published a book entitled *Laws of Media: The New Science* by "Marshall and Eric McLuhan." The subtitle, of course, is a nod to the *Scienza Nuova* by Joyce's great mentor, Giambattista Vico.)

In the early seventies McLuhan also revived his old dream of rewriting his Cambridge thesis. Tentatively titled "From Cicero to Joyce," the revision would amplify his early insights about logic, grammar, and rhetoric and apply them to everything he had learned about Pound, Eliot, Joyce, and other modern artists since the days he had toiled in the Cambridge University library under F.R. Leavis's tutelage.

As early as 1969 McLuhan was negotiating for the publication of this work, and for the next eight or nine years he would return to the thesis from time to time and write a few pages. Mostly, he talked at length to friends and visitors about the five divisions of rhetoric. It became another one of his intellectual pastimes, along with constructing tetrads for various artifacts, to detect the use of these five divisions in everything from the Lord's Prayer to virtually any classic work of literature, from Chaucer to Voltaire to Joyce. (McLuhan puzzled over the fact that his favorite writers, the early moderns, never let on that they were using the five divisions.) In fact, anything at all in five parts was guilty of being modeled on the five divisions, until proven innocent. Since these five divisions — inventio, dispositio, pronuntiatio, elocutio, and memoria — had, in McLuhan's mind, a highly elastic meaning, the thrill of the hunt, in this case, was not terribly useful to other literary scholars.[21]

Still another project McLuhan worked on throughout the seventies was a book he originally had planned to do with his publisher and friend William Jovanovich of Harcourt Brace, on the subject of the future of the book. Jovanovich had dropped out in 1969, but McLuhan persisted with the idea. In 1972 he reported in his diary the effort it was costing him to attempt, for the first time in many years, actually to *write* instead of dictate or "dialogue" a work into existence.[22] He did not succeed in this enterprise. Later in the decade the project metamorphosed into a book on the future of libraries when yet another collaborator, a University of Toronto physics professor named Robert Logan, hove into sight.

Logan, who was of a decided holistic bent, taught a course called The Poetry of Physics (alternately known as The Physics of Poetry) at one of the colleges of the university. He was the kind of intellectual asteroid more or less fated to fall into McLuhan's orbit, and he became McLuhan's unofficial "science adviser," fulfilling the Nevitt-like function of conveying to him the gist of various articles on science. (Whether because Logan filled this function very well or because McLuhan had a good instinct for the truly valuable and interesting, the articles McLuhan did seize upon and recommend to others in the scientific community were almost always worthwhile.)

In 1975, a year after they met, Logan and McLuhan negotiated a contract for a book on the future of the library with R.R. Bowker Company, a publisher specializing in works aimed at the library trade. The book never went beyond the first draft stage and never approached anything that Bowker would have found useful for its customers.

As the decade wore on, McLuhan's anxiety over some of these unfinished works grew, manifested chiefly in occasional bouts of irritability and crankiness.

Although constitutionally incapable of depression, McLuhan was sensible of the fact that none of his books since *Understanding Media* had enhanced his reputation or made much of an impression on the intellectual public. He rarely discussed the subject — it was a painful one — but this awareness, combined with his almost total absence of pride of authorship, helped undermine his will to finish any of his planned books.

He also knew that all his previously published books had good things in them. He knew that much of the criticism directed against them was wrongheaded. "They cannot read," he once complained to a reporter about his critics.

> These men literally cannot read what is written in front of them. Any time I have these people present and we talk about what I have written they say, "Ah, ah, now I see, ah yes, yes, yes, of course." They agree one hundred percent with everything I say. When they read it they cannot understand it. It isn't because what I write is difficult. What happens, I think, is that what I write requires that they re-arrange their own ideas about everything.[23]

McLuhan's plea to his readers was much the same as that of his old Renaissance hero, Francis Bacon, in the latter's preface to his *Novum Organon*. There Bacon urged his reader not to dismiss reflexively his "speculations." "Let him examine the thing thoroughly," Bacon wrote. "Let him make some little trial for himself of the way in which I describe and lay out; let him familiarize his thoughts with that subtlety of nature to which experience bears witness; let him correct by seasonable patience and due delay the depraved and deep-rooted habits of his mind."

To which McLuhan could only have added a heartfelt amen — painfully aware, at the same time, that few, if any, of his critics were prepared to make some little trial for themselves of his peculiar rhetoric, much less correct the depraved and deep-rooted habits of their minds.

On the other hand, McLuhan was too blithe in dismissing the inherent difficulty of his work. His readers did need some help. But as the editor of his first book, *The Mechanical Bride*, had realized years ago, McLuhan refused to bore himself by reworking his prose to make it easier for readers.

Alert and sympathetic interviewers seemed to elicit the most effective McLuhan prose. Realizing this, McLuhan never ceased looking for collaborators — in reality, one-man audiences — to help him get books written. If he had found collaborators who also possessed superb editorial powers and if he had trusted those collaborators to exercise such powers, McLuhan's later books might have been more successful.

As it was, McLuhan realized that his insights were being virtually ignored in the seventies. The times were partly to blame. In the sixties the world changed suddenly and everyone wanted to know why; in the seventies change continued, but in much grimmer and duller fashion, and everyone wanted only to cope.

Whatever the reasons, McLuhan could see that he would share the fate of Ezra Pound and Wyndham Lewis, who had also tried to wake up contemporaries with their insights and who had failed. McLuhan's insights were not negligible. His most important insight, perhaps, had been his realization that the introduction of a major medium of communication, such as print or television, involved a major shift in the mental posture of that medium's users, regardless of what was actually printed or broadcast. To indicate the specific dimensions of the shift, given its subtlety and complexity, was extraordinarily difficult. As soon as one aspect of the shift was pointed out, ten good arguments could be made that that aspect was exaggerated or was due to other causes.

But McLuhan was right to insist that unless the shift was at least recognized, intelligent appraisal of current events was impossible. Hence the need to wake people up. McLuhan tried to be more crafty than Lewis and Pound to avoid the disasters that had beset their careers, but ultimately he had no more success than they in making an appreciable impact on society. Because of this, throughout the seventies his public pronouncements were rarely free from a hint of anguish.

Robert Logan, meanwhile, continued to work with McLuhan. Perhaps his chief contribution was not his work for their book on libraries but the fact that he inspired McLuhan's interest in the hemispheres of the brain. McLuhan had heard before of the split-brain hypothesis — the theory that the two sides of the brain controlled different physical and mental functions. An admirer had sent him an essay on the subject as early as 1967. Now, under Logan's prodding, he took up the idea again.

In fact, as Logan puts it, McLuhan "went completely bonkers" over the theory. He had a diagram of the split brain blown up and put up on a wall of the center. Since the days of the charts he had devised for his "Report on Project in Understanding New Media," McLuhan had a fondness for visual aids, and this particular aid promised, finally, to bring home to people what he had been talking about all these years. To McLuhan, the right hemisphere — which was, according to his own interpretation of the hypothesis, the home of the acoustic, the simultaneous, the intuitive — was the hemisphere of the electronic age. The left hemisphere, attuned to the visual, the linear, and the quantitative, was the hemisphere of the phonetic alphabet and the printing press. The famous "dissociation of sensibility" that Eliot had talked about with regard to the poetry following Dryden and the seventeenth century, was really the splitting of the hemispheres and the rise of the left — already enhanced, in Western civilization, by the phonetic alphabet — to even further dominance through the invention of print.

McLuhan believed the split-brain hypothesis verified his life's work. It really added nothing to his basic insights; but it did provide a sort of neurological basis for the dichotomies he was constantly finding between acoustic and visual, analogical and logical, and so on. It also explained why so many people had rejected his ideas: he had been a lone right hemisphere talking to a crowd of left hemispheres. Finally, McLuhan seized upon the hypothesis because it was always a relief to have something new to use — in this case, an outsized diagram of the human brain — in talking before audiences.[24]

In theory, McLuhan maintained that both hemispheres were necessary and complemented each other and that no value judgments should be passed on their respective functions. This disclaimer was of the same nature as his constant protestations that he did not favor the acoustic over the visual. In fact, he prided himself on being a right hemisphere man, just as he prided himself on being at home among the resonating fields of the acoustic. "Left brain" became a handy term of dismissal in the McLuhan lexicon, along with "print-oriented," "visual," "linear," and so on. In one letter to a friend, written at the height of his obsession in 1976, he termed the left hemisphere, with a minimum of irony, "the villain."[25]

In his enthusiasm for the split-brain hypothesis, McLuhan tried to contact noted workers in this area, such as Roger Sperry of the University of California. None of the people he wrote bothered to get back to him, and this hurt McLuhan very much: he solaced himself as best he could by referring to Sperry and company as "left brain." Of course, given McLuhan's lack of patience with scientific research — as he sometimes put it, he didn't have to do ten years of work to confirm what he was able to perceive perfectly clearly — the indifference of these men of science was understandable. Moreover, McLuhan's interpretation of their work was not necessarily something they would have endorsed. He still could not refrain from giving his own interesting little twist to scientific reports, and the split-brain material was no exception.

This was clear at least to Marcel Kinsbourne, then professor of pediatrics at the University of Toronto Medical School and professor of psychology at the University of Toronto. McLuhan had consulted Kinsbourne as the local expert on the hemispheres of the brain. Kinsbourne listened patiently to McLuhan's explanation of the hemispheres, associating the left with the visual, the right with the acoustic. "That was his interpretation of the data," Kinsbourne recalls. "It wasn't mine. But it made for good conversation. I would never be so discourteous as to say, 'You're wrong.'" In Kinsbourne's interpretation, the right brain was perfectly logical and by no means intuitive, or receptive to what

McLuhan called the ESP of the acoustic world. Rather, the right hemisphere specialized in the perception of spatial relationships, especially those sequences of words we call written or spoken language. In simplest terms, according to Kinsbourne, the left hemisphere dealt with *what* a point was, and the right dealt with *where* it was.

It was easy to understand how McLuhan could use this distinction as a launching pad for his own distinctions between acoustic and visual, but what McLuhan meant by "visual" was not what Kinsbourne and his colleagues would have meant by "visual," and that made communication very difficult. In fact, Kinsbourne saw McLuhan, if anything, as a left hemisphere man par excellence. "He was one of the most linear people I've ever met," Kinsbourne recalls.

> He would filter out anything he didn't want to hear, like someone
> turning off his hearing aid, and he would simply talk past you if you
> said something contrary to what he had said. Nor would he see all the
> implications, the various multidimensional ramifications of something,
> the alternative interpretations. He would never say, for example, "Oh,
> I see what you mean," or "Oh, that's how that fits in." The right hemi-
> sphere, in this regard, is the discrepancy hemisphere. It says, "Ah ha,
> things aren't as I thought they were." The left hemisphere, on the
> other hand, is business as usual. And Marshall was certainly business
> as usual.[26]

But Kinsbourne was patient with McLuhan. "I enjoyed him as a terrific person-ality. After all, how many eccentrics do you meet in Canada?"

Left brain and right brain appeared like verbal tics in McLuhan's conversa-tions and articles after 1976. When he spoke at the American Psychological Association convention in Toronto that year, out came the blown-up diagram of the human brain. It was greeted by a wave of laughter. The diagram looked for all the world like the old phrenological charts that Elsie and Herbert McLuhan had found so fascinating in the 1920s. McLuhan played to the laughter and delighted his audience with a subsequent string of one-liners, not only on the subject of the split brain but on such distantly related topics as the way acade-mics dressed for conferences.

Left brain and right brain, in effect, became shorthand, a verbal nutshell, for the whole complex of ideas McLuhan had been working on since the 1950s. These terms joined two others that McLuhan employed throughout the 1970s — "figure" and "ground," borrowed from the world of painting. The part of a painting the viewer focuses on is the "figure," while the rest, falling into the background, is the "ground." The relationship between figure and ground would be dramatized in the kind of amusing sketch that looks like a lamp or two profiles, depending on which parts of the sketch the viewer focuses on. If the viewer sees the lamp, he does not see the profiles — and vice versa. Whatever the viewer does see is the figure; whatever he does not see is the ground.

McLuhan's point was that most people are trained not to look for the ground in any situation. They focus on one part and ignore the rest. If people consider the motorcar, for example, they focus on the car itself, rarely perceiving the network of gas stations, highways, neon signs, parking lots, and all the altered habits and perceptions that arise out of the existence of the car — the ground, in other words, of the automobile. True perception, according to McLuhan, is the ability to hold both figure and ground in one's attention, in a dynamic and resonating relationship.

True to his "dichotomania," McLuhan clustered a whole group of characteristics and associations around one or the other of these two terms. "Figure" was visual, conceptual, the ascribed cause of a thing. "Ground" was acoustic, perceptual, the perceived effect of a thing. Ground was a medium as McLuhan had always studied a medium — when he had noted the relationship between, say, print and nationalism. It was the environment, it was the source of all real change. Primitive man had possessed a keen, almost instinctive awareness of ground, but literate man had concerned himself almost wholly with figure. Now McLuhan was concerned to retrieve some of that primitive awareness. One way of encouraging awareness of the ground of a thing, instead of merely focusing on the thing itself, was to imagine what it would be like to be suddenly deprived of that thing. Or one could take a thing and put it in a different ground — imagine a car in the nave of a church.[27]

McLuhan first began to meditate fully on this dynamic while working on *From Cliché to Archetype* and considering the ways in which clichés flipped into archetypes and vice versa. He then altered the meaning of these same terms slightly, dubbed them "figure" and "ground" in *Take Today*, and used the new terms with increasing frequency from about 1974 onward. Once he hit on an idea like this, he could not let it go.

It was much the same with still another insight he promoted in the seventies, the idea that North Americans, virtually alone among the peoples of the world, sought privacy and the feeling of being alone outside their homes, while using their homes for socializing — the opposite of the European pattern. This idea came to him in the summer of 1971 when he spoke to a group of advertising executives in Great Britain and pondered the question of why Americans, unlike Britons or Europeans, disliked advertising in the movie theaters. The answer that came to him was precisely the existence of this habit of going out to be alone. Henceforth he used the idea as a key explanation for almost every

North American social phenomenon, from decreasing attendance at Mass to the American resistance to public transit.[28]

McLuhan's infamous tendency to repeat himself was even more marked, in the middle and late seventies, by the constant appearance in his correspondence and conversation of these particular phrases and ideas. A casual talk with him about almost anything might be pretext for a complicated argument involving figure and ground or left brain and right brain or North Americans going out to be alone. So obsessive and cranky was he about these few notions that it was almost as if he took pleasure, as Kinsbourne would say, in not taking note of alternative explanations.

The mind was still powerful, in short, but its marvelous acrobatic freedom of movement continued to be worn down by age and ill health, the effects of which were not mitigated by the McLuhans' hectic social life. Corinne was constantly on duty as hostess, and a dinner spent alone was a rare and cherished occasion for the couple. Moreover, the few moments of peace and quiet left over from work and socializing were often absorbed by worries over their children.

In 1973 Eric, after earning a B.Sc. in communication from Wisconsin State University, had become his father's full-time assistant, helping with the twentieth-century Baedeker and the laws of the media, assisting with McLuhan's university chores, and sitting in on seminars. In time, he became his father's most trusted surrogate, and McLuhan, for all his occasional rudeness toward, and impatience with, his eldest child, did genuinely feel that he contributed greatly to his work through their joint "dialogues."[29] At the same time, he was very aware that Eric was in the midst of a hostile environment, as it were, at the University of Toronto, in particular, and the world of academia generally. He could not help but worry what would happen to him after he was gone.

In the spring of 1974 McLuhan had the pleasure of seeing Eric wed in Hereford, England, to a young woman from that country who had been working in Canada, Sabina Ellis. (It was the only first marriage among McLuhan's children that proved durable.) After the ceremony McLuhan went to Paris, where he was interviewed on the occasion of the French publication of *From Cliché to Archetype*, and from there he traveled to a speaking engagement in Stockholm. He found the dull, provincial atmosphere of that Scandinavian city, where he had lunch with the king of Sweden, pleasingly reminiscent of Toronto.[30]

McLuhan's daughter Elizabeth was also wed in 1974, to a young man from Toronto named Sean O'Sullivan. McLuhan's joy on the occasion, however, was somewhat tempered by the fact that Elizabeth and Sean had been living together before the wedding.[31] This was one trend of the acoustic age he had no use for.

That same year he was closely involved with the troubles of another daughter. Mary had been the first child to marry — she settled in California — and had provided McLuhan with his first grandchild, Jennifer, born in 1970. McLuhan had been immensely pleased with attaining the status of grandfather; it had come, he thought, none too quickly.[32] In 1973, however, Mary separated from her husband and for more than a year afterward McLuhan was preoccupied with trying, long-distance, to help her negotiate financial and legal problems arising out of her divorce. He did all he could to buck up her morale and to advise her on career moves. Finally, in near desperation, he asked a friend in California if he might try to persuade her to come back to Toronto, with her prized Mercedes in tow if necessary.[33]

Mary stayed in California, however, phoning her parents from time to time with news of her plight. At one point McLuhan wrote her nearly every day — he felt that of all his children Mary was the least able to cope with the hard blows of life. Unfortunately, his paternal charity aroused the jealousy of his other daughters. It was almost as if they were still smarting from the days when their father, weary and distracted, had acted as if he did not know who these girls were. Now, it seemed, he was paying for this neglect. In 1975 he noted with sadness in his diary that his daughters were more trouble to him as adults than they had ever been as children.[34]

Part of the problem was that the daughters were not particularly friendly with each other. Mary's twin sister, Teri, once commented, "Rivalry is an innate part of twinship. It leads to a great urge to assert your own individuality."[35] She might have been speaking for all of her siblings, even the non-twins. In her own pursuit of individuality, she worked hard to save enough money to attend an out-of-town university (she went to the University of Ottawa instead of taking the expected course and attending the University of Toronto), taught English in Paris after university, and published an anthology of speeches and writings of the North American Indian, entitled *Touch the Earth* in 1971. The book, full of the haunting pictures of the great nineteenth-century photographer Edward Curtis, was a critical and commercial success.

The most talented as well as independent of McLuhan's children, Teri adopted her father's strategy of studying things she found particularly oppressive. In her case, it was twinship. In 1980, partly as a result of her long study of the subject, Teri made a dramatic feature film involving twins, called *The Third Walker*. Shot in Nova Scotia, the film never received wide distribution.

Her two younger sisters, Stephanie and Elizabeth, also engaged in highly creative pursuits. Stephanie made television documentaries; Elizabeth wrote poetry and became an authority on native Indian art. Despite their individual accomplishments, the spirit of rivalry persisted among the daughters, combined with a feeling of living perpetually under the shadow of their father. McLuhan told a friend that one of his daughters had complained to him about the difficulty of finding anyone to marry: who, she had asked, could compare with her father? Recognizing the problem, McLuhan hoped they would marry men of no intellectual pretensions.[36] In fact, the two daughters who married did choose husbands of no intellectual pretensions: unfortunately, this did not save their marriages.

The only child who seemed not to be in awe of his father and who felt no need to win his approval was the youngest, Michael. McLuhan watched the vagaries of this child's career, which included living in a commune and working as a chef in the commune's restaurant, with a mixture of bemusement and alarm. He lamented the fact that none of his children seemed happy. Their frustrations puzzled the man who considered that he himself had always lived a very happy life.[37]

As the years went by, attendance at McLuhan's regular Monday night seminars gradually tapered off, until he was reduced, in 1978, to asking his administrative assistant to try to drum up interest among students and professors. About this time, McLuhan appeared at the center for the first day of class of his graduate course in English and was unpleasantly surprised to discover that only six students had signed up for it.[38]

Occasionally he still worried that some of the students at the center, and even some of the people he worked with, were spies for his old enemies in the university or for the government of Ontario, whose office buildings were within sight of the coach house. He never revealed these fears to any except his closest friends. To those who came to the seminars simply out of curiosity, McLuhan usually appeared as an amiable, aging professor with a still dazzling verbal output and a habit of scribbling notes to himself, from his reading or conversation, on note pads or envelopes, which he would then stuff in his jacket pocket for further reference. A reporter who visited the center in 1976 noted, "If he has an exciting idea or interesting joke in mind he'll try it out on anyone around. He expects witty feedback and when he doesn't get it, he wanders off absent-mindedly with a vague look of annoyance."[39]

As the years went by, the "percepts" McLuhan generated in his seminars were more and more confined to certain standbys. The flow of his talk, while remaining a source of wonder to most visitors, was far from the pyrotechnic display of ideas generated by the McLuhan of the forties and fifties. Even the language of his ideas varied surprisingly little. Moreover, as technology changed, McLuhan did not appear interested in exploring the implications of the latest advances — not even that great technological wonder of the seventies, the microchip computer. For him, electronic technology was largely broadcasting

— instantaneous, global communications. He did ingest some information about computers from people like Barry Nevitt and retained a certain curiosity about their functions, but it was a limited curiosity at best. Computers were interesting to him as means for the storage and retrieval of information, as part of the happy hunting ground of discarnate man, but they were circumscribed by their binary, yes or no, mode of operation — what McLuhan called their "two-bit wit." Subtleties such as yes *and* no were beyond them, as well as the generation of any authentic percepts, for which one still had to look to artists.[40]

Of course, there were still occasions, in the Monday night seminar and elsewhere, when McLuhan would meet some fascinating individual and suddenly shut off his own flow of talk and listen, happily absorbing something truly new. One Monday night in the mid-seventies, he noticed a small, rather mousy-looking individual sitting in the seminar, with a pipe and a suit almost as dingy and academic-looking as his own. It was the third week the man had made an appearance. No one else had paid him the slightest attention. McLuhan, who disliked people sitting in his seminar week after week without saying a word, interrupted his talk to address the man. "You, sir, what are your perceptions on this matter? Incidentally, who are you?" ("Incidentally" was McLuhan's favorite word in conversation.) The man introduced himself as an anthropologist who was a visiting professor at the University of Toronto. McLuhan pulled his chin in and resumed his monologue.

The next week McLuhan once again noticed the little man with the pipe and once again interrupted his talk. "You, sir, you haven't said anything. Incidentally, who are you?" The man once again introduced himself. "I'm a visiting professor in the anthropology department." "Where do you hail from?" "France." "Oh yes," McLuhan murmured. "Lévi-Strauss." "As a matter of fact," the man replied, "I was his student."

McLuhan suddenly became very alert. A smile appeared. He began to ask the man questions about working with Lévi-Strauss, and that was the end of that seminar as far as the rest of the participants were concerned.

The next week the seminar was held at the home of one of the regulars, and the man with the pipe was again in attendance. As soon as McLuhan arrived, at 7:30, he picked up where he had left off the previous week, continuing the barrage of questions. How had Lévi-Strauss interacted with the natives he had studied in Brazil? How had Lévi-Strauss approached the question of sensory preferences among them? The man quietly answered all of McLuhan's questions. Moreover, he had a seemingly endless supply of fascinating anecdotes about Lévi-Strauss. McLuhan could not get enough of it.

At 9:30 Eric interrupted. "Dad, it's probably time for a break." (The seminar was supposed to end at 10:00.) McLuhan ignored him and fired off another question. At 10:00, Eric again spoke up. "I think it's really time for a break." Again, McLuhan seemed oblivious. The host of the seminar cast a look of despair at the table he had spread with cheese and crackers, clusters of grapes, and bottles of wine. McLuhan kept asking questions, and Lévi-Strauss's pupil kept answering them. At 11:00, Eric made a considerable show of placing an alarm clock directly in front of his father. McLuhan glanced up. "Well, it must be time for a break," he said. Later, he rode home in the car with the visiting professor and continued the interrogation. Even when the car stopped at the man's residence and he got out, McLuhan kept up the barrage through an open window until Eric, unable to stand it any longer, drove away in somebody's mid-sentence.

Few people were capable of feeding McLuhan's appetite for anecdotes, odd little facts, and so on in this manner. For the most part, he seemed curiously alone in his center, even though he was constantly writing to people, the famous and the unknown, forever hatching projects and schemes with anyone who seemed a remotely likely candidate for collaboration. It was not enough that he had a half dozen or so unfinished books in the works; he had to be thinking of more and more things to do: how to launch media centers in other countries, how to advise Nixon, how to advise Trudeau, how to rewrite *Take Today*, how to link media studies with psychology, with management consultants, with the problem of unemployment. Toward the end of the decade, McLuhan was even thinking about how to invent a television set that would not override the left brain so thoroughly.

Amid this vortex of hypothetical activity, McLuhan became increasingly disconnected from his old clerical colleagues at St. Michael's College. He still ate lunch in the faculty dining room, but he began to grow annoyed with what he considered petty sniping from the gentlemen in the reversed collars. He wrote one sympathetic Catholic philosopher that he had a feeling of being "among a group of spoiled priests" in a "rather stale and sick atmosphere"; the Basilians could not seem to comprehend a word of what he said.[41]

Especially galling were some of the Basilians in the philosophy department, the Thomists. McLuhan felt that, with one or two exceptions, they almost willfully misunderstood the object of their studies, St. Thomas Aquinas, and thereby hindered McLuhan from enlisting Aquinas in the ranks of the proto-McLuhanites. If Aquinas were living today, McLuhan declared, he would certainly not be a Thomist. To establish his point, he frequently dug up items from the *Summa Theologica* that, to his mind, demonstrated conclusively that he and St. Thomas were on the same wavelength.[42]

His theory of communications, McLuhan now insisted, was "Thomistic to the core." One support for this idea came from a book by his closest friend among the Thomists, Father Joseph Owens. "The cognitive agent itself," Owens wrote, "becomes and is the thing known." This confirmed McLuhan's idea that the "content" of any medium was the user of that medium.[43]

This formulation had long since replaced his old dictum, expressed in *Understanding Media*, that the content of a medium was another medium (for instance, the content of television was film). Ultimately, his new formulation owed less to Aquinas than to his old Cambridge mentors William Empson and I.A. Richards, who had first introduced McLuhan to the notion that poetry communicated its meaning by arranging impulses inside the reader's head. It also led directly to McLuhan's dictum that the "cause" of works of art and philosophy was not so much those who created the works but the audience or public that received them, another lesson derived from his Cambridge years, when he had read Q.D. Leavis's *Fiction and the Reading Public*.

Owens himself was not exactly ready to ratify McLuhan's use of Aquinas or of his own phrase — further evidence to McLuhan that his colleagues in the philosophy department were obtuse and obstructive. He was obliged to point out to them the significance of what they themselves had uncovered.[44]

But if McLuhan was less and less patient with his old friends at the college, he unexpectedly gained a new audience of fellow academics in 1975. That year an assistant dean in the university named Robert Painter invited the heads of the university's interdisciplinary centers and institutes to dinner for an evening of conversation. The evening proved so convivial that one participant suggested that the group hold regular meetings, dubbing it "the best club in town." Thus the Best Club was formed.

McLuhan rarely missed a meeting. In response to his jokes, which he brought on scraps of paper, and his prefab interpretations, he got arguments of course, but generally his ideas were greeted in a spirit of affection and tolerance that he had seldom encountered elsewhere in the university. Perhaps there was some special fellow feeling among these heads of institutes, all of whom were basically entrepreneurs trying to keep their little enclaves going by fishing for grants. In any case, late in McLuhan's career, this club was the closest thing to his adolescent dream of a fine circle of intellects.

Among themselves, the members of the Best Club engaged in the old debate: was McLuhan a phony or was he for real? They could not decide. For them, as for many of their colleagues, McLuhan was ultimately an enigma. But his extraordinary success beyond the confines of the university and even of his native country had to mean something. No other member of the Best Club had been asked to write articles for the editorial pages of the *New York Times* on the subject of Watergate or inflation.

Watergate was the kind of topic on which McLuhan could be at his best. He observed it with the contented curiosity of a man who had no moral investment in Nixon's future. He saw the scandal, in fact, as a conflict between the new acoustic technology and the old print-oriented technology. The technology of bugging and tape recording had amplified the oral, secretive side of the Nixon administration, which looked very sinister when translated into printed transcripts. Nixon, with his lawyer mentality, his commitment to specific goals and points of view, was out of his depth in the new acoustic technology. His desperate gesture, in a television broadcast, of pointing to the stack of printed transcripts was like Poe's sailor in the maelstrom trying to save himself by clinging to the sinking ship instead of studying the action of the waters.[45]

McLuhan had presciently observed in 1973 that the old secrecy and confidentiality of political confederates hatching plots was impossible under the conditions of the electronic age, in which information circulated at the speed of light, out of anyone's control. Everything was in the open. Discarnate man — his personal, private morality obliterated, his involvement in the political drama surrounding him total — was absorbed by what McLuhan in one article termed a new "austere public ethic."[46] (The austerity of this new ethic, as exacting as the specifications for a television talk show host, would bedevil another presiden, Bill Clinton, twenty-five years later.)

Privately, McLuhan had other theories about Watergate. He was certain that it involved internecine conflict between the CIA and the FBI, with the KGB reaping the rewards. Cynicism was not in his nature, but McLuhan had known since the forties that politics was a branch of show business. Now he saw in Watergate an almost ritual degradation of tribal leaders mounted for the delight of television viewers.[47]

If McLuhan observed Watergate with neither moral dismay nor gratification, he also refused to wring his hands and join in many of the high-minded crusades contemporaneous with the scandal. He had no use for the warnings of imminent doom from the Club of Rome, that informal assemblage of leading industrialists, academics, and futurists. When he met its most prominent spokesman, Aurelio Peccei, in February 1973, he told Peccei that his gloomy predictions of global shortages in oil, food, and other commodities were eagerly absorbed by the media because the media needed a lot of bad news to help sell its good news — its advertising. People did not mind bad news, McLuhan believed. Global shortages in oil merely consoled them for not being able to afford that new BMW. Genuinely good news, on the other hand, was upsetting because it almost always threatened people with change. Peccei was not thrilled by these observations.[48]

McLuhan's attempts to participate in ventures that did seem to support the common good were no more fortunate. Serving as a consultant for the Pontifical

Commission on Communications in 1973, for instance, proved to be an exercise in futility. Instead of setting an example of communicating in the electronic age by using, say, telephone conferences among the members, as McLuhan suggested, the commission laboriously circulated documents, briefs, and so on.[49] Not a single member of the commission, McLuhan commented, understood a word of what he had to say. But, as he had remarked of another commission (the Presidential Commission on the Causes and Preventions of Violence established in 1968 by Lyndon Johnson) years previously, such bodies were "deeply committed to avoiding any understanding" of the issues they were set up to explore.[50]

In the spring of 1975 McLuhan began a sabbatical year by spending a month at the University of Dallas, a small Catholic college in a city where McLuhan had a few close friends. McLuhan was favorably impressed by the students there, attributing their acuity partly to the small size of the institution, which he believed encouraged dialogue. Eric McLuhan had chosen to spend his next two years at this university obtaining his M.A. in English. While he was there, he received a steady stream of advice from his father, advice that echoed Erasmus's superbly knowing letter to the young man summoned to the court of a Renaissance prince. McLuhan reminded his son that occupants of a university did not love each other. Eric must never say anything without being aware that it would be repeated and would be embellished in the retelling. Therefore, only honeyed remarks about others should pass his lips. Plainly speaking from experience, McLuhan added that anyone who was self-assured was always in danger of stepping on other people's toes; most people "are naturally diplomatic from sheer timidity and uncertainty."[51]

After his return from Dallas, McLuhan launched into the projects he had planned to undertake while on sabbatical. These included the revision of *Understanding Media*, which he and Eric had been taking stabs at during the previous two summers. The revised version would be the beneficiary of his recently evolved theory of tetrads; it would also take cognizance of such new and trendy academic products as structural linguistics and semiology. In the fall of 1975 he was introduced to a young woman named Nina Sutton, who offered to interview him over an extended period and then write up the interviews — in French, which was her first language — as the substance of this revised version. The resulting manuscript would be published, she assured McLuhan, by Éditions Stock of Paris.

McLuhan was persuaded to accept this arrangement, since it seemed to offer a convenient way of getting the book written. That fall he spent more than a month dictating. Although he was exhausted, at the end of it he had at least thirty-nine ninety-minute tapes, which amounted to about six or seven hundred

pages of transcription, to show for his efforts. Sutton took the tapes and transcripts with her to Paris, and that was the last McLuhan ever saw of them. Two
years later, he received a phone call from her in France. She informed McLuhan
that she had taken on a job in journalism and had given up her attempts to complete the manuscript and publish the book.[52]

Meanwhile, McLuhan kept telling himself that during his sabbatical year he
would totally revise his old doctoral thesis for publication. As usual, however, it
was easier to continue thinking of new projects. In the spring of 1976 he saw *A
Chorus Line*, which revived his dream of writing a musical comedy, specifically
a musical comedy about media. He thought the format of *A Chorus Line*, in
which dancers auditioning for parts in a musical more or less gave the audience
their life stories, might do very well for his media play: the various media would
take the role of the dancers and explain themselves to the audience. Television,
for example, could explain how it nearly destroyed baseball and then revived it
through the technique of instant replay.[53]

McLuhan appeared briefly in public in his own role of explainer of the
media when he testified in a Toronto courtroom in June 1976 as a witness for
the prosecution against *Show Me*, a sex education book whose publishers were
on trial for distribution of obscene materials. The book had provoked court
action by featuring nude photographs of small children. McLuhan explained to
the judge that the photographs represented an isolation and a heightening, in a
fragmented, specialist manner, of certain aspects of the human body. He tried
to clarify his point by talking about figure and ground, with the photographs
representing figures (bodies) without grounds (personalities). He succeeded in
thoroughly baffling the court. He could hardly have done otherwise. McLuhan
detested, in every fiber of his being, a book like *Show Me*; true to his form, however, he attacked it not on moral grounds but on metaphysical grounds. He realized, accurately, that this approach was the only possible way to get people to
look at such material in a fresh light. It was a pity that his metaphysics were so
impenetrable to most people.[54]

The chief event of McLuhan's sabbatical year took place shortly after this
trial. In July he received a phone call from filmmaker Woody Allen asking him
to play a cameo role, as himself, in Allen's new movie, *Annie Hall*. McLuhan
protested that he was no actor — an uncharacteristically modest reply. Allen
told him that the part required somebody possessing what one might call a
slightly...idiosyncratic temperament. Was there anyone, in that regard, more
suitable than McLuhan? All he had to do, Allen assured him, was play himself,
perhaps use one of his favorite one-liners. McLuhan agreed.

In fact, he did not need much persuading. Robert Logan was with McLuhan
when he spoke to Allen and recalls, "He came back from the conversation like a

kid who had just raided a candy store."[55] Allen's offer appealed both to the playful side of McLuhan's character and to his need to be taken seriously. Two or
three weeks later, dressed in his faithful reddish-brown tweed suit, McLuhan
was in New York to do the scene, which involved Allen, in his character as Alvy
Singer, hauling McLuhan out from behind a signboard to deflate a man in line at
a movie theater who had been pontificating about McLuhan's work.

McLuhan wanted to include in this scene his favorite put-down for hecklers:
"You think my fallacy is all wrong?" The phrase, he once advised Trudeau,
should be uttered "with a certain amount of poignancy and mock deliberation."[56] Allen, however, did not seem to find the phrase as scintillating as
McLuhan did. McLuhan was determined to keep it in, and the two men had
a sharp exchange. A few days after the scene was shot, McLuhan was called
back to redo the dialogue. Although the dubbing was probably required because
of poor sound levels at the outdoor location where the scene was filmed,
McLuhan concluded that the exercise was the result of Allen's pique. He was
miffed, McLuhan thought, because the line was too funny and stole the show
from him.[57]

The episode was fatiguing, and the morning McLuhan returned, Scott
Taylor, then doing an M.A. thesis under his supervision, knew there was something wrong with McLuhan the moment he walked into the center. He kept rub-
bing the back of his neck as if the muscles in his neck and shoulder were giving
him pain. At the same time he peered around the room, apparently looking for
something. Then he shook his head gently.

Taylor handed him part of a manuscript, which, he explained, was the first
few chapters of his thesis. McLuhan looked blank. He asked Taylor what the
topic of the thesis was. Taylor told him. McLuhan said, "Oh...who is your
supervisor?" Taylor, reaching for an answer that would not embarrass McLuhan,
said, in as light a tone as he could manage, "As far as I know, it's some guy called
Marshall McLuhan. Have you heard of him?" A pained smile appeared on
McLuhan's face; suddenly he seemed to focus again. "How long has he been
your supervisor?" he asked Taylor. Then he looked at the thesis, as if it were an
entirely unknown object, and said, with a sad sigh of recognition, "Well, then,
I guess I'll have to read it."[58]

The minor stroke McLuhan suffered that morning in 1976 may or may not
have been caused by the strain of filming *Annie Hall*. On more than one occasion during that year he had suddenly lost his memory and his ability to focus
on what was in front of him.[59] He always tried to work through these episodes.
The morning he returned from New York was no exception. Ignoring what
happened, he amused his friends with the tale of the sixteen or seventeen takes.

When the movie was released, McLuhan found it not altogether to his taste, if for no other reason than its explicit emphasis on the hero's sex life. But the movie did prove to be a good PR move for McLuhan, fixing an image of him in viewers' minds to a surprisingly lasting degree.

A few weeks after filming the movie, McLuhan received public exposure of a different sort when he appeared on NBC-TV with Tom Brokaw and Edwin Newman to discuss the first Carter-Ford debate. McLuhan watched that debate carefully, noting the body language of both participants, and observed, with some amusement, that moderator Frank Reynolds projected far more authority than either of the two candidates. He was also intrigued by a technical breakdown in the television studio, which resulted in the two men standing side by side, stiffly and silently, for several minutes. As far as he was concerned, this technical breakdown was the most exciting part of the debate, allowing for some unexpected drama and audience participation in an otherwise stilted format.[60]

When Carter won the election, McLuhan was delighted. He noted with pleasure that Carter was a religious man, and his southern roots had a sentimental appeal to both McLuhan and Corinne. All of McLuhan's old fondness for the cavalier, rhetorically inclined South surfaced again (although now it was cast in different terms: the South was acoustic and right brain, whereas the North was visual and left brain). McLuhan also noted, with considerable justification, that Carter's election was a major step toward the election of a black president. He sent Carter a congratulatory telegram and no doubt hoped to obtain his ear the way he had obtained Trudeau's. Carter, however, evinced no interest in making contact with McLuhan.[61]

McLuhan still loved the idea of influencing the thinking of the world's great minds; he was thrilled when André Malraux sent him a copy of his latest volume, *La tête d'obsidienne*, in 1974, with a flattering inscription. It was loss of potential influence, as much as any merely personal vanity, that made McLuhan regret the passing of fame. At times he professed actually not to mind its passing at all. All the "ballyhoo and notoriety," he claimed, interfered with his work. "What's worse," he told one reporter in the mid-seventies, "all the publicity never helped in getting people to understand what I say."[62]

Regardless of how he viewed fame in the abstract, McLuhan still complained of bad press, especially from local papers like the *Toronto Star* and the *Globe and Mail*. But no local paper handled him as roughly as the Australian paper that began an article covering his arrival on that continent in June 1977: "Media mystagogue Marshall McLuhan, wearing reading glasses to hide a black eye and claiming the clicking sounds of the press cameras were forcibly punctuating his talking, said yesterday unemployment did not exist."[63]

However much he may have smarted from such comment, McLuhan never resorted to blaming the press for his fading appeal. At one point in the seventies he suggested that it might be due to the hasty retreat of the TV-era kids of the 1960s to law school and corporate employment. No longer were they outraging their elders by barricading the dean in his office and quoting McLuhan.[64] There was some degree of truth in this analysis. Whatever the explanation, however, it is certain that McLuhan made no real effort to perpetuate his insights or to ensure that the work of the center would carry on after him — even though he feared for the fate of a society careless about the effects of technology.

His refusal to really settle down in the mid to late seventies to work on his books was an indication of something approaching unworldliness in his character, at least in regard to his own reputation. As one associate put it, McLuhan was more like "an adult playing in a garden of delightful ideas" than a man making a serious bid for immortality by preparing a body of work that would survive him. When his obsolescence came, he was unprepared for it.[65]

That insouciance was part of his character, but now he was, for the first time, hindered by increasing lapses in his ability to prepare a face to meet the public. He still addressed various audiences, but as the seventies progressed he was unable to put his audiences on with the deftness of previous years. Canadian author Peter Newman recalls an address McLuhan delivered to a group of advertising people in the late seventies. Almost compulsively, he spewed out one joke after another, whole strings of amusing one-liners, until his listeners lost their willingness to take him seriously on any level. "The audience was laughing as much at him as with him," Newman recalls, "and I don't think he even knew."[66]

At a talk at New York University when he received NYU's Creative Leadership in Education Award in 1977, McLuhan went to the opposite extreme. He trotted out his brain diagram and plunged into one of his more incomprehensible and rote performances. Neil Postman, who was in the audience, recalls, "He talked and he talked for about an hour without ever making reference to this diagram. The audience began to get really annoyed. I remember one man saying, 'Oh my God, he's talked for an hour and he hasn't even gotten to his chart yet.'"[67]

The reality was that McLuhan's physical, if not mental, powers were wearing thin. The body that he had pushed so mercilessly throughout his life was beginning, finally, to fight back. In October 1976, after a severe bout of the flu, he experienced a heart attack, which he usually referred to afterward as a "wee heart attack" but which was severe enough to require hospitalization in the cardiac care unit for almost a week. Corinne solaced herself with the thought that at least he was getting some rest in the hospital. (Even when suffering chills and fever from the flu, McLuhan hated to slow down. Friends were perpetually trying to get him to take it easy, to lie down on the couch from time to time. When he did lie down, McLuhan would pick up a book and read, or he would lecture and discourse nonstop if anyone else was present.)[68]

Back on his feet a few weeks after his heart attack, McLuhan told everyone he had been temporarily sidelined by the flu (no sense in rousing the blood lust of his enemies by advertising the heart attack) and resumed his teaching and "writing." He worked with Logan on their book on libraries, in which McLuhan articulated one of his more recent insights: whereas Gutenberg's invention made everybody a reader, the invention of the photocopy machine made everybody a publisher. (This idea would, of course, acquire added force with the advent of the personal computer.) McLuhan and Logan also wrote an essay, entitled "Alphabet: Mother of Invention," on the effects of the phonetic alphabet on ancient and modern science, sent off an excerpt from their library book to *Harper's*, which rejected it, and concocted a scheme for a television show, to be launched on the CBC, in which McLuhan and Logan would analyze the television images of various Canadian politicians and predict their political futures. (This last idea was prompted by the introduction of televised sessions of the House of Commons.) The CBC was unreceptive.

Meanwhile McLuhan kept generating more ideas for books, as if the mere multiplication of ideas might ensure that at least one of them would become a reality. In February 1977, Harper & Row offered him and Edward Wakin, co-author of the book *The De-Romanization of the Catholic Church*, a contract for a book the two had proposed on Catholicism. At the same time, McLuhan was planning a book with his old friend Lou Forsdale of Columbia Teachers College called *Art as Survival in the Electric Age*. McLuhan never actually worked on this book, but from time to time he inspected Forsdale's work in blocking out the various parts of the book, selecting illustrations, and so on.

Later in 1977, McLuhan once again attempted to play the role of intellectual guide to the powerful — this time to California governor Jerry Brown. An ex-Jesuit seminarian with a Zen Buddhist orientation, Brown seemed just the man to take the Guru of the Electric Age seriously. In July, McLuhan flew to Sacramento and spent three and a half hours with Brown and his cabinet, discoursing on satellite information systems, discarnate man, the dyslexic television child, Christianity, and the twin hemispheres of the brain. As it turned out, McLuhan felt no rapport whatsoever with Brown, who barely got a word in edgewise, but he was happy to take on the part, contrary to his popular reputation, of the champion of literacy. After the meeting he wrote Brown to suggest a rationing of television for the population in general. Brown's assistant superintendent of education replied to McLuhan, gently suggesting that such rationing was somewhat unrealistic, politically speaking.[69] McLuhan was hurt that his suggestion was considered impractical, for he prided himself on being clued in to the secrets of managing affairs, public and private: he was no ineffectual academic. Like Ezra Pound, hoping that once he got the ear of Stalin or Roosevelt he could make him see the absolute necessity of saving the world through monetary reform, McLuhan hoped that he might yet wake up a politician and help avert disaster for discarnate man.

In 1977 he did make a further contribution to countering the effects of electronic technology by publishing a textbook called *City as Classroom*, which he had completed with Eric and a schoolteacher named Kathryn Hutchon the previous fall. This book, the last McLuhan book published in his lifetime, brought his career full circle, harking back to the work of F.R. Leavis and Denys Thompson, *Culture and Environment*. That work, which he had seized upon as a young college instructor fresh out of Cambridge, had been an attempt to enlist the school system as a prophylactic against the effects of advertisements and best sellers. *City as Classroom* was an attempt to moderate the effects of electronic living by sharpening the perceptions of high school students concerning the world, or the city, they lived in. McLuhan wanted to turn high school students into fellow sleuths, probing and exploring their surroundings for clues to the nature of the times they lived in, seeing worlds of significance in street lamps and automobiles.

The completion of the work was a severe trial to his patience, and it would doubtless have ended like all his other unfinished projects had it not been for the determination of Eric — who had been working on the project since the mid-sixties — and Hutchon to see it published. It began with a discussion of McLuhan's figure/ground approach to analysis and proceeded to treat, in the manner of *Understanding Media*, various specific media such as money, clocks, television, and so on. The text was largely in the form of questions and class projects for students; these questions represented a grab bag of McLuhan's insights from the last twenty years or so.

Again, McLuhan groaned under the demands of his publishers. Much rewriting had to be done to make the text reasonably accessible to high school students; as Lou Forsdale had noted long ago, McLuhan had highly inflated views of the capabilities of students. He was forced to eliminate actual examples of advertisements and to make the tone of the book less lighthearted than he originally intended. As in the case of his first published book, so with his last published book he nurtured suspicions that the publisher was covertly antagonistic to the whole project.

McLuhan hoped the book might be an answer to radical educators like Ivan Illich, who McLuhan believed had purloined almost all of his insights.[71] Instead of junking the schools, as Illich advocated, McLuhan believed the schools might still be useful for examining the *real* education children were receiving from the electronic media. Unfortunately, the book, once published, disappeared from view, a fate that seemed to await all of McLuhan's books published in the seventies. It had no effect whatsoever on school systems. A few years after its publication, few educators had even heard of the book, even when it was republished in the United States in 1980 under the title *Media, Messages & Language: The World as Your Classroom*. Doubtless, *City as Classroom* was too patently a bouquet from McLuhan's garden of ideas and would not have fit in easily with existing course structures (exactly the point of the book, McLuhan might have added). Nevertheless, it certainly deserved more attention from teachers, especially those eager to show their more alienated and restless students that schooling could at least sharpen their awareness of their daily lives.

The failure of the book saddened McLuhan but did not abate for a moment his willingness to continue with new projects and new collaborators. One of the more interesting of these new collaborators was Gerald Goldhaber, a professor in the communication department of the State University of New York at Buffalo. McLuhan met Goldhaber at a conference on communications in Mexico in October 1975. Before the year was out, Goldhaber was addressing a Monday night seminar. The night that he spoke, Goldhaber recalls, McLuhan was particularly interested to determine if he was speaking from the left or the right side of the brain.

For the next few years, Goldhaber and McLuhan maintained a relationship more or less confined to the exchange of ideas. Then, in 1978, Goldhaber, who had his own polling and market research company, met a political consultant in Nevada named Don Williams. Goldhaber had been hired as a consultant in a political campaign in that state, and he was bowled over by some of the television advertisements that Williams, working for the other side, had mounted. He discovered that the inspiration for these stunningly effective political ads was none other than the work of Professor H. Marshall McLuhan.

Goldhaber subsequently set up a meeting of Williams, McLuhan, and himself in McLuhan's home in Wychwood Park. The result was a new company — McLuhan, Goldhaber and Williams — whose business was "communications research," particularly in the area of politics and entertainment, Goldhaber's specialty. Goldhaber and Williams underwrote the company, gave some stock to McLuhan, and promptly set up offices in Buffalo and Las Vegas. McLuhan's part was to supply contacts — people he had met in film and television — and, of course, ideas.

Foremost among those ideas was the left brain/right brain dichotomy. In response to this notion, Goldhaber came up with a technique of polling people that would register both their left brain responses — what they said they thought about certain issues — and their right brain responses — their emotional, instinctive responses to the issues. "I'm not going to reveal our trade secret," Goldhaber comments, "but this technique was based directly on conversations with McLuhan."[72]

Another project Goldhaber pursued under McLuhan's guidance was a proposed study for the Toronto Transit Commission to determine whether redesigned bus and subway seats would attract more passengers. This project was inspired by McLuhan's cherished notion that North Americans go out to be alone. McLuhan told Goldhaber that he had once passed on this observation to Henry Ford II, who seemed wholly incurious about it. The next thing McLuhan knew, the Ford Motor Company was running television ads showing a man leaving a house full of people screaming at each other, getting into his car, rolling up his windows, and driving off with a look of ineffable contentment on his face. Obviously, McLuhan told Goldhaber, his idea had been ripped off. This time he wanted due credit — and some money — for his insight.[73]

In June 1979, Goldhaber and Williams lured McLuhan to Las Vegas, where he spoke to a convention of cable television operators. The occasion was memorable not so much for McLuhan's speech or for any business transacted at the convention — McLuhan's involvement with the small firm remained very slight in the year or so between its launching and his last, incapacitating illness — but for a rare display of tipsiness on McLuhan's part. He exceeded his usual quota of alcohol — one drink — before dinner upon hearing the dismaying news of Trudeau's defeat in the Canadian federal elections.[74]

The previous year McLuhan had worked with some of Trudeau's aides, including Jim Coutts, Trudeau's principal political adviser, in the 1979 throne speech, the statement of government policy initiatives read before the Canadian Parliament by the Queen's representative, the governor general of Canada. The 1979 speech was basically Trudeau's campaign platform. McLuhan had tried to convince Coutts (who was dating McLuhan's daughter Teri at the time) that the speech should not list the usual government intentions but should create the "emotion of multitude." This was a favorite theme of McLuhan's, derived from his study of the epyllion (the little epic) and an essay by Yeats in which the poet discussed the use of plot and subplot by Shakespeare in *King Lear*. The reverberating and suggestive parallels between plot and subplot, Yeats maintained, created, almost as if by magic, a sense of universality — the "emotion of multitude." Such emotion drew the spectator ever more deeply into the world of the play. Using this idea, McLuhan recommended that the throne speech should not be a connected progression of items but should somehow juxtapose those items, almost at random, to create all sorts of implicit parallels and suggestions of swarming multitude. In other words, the speech should sound very much like an article by McLuhan. Coutts listened politely, but the final product had very little to do with McLuhan's approach.

The last time McLuhan ever advised Trudeau was September 1979, when Trudeau, then leader of the opposition in the House of Commons, sported a beard. The beard, McLuhan wrote him, "has cooled your image many degrees!"[75] (in other words, it made Trudeau more of an enigmatic, playful, absorbing figure). McLuhan suggested that such an effect was not desirable at this point in Trudeau's career. Trudeau shaved off his beard (reportedly because he disliked the fact that it was almost entirely gray), and McLuhan lived to see him win still another election, in 1980.

McLuhan's never-ending search for collaborators found him working in 1979 with Bruce Powers, a professor of communications for Niagara University. Powers was given the title Associate for Educational Research at the center in June 1979, with the specific charge of helping McLuhan develop instructional materials on the media. Powers was knowledgeable about new information technologies such as fiber optics and microwave transmissions, which McLuhan was largely ignorant of; together the two planned to write a book called *The Social Impact of New Technologies*. Such a book, Powers suggested, might reestablish McLuhan's reputation as a media guru in academic circles. McLuhan also hoped that Powers would help him do something with the laws of the media, which seemed further than ever from publication.[76]

Powers was a forceful individual, but it is doubtful he could have had more success than others in nudging McLuhan into more businesslike and persevering ways. Much as he craved their presence, the truth was that McLuhan had never really known how to use collaborators or how to exercise much judgment in choosing them. His attitude toward them remained ambivalent. If he seemed to need them, he almost invariably came to resent them.

McLuhan had a dream in 1975 that made a profound impression on him and that seemed to dramatize his ambivalence about working with collaborators. In the dream, a waiter in a Victorian railway dining car spilled a tray of breakfast food in McLuhan's hair. McLuhan tried to shake the crumbs out of his hair. Railroad officials then launched a series of tedious bureaucratic inquiries to determine if McLuhan had, in fact, got everything out of his hair.

When he told the dream to Nina Sutton, who was interviewing him at the time for the planned revision of *Understanding Media*, she interpreted it to mean that McLuhan felt that she was in his hair. The railroad setting meant that she was trying to put McLuhan on the rails, as it were — forcing him both to get on with his work and to cast it in a reasonably connected, linear form.[77]

McLuhan often felt that collaborators, besides trying to put him on the rails, were merely fussing over him and picking his brain. While he did not mind having his brain picked, to a certain extent — that's what it was there for — he sometimes felt a certain exasperation that people didn't pick their own brains more. Years earlier, McLuhan had remarked that James Joyce had paid for his "high vitality" by incurring the resentments, sometimes unconscious, of men less well endowed. Even friends could harbor such resentments. Doubtless McLuhan was also thinking of himself. As a person of high vitality he constantly felt thwarted even by reasonably well intentioned people who happened to possess less energy and drive than he did. This usually meant collaborators.[78]

From time to time in the late seventies, McLuhan wondered whether he had, throughout his life, squandered much of that vitality on projects that had gone nowhere. It is not unusual for those who make their living teaching or writing to have doubts toward the end of their careers about the directions they took — to wonder what it would have been like to explore different areas. There were times when McLuhan felt proud that he had been an intellectual pioneer, almost the first person in the West since Plato, as he sometimes put it, to study effects rather than to talk about causes. At other times, however, he felt that his equivocal intellectual standing in the world was a sort of judgment on him, a sign that he had been careless in the application of his vast mental energies.

At any rate, he fully intended to continue his work as long as he was physically and mentally able to do so. Unfortunately, in the last year in which he was able to work, he lost his faithful secretary, Margaret Stewart, the woman who had taken dictation for so many of his books and who had spared him much of the practical and administrative trouble of running the center. In May 1978, she fell from a ladder and broke her hip and never recovered sufficiently to return to her post.

Her replacement was Eleanor Austen, who came from her position as alumni secretary at St. Michael's College. Austen had always been friendly with McLuhan, but she was not quite as worshipful toward him as Stewart had been. After she began working for McLuhan, even the friendliness quickly disappeared. For one thing, she was not happy at receiving phone calls from him at all hours of the night asking her to take down his latest breakthrough. For another, working in the center required a high degree of tolerance for disorganization and chaos, which Austen hated. Occasionally she was caught throwing out one of the scraps of paper that McLuhan had scribbled on and that littered the center. Evidently she did not realize just how valuable these scraps of paper were — at the very least, McLuhan's friends and assistants tried to point out, they could be sold to somebody, someday, for good money.

Finally, Austen made the mistake of taking issue with McLuhan over his familiar percepts. She argued strongly against his notion, for example, that jokes were based on grievances. Her arguments may have been sound, but McLuhan was not interested in having to answer them, and Austen soon came to be regarded by most of the regulars at the center as an enemy, subtly undermining McLuhan's work.

The atmosphere of the center was also soured by the possibility, horrible for McLuhan to contemplate, that he might have to pay back his Doubleday advance, since neither his "twentieth-century Baedeker" nor the book on the laws of the media was going anywhere. In a bid to restore some impetus to the latter project, McLuhan applied in 1979 for a grant of $26,000 from the Social Sciences and Humanities Research Council of Canada. (In the application McLuhan stated that the book was, in fact, already written and that the grant would enable him to further his investigation of the tetrads.)

McLuhan told the council that the grant would cover a full-time salary for one research assistant, his son Eric, and a part-time salary for another, his administrative assistant, George Thompson. Since Eric lacked a Ph.D., the council was somewhat dubious about his appointment. McLuhan was obliged to write a letter to the council defending his choice, pointing out that Eric had been at his side for virtually the past fifteen years. He did not say this exactly in the letter, but Eric was his chief listener, and he played back what he heard with almost inspired comprehension. McLuhan needed this kind of response, more than anything else, to do his "research."[79]

McLuhan's choice of assistants was not the only problem with the application. The council also clucked over the lack of academic standards in McLuhan's analysis of the tetrads. As one of the council's assessors put it, "there is little point in trying to palm this off as significant science, or social science. It is interesting intellectual conversation."[80] (How often McLuhan's colleagues dismissed his work as merely "interesting speculation" or "stimulating conversation" — in a tone that suggested that they, too, could be as interesting or as stimulating were it not for some vague moral principle.)

To gain support for his application, McLuhan wrote urgent letters to Trudeau and Coutts.[81] It was bad behavior on his part — powerful politicians were supposed to have no influence whatsoever on the research council's decisions — but the end run may have had some effect: in April 1979, he received the full $26,000 from the council.

McLuhan was relieved, if for no other reason than the grant provided a year's salary for Eric. As he approached retirement, the question of Eric's employment took on increasing importance. As in the past, McLuhan did not hesitate to approach even those who had never shown him warmth or favor if it meant a chance to advance his son's interests. Dennis Duffy, a University of

Toronto English professor who had published mildly hostile criticism of McLuhan's work, received a lunch invitation from him in the late seventies, when Duffy was principal of one of the colleges. "I thought, oh no, here I am, going to be monologued out for forty-five minutes," Duffy recalls. "But during the lunch he said, 'Have you met my son Eric? He's looking for a job and you're principal of Innis College and maybe you could give him one.' It was all so clumsy and innocent and nice that I didn't come away mad at all."[82]

McLuhan also began to think, finally, of how the center might continue without him. He hoped, of course, that Eric would succeed him as its head. Whatever decision was made about the future of the center, it would have to be made quickly. In 1976, when McLuhan was approaching the mandatory retirement age for University of Toronto professors (he was supposed to retire in 1977), he had made a deal with the administration that allowed him to stay for three more years as director of the center, continuing to teach one graduate course. June 30, 1980, was to be his last day on the payroll.

As his final academic year began in the fall of 1979, McLuhan began to hope that the deadline might be postponed for another year or that he could somehow run the center independently of the university. Once again, he wrote to his friend Trudeau asking for a chance to talk with him about the matter.[83] The center had been the focus of his life for so long that he could hardly imagine living without it.

13. Silence (1979–1980)

Thought is a garment and the soul's a bride
That cannot in that trash and tinsel hide.
— William Butler Yeats, "Ribh Considers Christian Love Insufficient"

On September 25, 1979, Father Armand Maurer of the St. Michael's College philosophy department dropped by the center to pick up a photocopy of McLuhan's now famous diagram of the split brain. "He came out of his office and he looked terrible," Maurer recalls. "It struck me at the time that the man shouldn't be there, he should be on vacation. He looked so drawn. That worried me. He was obviously very, very tense. But he was always so glad to see you."[1] McLuhan's delight was by no means affected, despite his grumbling about spoiled priests and obtuse philosophers. Until the day of his death, he remained at heart a basically cheerful and affectionate man.

The next day McLuhan sat at his desk working on the proofs for an article he had written for *Maclean's* magazine. Eleanor Austen noticed that he had trouble holding the pencil in his hand and that his hand was shaking. He rose from his desk unsteadily and made his way downstairs, where he collapsed on the floor. George Thompson helped him to the couch and then phoned for help.

McLuhan had suffered a massive stroke. Ten days after he entered St. Michael's Hospital he underwent surgery to help restore circulation of blood to his brain. Two weeks after the operation he walked out of the hospital. He eventually regained almost complete physical mobility — but his ability to read and write had been annihilated. Worst of all, he could no longer speak, except for a few odd phrases. He would repeat "oh boy, oh boy, oh boy," for hours; sometimes he managed to say "yes" and "no." When he was anxious to express something, the best he could usually do was a sort of "wuh, wuh," like a deaf-mute in an old Hollywood movie. (Like many stroke victims, however, he could sing, and at church he would stand up and belt out the hymns he knew by heart.)

So began McLuhan's stay in an earthly purgatory in which he understood everything but could say nothing. For a man who lived to talk, it was the ultimate torment. Corinne tried to ease it as best she could. She hired one of his disciples, Carl Scharfe, to attend him. A few friends visited regularly to read and talk to him. Some, like Barry Nevitt, entertained the hope that McLuhan's brain functions might be reactivated through sign language or some other symbolic form of speech. Since McLuhan had worked all his life at integrating his left and right brain functions — he had been, for example, left-handed in childhood but was subsequently compelled to use his right hand — both hemispheres might still be operative in some undamaged area of his brain. The idea was very comforting, but unfortunately no one really had a clue how to make it work. "It was pathetic," Marcel Kinsbourne recalls. "When one of his staff asked me about this, I realized they were trying to use McLuhan's ideas to retrain him. With the best will in the world I couldn't see how it could be done. But these people believed so strongly in him."[2]

Family and friends eventually resorted to more prosaic approaches. They presented McLuhan with a Speak & Spell, a device used to teach children to read and spell, which reproduced the sounds of human words. McLuhan looked at it wistfully, as if pondering the curious fact that computer chips could speak and he could not.

His daughter Teri was particularly diligent in her attempts to help him regain his speaking abilities. She used a large-print Bible (her brother Michael had suggested the Bible as the text most familiar to their father) and placed McLuhan's index finger under a word, not letting him move it until he said the word aloud. Often McLuhan became frustrated with the process, pushing her away and displaying a childlike temper understandable in a man whose brain had been so assaulted. Teri, who had been impressed by Norman Cousin's book *Anatomy of an Illness*, in which the author recommended the therapeutic powers of laughter, also arranged for the screening of movies by Chaplin, Buster Keaton, the Marx brothers, and W.C. Fields at 3 Wychwood. McLuhan's laughter on these occasions was so ebullient it alarmed his nurses.

Most of his visitors came simply to read to him. His favorite book at this time was, ironically, *Caught in the Web of Words*, an account of the compilation of the *Oxford English Dictionary* in the nineteenth century. It was a book that seemed made to order for McLuhan, who had always maintained that words were the most potent and unfathomable of all human artifacts. He also appreciated lighter fare, such as *Jennie*, the biography of Winston Churchill's mother, and cried helplessly at some of the more moving passages. Like many stroke victims, he could no longer keep up his emotional guard.

Nonetheless, he was still capable of great intellectual excitement. On one occasion, he became extremely agitated after hearing two or three pages from *Gödel, Escher, Bach*, a witty and poetic work about mathematics, music, and computer intelligence, among other things. He reached out for the book, clutched it in his hands and hurried, as fast as he was able to walk, over to a neighbor's house. When the neighbor appeared at the door, McLuhan thrust the book at him. McLuhan could say only "Wuh, wuh, wuh," but the neighbor understood. As in the days of his prime, McLuhan was eager to share a bit of intellectual gold he had just discovered in his endless prospecting.[3]

He maintained, as well, his habit of closely observing the strange and some-times beautiful coherence of the world, which had so often recompensed him for personal pain and difficulty. Walking about in Wychwood Park, he pointed to the leaves moving in the wind, wordlessly expressing his delight at the scene. He also watched television from time to time. In 1980 he became fascinated by the televised meetings of Prime Minister Trudeau and the premiers of the ten Canadian provinces haggling over Trudeau's proposed addition of a charter of rights to the Canadian constitution. McLuhan followed these meetings as if they gave promise of revealing some tremendous secret. It was a chance, at any rate, for him to cheer on his old friend and hero Trudeau.

Barry Nevitt tried to maintain with the speechless man something resem-bling their old dialogues. McLuhan had articles and books he was interested in ready for Nevitt to read when he visited, once or twice a week, and Nevitt dis-cussed them as he read. Nevitt tried to present both the way he would look at the material and the way McLuhan would look at it. When he said something that struck his listener, McLuhan would nod his head and say, "Oh yes, oh yes, oh yes" and take hold of Nevitt's arm. Conversely, Nevitt sensed when he said things that bored or irked McLuhan and would immediately take another tack. Sometimes he tried three or four approaches to spark in McLuhan something of the old intellectual excitement of their dialogues.

Other friends who dropped by were not always so patient. Tom Easterbrook, who had once wandered the streets of Winnipeg with McLuhan arguing till dawn, now sometimes found himself almost inadvertently punching away at his old friend's positions until his defenseless listener resorted to turning on the radio full blast to drown out the words. Neither of these men — both of whom had burned in youth with an ambition to make a difference in the great matters of the world — had yielded to the other in their arguments. Now Easterbrook was left alone to carry his part of the argument, which McLuhan could rebut only by not listening to it.

It disturbed Corinne that so few of McLuhan's colleagues from St. Michael's College came to visit him. For their part, many of those colleagues simply didn't know how welcome they were at Wychwood Park. The relationship between the McLuhan family and St. Michael's College and the University of Toronto in general had become increasingly strained. Now Corinne was desperately anxious to keep the news of her husband's disability from becoming general knowledge. The day after his stroke, she had made Eleanor Austen promise not to say anything about it to anyone. At first, she had hoped that McLuhan might yet recover and be able to fulfill some of his speaking engagements and book contracts. There was also the center to protect.

But the news could hardly be suppressed for long. In November Austen received a phone call from a local reporter asking her pointblank if McLuhan had been incapacitated. She asked him to speak to a member of the family but admitted that the professor had suffered a debilitating stroke. The word spread quickly that the expert in communications could now hardly speak a word. The university administration came under pressure to make some decision about the future of the center. John Leyerle, then dean of the School of Graduate Studies, recalls, "We delayed and delayed this decision hoping that McLuhan would recover. We kept getting stories that McLuhan was recuperating and starting to speak and it would just be a couple of weeks before he recovered. We were being told what people wanted to believe, not what was the case."[4] Eventually Leyerle and others in the administration confronted the fact that McLuhan was not going to recover. In a university hard pressed for cash, the temptation to save $75,000 by cutting off funds to a center that was no longer, in effect, oper-ating was irresistible. In March 1980, a review committee was formed, headed by McLuhan's friend and admirer Ernest McCulloch. In May the committee issued a report stating, correctly, that "Marshall McLuhan was the Centre, and the Centre was Marshall McLuhan." Under the circumstances, the committee recommended that the center be replaced by a McLuhan Program in Culture and Technology, whose main function would be to sponsor lectures on the work of McLuhan and related matters.[5]

In June the university announced it was indeed going to dismantle the cen-ter. In its place, the McLuhan Program, with a budget of $20,000, would attempt to carry on the legacy of his work at the university. Teri McLuhan then launched a letter-writing campaign to influential friends and admirers of her father, which elicited protests from men like Buckminster Fuller, Woody Allen, and Tom Wolfe. From California, an aide to Governor Brown called Dean Leyerle demanding to know what was going on with the center ("A real piece of straight-arm tactics," Leyerle terms the phone call), but this bit of pressure merely added to the irritation of administration officials. Pierre Trudeau him-self wrote McLuhan a letter in April, commiserating with him over the univer-sity's plan. He had also written to the university itself, but even he could do nothing to save the center.[6]

For many of McLuhan's friends, the death of the center symbolized and capped the long history of the University of Toronto's antipathy to McLuhan and his works. But if it was true that Leyerle, for example, was no McLuhanite — he still calls the center the home of "an army of mind ticklers and syco-phants"[7] — nonetheless the decision to close the center was by no means a spiteful one. Centers like McLuhan's were supposed to produce the kind of research that finds its way into the pages of academic journals or to train grad-uate students through full-fledged graduate programs. McLuhan's center had done very little of the first and none of the second. The review committee was right: the center was McLuhan; after that, there was very little to be said for it. And certainly, in 1980, there were no very promising candidates for filling McLuhan's shoes.

McLuhan paid occasional visits to the coach house, like a sea captain view-ing the remains of his shipwrecked vessel. Margaret Stewart received a phone call from Corinne one night, asking if she could help McLuhan find some mate-rial at the center. She grabbed her crutches and drove to the coach house, where she was met by McLuhan in the doorway. After she retrieved the document he wanted, McLuhan raised his open hand to signify that he wanted five copies made. This done, Stewart looked about the center. The place was in chaos, as if somebody had deliberately trashed the files. "It was just a complete garbage dump," she recalls.

> All of the records I had made for McLuhan were in a garbage heap. I
> don't know what happened to all his stuff, but when I saw it I just felt
> my whole life fall apart. I said to Marshall, "Do you want me to help
> you?" and he put his arms out and then he came over and he hugged
> me. It was pitiful. It nearly tore me apart.[8]

Later, as he was going through his papers before leaving the coach house for-ever, McLuhan wept.

He was defenseless now in almost every respect. The essential McLuhan, the man at his most characteristic, was the man who in the forties and fifties had held forth at parties unself-consciously, with a stream of brilliant conversa-tion, dizzying his listeners with insights that, when grasped, opened up worlds for them. No gossip, no sour academic laments about the state of things, no alcohol-inspired silliness ever tainted the atmosphere of a social gathering around Marshall McLuhan. Even in the latter years of his career, even when he was racked with pain, he was capable of observations and insights that stayed forever in the minds of his students. They were statements that, in spite of all McLuhan's talk about "putting on" an audience, were never calculated to achieve any effect; they seemed, instead, to be the overflowing excess from some bottomless reservoir of perception.

McLuhan wept freely in his stricken condition, but he also laughed, and laughed heartily. In the last week of his life, he was watching a sitcom on television in which a figure representing the devil talked about the eighth deadly sin, which had cost him much more effort to invent, he said, than the first seven. That eighth deadly sin was advertising. When McLuhan, who had had more to say about advertising than anyone else in this century, heard the line, he roared with laughter. Barry Nevitt, who was watching the program with him, had never seen him so struck with hilarity.[9]

Still, McLuhan continued to suffer greatly from his affliction. Nevitt and a few other friends tried to comfort him by suggesting that, when he recovered, he would have an extraordinarily interesting tale to tell of the experience. The suggestion only caused the look of pain in McLuhan's face to deepen. He knew he would never recover. He also knew that even if he did recover he would never want to speak of what had happened to him. His old friend Sheila Watson was more honest when, in November 1980, she said to him, quietly and with no attempt to soothe or placate but simply to afford McLuhan the relief of acknowledging his state, "It's hell, isn't it?" McLuhan, with immense and simple gravity, nodded his head.

On December 24, 1980, in the company of Corinne and Teri, McLuhan visited an exhibition of sculpture by Sorel Etrog at a local gallery. Etrog, an admirer of the works of Samuel Beckett as well as of McLuhan's writings, had infuriated McLuhan earlier that month by comparing him to Beckett. McLuhan, who regarded the absolute godlessness of Beckett's work with something approaching horror, grew so red in the face that one of his veins stood out.

On this occasion, however, McLuhan was laughing and in good spirits. At lunch time, Etrog accompanied him to one of his favorite haunts in Toronto, the lounge on top of Sutton Place, a high-rise apartment building. During the lunch, McLuhan took Etrog's arm and walked around the four glassed-in walls of the lounge, overlooking the city below. "He knew he was dying," Etrog maintains. "He knew that he was looking at the city for the last time." The speechless man was, in effect, taking his farewell of the city that had been his home for more than thirty years. He had come to Toronto desperately hoping that the new environment might signal a change of luck for him, might somehow facilitate his great task of making the world see what he saw. Strangely enough, it had.

Five days later, a Jesuit priest from New York, Frank Stroud, who had met McLuhan while doing research in media and education in 1974, dropped by the McLuhans' house for a post-Christmas visit. Teri had warned Stroud that her father might tire easily, but McLuhan seemed to welcome his visit. He proudly showed Stroud how his Speak & Spell worked. "Teri kept saying to me, 'Just keep encouraging him, do anything you can to keep him going,'" Stroud recalls. That night they all went out to a restaurant for dinner. During the meal, Corinne pointed to a badge on McLuhan's lapel, a replica of the Order of Canada, an honor conferred on Canadians for exemplary achievement in major fields of endeavor. "She said, 'Do you see that little emblem?'" Stroud remembers. "'He's proudest of that of any of the things he has.' And then Marshall sat back and said, 'Mmmmmm, mmmmmm.' He couldn't say anything, but you could see he was taking pride in our saying things like that."[10]

Stroud was invited back to the McLuhans' the following evening, December 30. He said Mass in the living room, using a bottle of fine Burgundy a colleague of McLuhan's had brought back from France. Everyone was in convivial spirits as champagne was poured, toasts proposed, and the rest of the Burgundy consumed at the dinner table. McLuhan and Stroud lit up cigars after the plates were cleared and watched the evening news on television together. At one point, McLuhan's daughter gently suggested that he might want to go to bed, but he shook his head vigorously, as if he didn't want to miss a moment of the evening's fun. When Stroud finally left, the two men embraced.

McLuhan never lived to see the morning. At some point in the early hours of the last day of 1980, he died peacefully in bed.

McLuhan's friends and students, even if they never read a word he published, were more fortunate than those who knew him only through his books. Those books, of course, presented difficulties even for well-disposed and perceptive readers. It is entirely conceivable that some of them will never be reprinted. Unlike writers who put the best part of themselves in their books, McLuhan could never quite convey in print his own vitality and the free play of his mind.

Nevertheless, as a body of work, his books will probably be mined for years to come by clever prospectors hunting, as he did, for bits of invaluable ore. Such prospectors will know how to disregard the flagrant exaggerations, the wild generalizations, the off-target "percepts." They will pass over such material as the hopeful prophecy that cars would disappear.[11] They will even pass over some correct predictions — McLuhan's remark, for example, in *Understanding Media*, uttered two decades before the videocassette explosion, that "at the present time, film is still in its manuscript phase, as it were; shortly it will, under TV pressure, go into its portable, accessible, printed-book phase."[12] (McLuhan's attitude toward predictions was, predictably, idiosyncratic: he insisted that he predicted only what had already happened. Anyone who truly perceived the present, he maintained, could also see the future, since "all possible futures are contained in the present."[13] If something occurred to him, he further reasoned, sooner or later it would occur to others — and then it would come to pass.)

His predictions, true or false, were in any case not the true McLuhan ore. That ore was McLuhan's ability to stimulate, in the phrase of Matthew Arnold, "a stream of fresh and free thought upon our stock notions and habits." Such ability has been sufficiently uncommon, and remains sufficiently uncommon, to keep McLuhan's work alive for a very long time. It was an ability that was not simply, as McLuhan's enemies always maintained, the result of freewheeling speculation. It showed itself in insights that offered some genuine understanding of a society in which the extensions of man had become almost more real and more bizarre than mankind itself.

McLuhan had his period of unique celebrity, and he paid for it. The social historian could henceforth mark him merely as a phenomenon of the sixties, as a guru of the young — who seemed to have a large appetite, in that decade, for such figures. Measured by this role, McLuhan at least compared well with other claimants. It was obvious that he was in a different category altogether from such figures as Timothy Leary, Carlos Castaneda, Ken Kesey — men whose appeal to the sixties generation had little to do with their powers of intellect. And certainly he was a match for the more cerebral heroes of the period — Herbert Marcuse, Norman O. Brown, R.D. Laing, though none of these men could equal the incandescence and liveliness of McLuhan's intelligence. None of them possessed his sheer high spirits. The work of these others has an underlying dourness of outlook, a hint perhaps — and more than a hint, in some cases — of despair. McLuhan, on the other hand, lived and worked as if he really believed — in the most strictly orthodox Christian sense — the words of the medieval mystic that all would be well, and all manner of things would be well.

Notes

Complete bibliographical information about works cited in these notes can be found in the Bibliography. Each note contains a shortened form with the author's last name and a short title. Most of McLuhan's papers have been collected in the National Archives of Canada (cited as NA). These papers contain thousands of letters written by McLuhan, a selection of which is included in *Letters of Marshall McLuhan* (cited as *Letters*). All letters cited in these notes were written by McLuhan unless otherwise noted. All letters, including those written by others, are located in the National Archives of Canada (NA) unless otherwise noted.

The National Archives of Canada also holds copies of diaries written by McLuhan at various periods in his life. A one-year diary, with extensive entries, covers the year 1930. A five-year diary, with shorter entries, covers the period 1936–1939. A series of diaries with short entries covers most of the decade of the seventies. All these are cited as Diary.

Other important sources of letters and documents written by or pertaining to McLuhan are located in the Ford Foundation Archives in New York City, the Wyndham Lewis collection in the Cornell University Library, the Pound collection in the Lilly Library of Indiana University, and the archives of St. Michael's College. Among documents in private hands, one of the most extraordinary records of the development of McLuhan's thought is contained in the series of letters, written at frequent intervals by McLuhan over a period of more than two decades — from the late fifties to the late seventies — to Sheila and Wilfred Watson, then professors of English at the University of Alberta. An unpublished history of the McLuhan family by Stuart Douglas Mackay of Edmonton, Alberta, is an invaluable source of information about McLuhan's immediate progenitors.

All interviews, except three, were conducted by Philip Marchand in the years 1985–1988.

1. Childhood on the Prairies

1. Interview with Ernest Raymond Hall.
2. Creighton, *The Story of Canada*, p. 192.
3. Interview with Ernest Raymond Hall.
4. Interview with Maurice McLuhan.
5. Interview with Ernest Raymond Hall.
6. Interview with Chester Gamble.
7. John Kenneth Galbraith, *The Scotch* (New York: Macmillan, 1964), p. 105.
8. Quoted in Mackay, "The Ancestors and Descendants of William McLuhan Jr."
9. *Times Literary Supplement*, November 9, 1967. Quoted by Tom Nairn, "McLuhanism: The Myth of Our Time," in Rosenthal, *McLuhan: Pro and Con*, p. 152.
10. Letter to Jerry Royer, June 16, 1975.
11. Interview with Harley Parker.
12. Finkleman, "Marshall McLuhan," p. 23.
13. McLuhan, "Autobiography: For the H.W. Wilson Co.," NA.
14. Letter to Ray and Pearl Hall, August 12, 1977.
15. Letter to Elsie McLuhan, [September 5, 1935], *Letters*, p. 73.
16. *Speaking of Winnipeg*, p. 32.
17. Interview with Maurice McLuhan.
18. Interview with Una Johnstone.
19. Interview with Peggy Morton.
20. Letter from Herbert McLuhan to Elsie McLuhan, n.d. Letter from Herbert McLuhan to Elsie McLuhan, September 17, 1933.
21. Cooper, "Pioneers," p. 78.
22. Diary, July 1, 1930.
23. Diary, March 31, 1930.
24. Marshall himself retained a serious interest in phrenology at least until his early twenties. At the age of eighteen he disparaged in his diary the ignorance of psychologists about phrenology and insisted on the basic validity of its fundamental principles. When he fell in love with Marjorie Norris at the age of twenty-one, he noted with approval her "Celtic head" and included in his diary a clinical description of its shape. Less happily, on another occasion he noted the severe intellectual and spiritual deficiencies of many Polish and Czechoslovakian Canadians (great numbers of whom lived in Winnipeg) from a phrenological point of view. Interview with Maurice McLuhan. Diary, January 15, May 25, 1930; January 30, 1933.
25. Diary, June 5, 1930.
26. Diary, August 7, 1930.
27. Interview with Maurice McLuhan.
28. Letter from Herbert McLuhan to Elsie McLuhan, September 17, 1933.
29. Letter from Herbert McLuhan to Elsie McLuhan, n.d.
30. Diary, August 5, 1930.
31. Interview with Maurice McLuhan.
32. Interview with Una Johnstone.
33. Diary, July 1, November 22, December 1, December 2, 1930.
34. Letters from Herbert McLuhan to Elsie McLuhan, n.d.; September 17, 1930.
34. Diary, November 6, 1930.
36. Diary, April 3, March 3, March 30, February 24, May 27, 1930.
37. Diary, January 11, 1930.
38. Diary, November 29, 1930.
39. Diary, September 10, 1930.
40. Interview with Maurice McLuhan.

41. Diary, October 2, 1930.
42. Callwood, "What Are You Doin'."
43. Interview with Maurice McLuhan.

2. University of Manitoba

1. "Understanding Canada," *Mademoiselle*.
2. Cooper, "Pioneers," p. 93.
3. Diary, September 24, 1930.
4. Diary, October 24, 1930.
5. Diary, March 8, April 2, November 9, 1930.
6. Diary, October 31, 1930.
7. Letter to E.K. Brown, December 14, 1935, *Letters*, p. 79.
8. Letter to Noel Fieldhouse, February 4, 1970.
9. Interview with Saul Field.
10. Letter to Catherine Court, September 25, 1970.
11. Interview with Scott Taylor.
12. Letter to A.E. Safarian, April 26, 1973.
13. Interview with Scott Taylor.
14. Diary, February 3, May 13, 1930.
15. Diary, October 24, February 23, 1930.
16. Letter from Herbert McLuhan to the McLuhan family, December 10, 1945.
17. Diary, January 9, 1930.
18. McLuhan, "Autobiography: For the H.W. Wilson Co."
19. Diary, January 27, 1930.
20. Diary, October 14, 1930.
21. Diary, August 11, 1930.
22. Diary, August 14, 1930.
23. Diary, July 15, 1930.
24. Diary, January 24, July 27, 1930.
25. Diary, March 6, 1930.
26. Diary, October 4, October 13, 1930.
27. Diary, June 8, 1930.
28. Diary, March 3, April 30, 1930.
29. Diary, October 13, 1930.
30. Diary, January 6, 1930.
31. Diary, March 12, June 4, February 23, August 9, May 8, 1930. Letter to Elsie, Herbert and Maurice McLuhan, November 10, 1934, *Letters*, p. 38.
32. Diary, January 13, January 16, July 6, 1930.
33. Diary, April 18, June 15, October 26, November 26, November 27, 1930.
34. Diary, November 27, 1930.
35. Interview with Mrs. Andrew Currie.
36. Diary, June 24, February 28, May 23, 1930.
37. Diary, January 8, April 17, 1930.
38. Diary, December 3, March 19, 1930, January 30, 1933.
39. Diary, March 11, 1930.
40. Diary, January 30, June 19, 1930.
41. Diary, April 8, 1930.
42. Diary, January 16, 1930.

43. Diary, December 27, January 18, 1930.

44. Diary, April 10, January 14, April 5, 1930.

45. McLuhan, "Dagwood's America."

46. Diary, April 10, 1930.

47. Diary, February 9, 1930. Letter from Marshall McLuhan to Elsie McLuhan, n.d. [1933].

48. Diary, December 4, 1930.

49. Finkleman, "Marshall McLuhan," p. 34.

50. Finkleman, "Marshall McLuhan," p. 260.

51. Diary, September 18, 1930.

52. Letter to Elsie, Herbert, and Maurice McLuhan, [June 17], 1932, *Letters*, pp. 12–14.

53. Ayre, "McLuhan Revisited."

54. Letter to Robert Hittel, February 24, 1976.

55. Letter to Malcolm Ross, July 27, 1976.

56. Letter to Robert Hittel, March 4, 1974.

57. Chesterton, *What's Wrong with the World*, p. 33.

58. Introduction to *Paradox in Chesterton*, p. xix.

59. Horace Bridges, cited in *Paradox in Chesterton*, p. 14.

60. G.K. Chesterton, *William Blake* (London: Duckworth, 1910), p. 58.

61. Diary, January 30, 1933.

62. Diary, January 30, 1933.

63. Diary, April 9, 1933.

64. Cooper, "Pioneers," p. 101.

65. McLuhan, "George Meredith," p. 8.

66. McLuhan, "George Meredith," p. 2.

67. McLuhan, "Macaulay."

68. Kettle, "Marshall McLuhan."

69. *The Manitoban*, n.d.

70. *The Manitoban*, n.d.

71. Diary, July 1, 1930.

72. Interview with Maurice McLuhan.

73. Letter to Elsie McLuhan, January 18, [1935], *Letters*, p. 51.

74. Letter to Elsie McLuhan, January 18, [1935], *Letters*, p. 51.

75. Diary, June 7, 1936.

76. Diary, August 26, 1976.

3. Cambridge

1. T.E.B. Howarth, *Cambridge Between Two Wars* (London: William Collins), p. 195. Lucas's own pose of a man refreshingly free from academic pretentiousness McLuhan considered to be a gross affectation.

2. Dobbs, "What Did You Say?"

3. McLuhan, "'Colour-Bar' of BBC English."

4. McLuhan, "Cambridge English School."

5. Letter to Elsie McLuhan, [December 17, 1934], *Letters*, p. 47; letter to Peter Drucker, April 18, 1960, *Letters*, p. 268.

6. Interview with John Wain.

7. Letter to Elsie, Herbert, and Maurice McLuhan, [February 1(?), 1935], *Letters*, p. 54.

8. Interview with Lionel Elvin.

9. Interview with Northrop Frye.

10. Letter to Elsie, Herbert, and Maurice McLuhan, October 14, 1934, *Letters*, p. 25.
11. Letter to Elsie, Herbert, and Maurice McLuhan, February 7, 1935, *Letters*, p. 58.
12. Richards, *Practical Criticism*, p. 206.
13. Richards, *Philosophy of Rhetoric*, p. 11.
14. McLuhan lecture notes, NA.
15. Letter to Muriel C. Bradbrook, January 12, 1973, *Letters*, p. 462.
16. Empson, *Seven Types of Ambiguity*, p. 274.
17. Howard, "Oracle."
18. Letter to Rene Wellek, January 3, 1973.
19. Empson, *Seven Types of Ambiguity*, p. 72.
20. Letter to Elsie, Herbert, and Maurice McLuhan, May 16 [17, 1935], *Letters*, p. 67.
21. Interview with Walter J. Ong., S.J.
22. Cooper, "Pioneers," p. 121. The remark was quoted to Cooper by Muriel Bradbrook, McLuhan's Ph.D. thesis adviser at Cambridge.
23. Leavis, *Culture and Environment*, p. 6.
24. Diary, March 26, 1930.
25. Diary, December 24, 1930.
26. Letter to Elsie, Herbert, and Maurice McLuhan, December 6, 1934, *Letters*, p. 41.
27. Eliot, *Use of Poetry*, p. 151.
28. McLuhan, *Understanding Media*, p. 18.
29. Eliot, *Use of Poetry*, p. 118.
30. Letter to Ralph Cohen, November 30, 1978.
31. Letter to Elsie, Herbert, and Maurice McLuhan, October 14, 1934, *Letters*, p. 24.
32. Harding, "Trompe l'oeil."
33. Robin Maugham, *Escape from the Shadows* (London: Hodder and Stoughton, 1972), p. 89.
34. Interview with A.W. Loveband.
35. Interview with J.R. Ellis.
36. Diary, May 8, December 4, 1935.
37. Maugham, *Escape from the Shadows*, p. 90.
38. Interview with H.L. Elvin.
39. Maritain, *Art and Scholasticism*, p. 25.
40. Letter to Jane Scheuerman, November 16, 1971.
41. Diary, October 8, 1935.
42. Letter to David Balcon, December 10, 1969.
43. Diary, March 24, 1935.
44. McLuhan, "Autobiography: For the H.W. Wilson Co."
45. Interview with Scott Taylor.
46. Letter to Elsie, Herbert, and Maurice McLuhan, [October 4, 1934], *Letters*, p. 20.
47. Letter to Elsie, Herbert, and Maurice McLuhan, [October 4, 1934], *Letters*, p. 20. Diary, April 19, April 20, April 21, 1935.
48. Interview with Lionel Elvin.
49. Letter to Elsie, Herbert, and Maurice McLuhan, [October 4, 1934], *Letters*, p. 20.
50. Interview with Morton Bloomfield.

4. Apprentice Professor

1. Letter to E.K. Brown, December 12, 1935, *Letters*, p. 79.
2. Interview with Kenneth Cameron.
3. Diary, September 23, 1936.
4. "A Day in the Life," *Weekend*.

5. Schickel, "Marshall McLuhan."
6. "On the Scene," *Playboy*.
7. Letter to Edward Wakin, July 27, 1978. Letter to Nicholas Johnson, chairman, Federal Communications Commission, August 29, 1973.
8. Diary, October 11, 1936.
9. Interview with Kenneth Cameron.
10. Interview with Morton Bloomfield.
11. Interview with Kenneth Cameron.
12. Interview with Kenneth Cameron.
13. John Henry Newman, *An Essay in Aid of a Grammar of Assent* (South Bend, Ind.: University of Notre Dame Press, 1979), p. 169. Diary, October 17, 1936; February 19, 1937.
14. Callwood, June. "The Informal Mr. McLuhan," *Globe and Mail*, November 25, 1974.
15. Interview with Maurice McLuhan.
16. Interview with Marcel Kinsbourne. Also letter to Sidney Halpern, January 9, 1974.
17. Howard, "Oracle." Letter to Walter G. Buckner, February 23, 1971.
18. "McLuhan Asks for Shakeup in Church."
19. Interview with Ruth Kates.
20. Interview with Sheila Watson.
21. Letter to Tony Hodgkinson, November 4, 1964.
22. Diary, October 28, 1936; January 20, 1937; October 4, 1936.
23. Interview with Vernon Bourke.
24. Interview with Pauline Bondy.
25. Letter to Bernard and Mary Muller-Thym, June 11, 1974, *Letters*, p. 498.
26. Kostelanetz, *Master Minds*, p. 101.
27. Interview with Maurice McNamee, S.J.
28. *The Modern Schoolman*, March 1942.
29. Editorial, *Fleur de Lis*.
30. McLuhan, "Peter or Peter Pan."
31. Memorandum of Helen White re: McLuhan, n.d., University of Wisconsin Archives.
32. Interview with Vernon Bourke.
33. Interview with Karl and Addie Strohbach.
34. Interview with Karl and Addie Strohbach.
35. Diary, August 10, 1938. Letter to Bill and Una Johnstone, August 24, 1938. Letter in author's possession.
36. Diary, October 12, 1937.
37. Letter to Corinne McLuhan, January 31, 1939, NA.
38. Diary, September 24, November 1, November 8, November 9, 1938.
39. Diary, November 14, November 30, December 7, December 30, 1938.
40. Interview with Corinne McLuhan.
41. Diary, March 17, 1939.
42. Interview with Corinne McLuhan.
43. Letter to Elsie McLuhan, August 31, [1939], *Letters*, p. 116.
44. Letter to Wyndham Lewis, December 2, 1943, *Letters*, p. 139. Wilson revised the definitive edition of Nashe's work in 1958.
45. Interview with Muriel C. Bradbrook.
46. Interview with Muriel C. Bradbrook.
47. Letter to Felix Giovanelli, n.d. McLuhan, whose daughter Elizabeth's five cats were named after the five divisions of rhetoric — inventio, dispositio, pronuntiatio, elocutio, and memoria — retained his fascination with the dynamics of a speaker in front of an audience and with classical rhetoric for the rest of his life.

48. McLuhan, *Understanding Media*, p. 32.

49. Interview with Patrick Watson.

5. In Search of a Home

1. *The Rambler*, no. 21, May 29, 1750.

2. At the University of Toronto McLuhan kept huge stacks of index cards in Laura Secord chocolate boxes, each card containing an idea for a thesis. Students who inspected these cards were amazed to see hundreds of ideas in subject areas other professors thought had been exhausted.

3. Interview with Maurice McNamee, S.J.

4. Schickel, "Marshall McLuhan."

5. Howard "Oracle."

6. Letter from Hugh Kenner to Philip Marchand, March 18, 1987.

7. Interview with Maurice McNamee, S.J.

8. Letter to J. Stanley Murphy, C.S.B., March 9, 1944, *Letters*, p. 517.

9. Unpublished note, NA.

10. Letter to Elsie McLuhan, April 12, [1936], *Letters*, p. 82; letter to Corinne Lewis, [February 1, 1939], p. 109.

11. Letter to Felix Giovanelli, May 10, 1946, *Letters*, p. 183.

12. Wain, "Incidental Thoughts."

13. Letter to Eric McLuhan, March 15, 1976.

14. Interview with Marshall McLuhan by Jeffrey Kirsch, KPBS-TV, September 13, 1976. Grady, "They Loved Her."

15. Grady, "They Loved Her."

16. "An Interview," *Miss Chatelaine*.

17. Interview with Maurice Farge.

18. Letter to Felix Giovanelli, n.d.

19. Letter to Wyndham Lewis, January 17, 1944, *Letters*, pp. 146–47.

20. Letter from Felix Giovanelli to Marshall McLuhan, August 13, 1945.

21. Letters from F.P. Wilson to Marshall McLuhan, n.d.; June 13, 1944.

22. Letter from John H. Randall, Jr., to Marshall McLuhan, May 17, 1944, St. Michael's College Archives.

23. Diary, June 15, 1938.

24. "Pound's Critical Prose," in McLuhan, *Interior Landscape*, p. 80.

25. McLuhan, Review of *William Ernest Henley*.

26. Letter to Wyndham Lewis, November 27, 1943, *Letters*, p. 138.

27. Letter to Felix Giovanelli, n.d. An alternative title of this projected work was *Typhon in America*. Typhon was the monster born of earth (mom) who was destroyed by Zeus (dad) in Greek mythology.

28. Letter from Felix Giovanelli to Marshall McLuhan, January 22, 1949.

29. Note of Marshall McLuhan, n.d., NA.

30. Thomas Daniel Young, *Gentleman in a Dust Coat*, p. 381. The other name Ransom cited, as "a still better man" than McLuhan, was Austin Warren of the University of Iowa. Neither became editor of the *Sewanee Review* (p. 382).

31. Mumford, *Technics and Civilization*, pp. 134–36.

32. Letter to Kamala Bhatia, April 6, 1971.

33. Interview with Scott Taylor.

34. Letter to Wyndham Lewis, June 2, 1944, *Letters*, p. 160.

35. Letter to Wyndham Lewis, January 17, 1944, *Letters*, p. 147.

36. Letter to Wyndham Lewis, January 17, 1944, *Letters*, p. 147.

37. Letter to Wyndham Lewis, September 2, 1943, January 4, 1944, January 17, 1944, *Letters*, pp. 133, 142, 148.
38. Lewis, *Self Condemned*, pp. 323, 315, 362, 350.
39. Letter to J. Stanley Murphy, C.S.B., May 9, 1944, Assumption College Archives.
40. Letter to J. Stanley Murphy, C.S.B., [March 1944], *Letters*, p. 158.
41. Letter to Wyndham Lewis, December 13, 1944, *Letters*, p. 165; to Wyndham Lewis, October 11, 1944, Cornell University Library.
42. Letter to Wyndham Lewis, October 2, 1944, Cornell University Library. Lewis, *Self Condemned*, p. 382.
43. Letter to Wyndham Lewis, December 13, 1944, *Letters*, p. 165.
44. Letter to Marius Bewley, n.d. [1944].
45. Letters to Wyndham Lewis, June 2, 1944, January 17, 1944, *Letters*, pp. 160, 147.
46. Letter from Wyndham Lewis to McLuhan, February 4, 1945. Cornell University Library.
47. Lewis, *America and Cosmic Man*, p. 21.
48. McLuhan, "Wyndham Lewis."
49. McLuhan, "Footprints."
50. Howard, "Oracle."
51. Cooper, "Pioneers," p. 24.
52. Quoted in Jeff Myers, *The Enemy* (London: Routledge & Kegan Paul, 1980), p. 148.
53. Letter to Wyndham Lewis, [December 1955], *Letters*, p. 248.
54. Interview with Pat Flood.
55. Letter to Wyndham Lewis, December 13, 1944, *Letters*, p. 165.
56. Letter to Walter J. Ong, S.J., May 18, [1946], *Letters*, p. 186.
57. Interview with Scott Taylor.
58. Interview with Jim Peters.
59. Interview with Pauline Bondy.

6. Twilight of the Mechanical Bride

1. Letter to Gerald Phelan, C.S.B., February 4, 1952, St. Michael's College Archives.
2. Letter to Gerald Phelan, February 4, 1952, St. Michael's College Archives.
3. Diary, May 2, 1938.
4. Interview with Edward Synan.
5. Interview with Armand Maurer, C.S.B.
6. Thomas Dilworth, "Remembering Marshall McLuhan for Philip Marchand," unpublished essay, n.d.
7. Interview with F.E.L. Priestley.
8. Letter to Sheila Watson, September 27, 1963.
9. Interview with Louis Forsdale.
10. Interview with Lawrence Shook, C.S.B.
11. Interview with Edward Synan.
12. *Letters*, p. 92.
13. Interview with Robert Madden, C.S.B.
14. Interview with Frederick Flahiff.
15. Interview with John Madden, C.S.B.
16. Interview with Mac Hillock.
17. "Oracle of the Electric Age."
18. Interview with Edward Synan.
19. Letter to Elsie McLuhan, March 18, 1951.
20. Interview with Pauline Bondy.

21. Kritzwiser, "McLuhan Galaxy."

22. Letter to Lyman C. Johnston, November 9, 1964. Wain, "Incidental Thoughts of Marshall McLuhan."

23. Letter to Corinne Lewis, [January 21, 1939], *Letters*, p. 103.

24. Interview with Cleanth Brooks.

25. Interview with Ernest Sirluck.

26. Letter to Felix Giovanelli, n.d.

27. Letter to Clement McNaspy, S.J., [January 15, 1946(?)], *Letters*, p. 180.

29. Letter to Felix Giovanelli, May 5, 1946.

30. Interview with Edmund Carpenter.

31. Interview with Seon Manley.

32. *Letters* with Seon Manley.

33. Letter from Felix Giovanelli to Marshall McLuhan, May 4, 1949. So bitter was the public controversy over the prize for Pound's *Cantos* that the Bollingen Foundation canceled its sponsorship of future prizes (subsequent Bollingen Prizes have been awarded by Yale University Library). Tate, Brooks, et al., by discriminating, more intelligently than their opponents, between the political views expressed in a work of art and the value of that work, ultimately had the better of the debate.

34. Letter from Felix Giovanelli to Marshall McLuhan, November 12, 1948.

35. "Defrosting Canadian Culture."

36. Letter from Hugh Kenner to Philip Marchand, March 18, 1987.

37. Letters to Felix Giovanelli, [September 1948], March 4, 1949, *Letters*, pp. 202–3, 211.

38. Letter from Hugh Kenner to Philip Marchand, March 18, 1987.

39. Letter from Hugh Kenner to Marshall McLuhan, July 14, 1948.

40. Interview with Mary Parr.

41. Letter to James Striegel, April 27, 1979.

42. McLuhan, "James Joyce: Trivial and Quadrivial."

43. Letter to Ezra Pound, July 7, 1948, Indiana University Library.

44. Letter to Felix Giovanelli, n.d., Indiana University Library.

45. Interview with Cleanth Brooks.

46. Interview with Cleanth Brooks.

47. Letters from Felix Giovanelli to Marshall McLuhan, December 28, 1948, December 2, 1948.

48. Letter to University of California, January 27, 1967.

49. Kenner's fund of useful McLuhan insights has not been exhausted by any means. A recent work, *A Sinking Island* (New York: Knopf, 1988), contains the following sentences: "The overdecorated William Morris book may remind us how obsolete processes claim survival as Art. As TV has turned old movies into 'cinema,' so the rotary press and the monotype (an 1887 patent) had just made hand presswork an art form" (p. 37). Although his name appears nowhere in the text, these observations are vintage McLuhan, circa 1964.

50. Letter to Ezra Pound, July 15, 1948, *Letters*, p. 197.

51. Ayre, "McLuhan Revisited."

52. Interview with Reg O'Donnell.

53. Interview with Ann Bolgan.

54. Interview with Joseph Kertes.

55. Interview with Don Gillies.

56. Interview with Lawrence Lynch.

57. Letters to Walter J. Ong, S.J., May 18, [1946], [December 1947], *Letters*, pp. 187, 191.

58. Letter to Clement McNaspy, S.J., [January 15, 1946(?)], *Letters*, p. 180.

59. Years later, when Hagon was dying from a long illness that eventually robbed him of speech, McLuhan visited him regularly — until the very latter stages of the illness, when the spectacle

of his speechless friend was more than he could bear. "It just broke his heart to come," one of the Hagon children recalls. "He came to visit Dad out of a religious sense of duty, and when that duty was pushed to the limit, he stopped coming." Hagon's speechlessness was a ghastly foreboding of McLuhan's own fate and, in a way, a vivid embodiment of one of his worst fears.

60. Interview with Hugh Kenner.
61. Interview with Louis Forsdale.
62. Interview with Joan Theall.
63. Interview with Northrop Frye.
64. *Chicago Daily News*, November 21, 1972.
65. McLuhan, "A Day in the Life."
66. Diary, January 31, 1930.
67. Letter to Ezra Pound, February 18, 1953, Indiana University Library.
68. Interview with Patricia Bruckman.
69. Letter to Ezra Pound, December 3, 1952, *Letters*, p. 233. Letter from Ezra Pound to Marshall McLuhan, n.d.
70. As early as 1934, McLuhan had noted Aristotle's affinity to Christianity, in contrast to Plato's affinity to Oriental religion, chiefly Buddhism. Shortly after, McLuhan wrote that "Aristotle is the soundest basis for Xian doctrine." See letters to Elsie, Herbert, and Maurice McLuhan, November 10, 1934, [February 1(?), 1935], *Letters*, pp. 39, 53.
71. Interview with Ann Jones. Interview with Joseph Keogh.
72. Interview with Ann Jones. Interview with Joseph Keogh. Letter to Joseph Keogh, April 18, 1972. Interview with Lawrence Lynch. Interview with Reginald O'Donnell.
73. Interview with John Kelly, C.S.B. Interview with Joseph Keogh.
74. Letter to Sheila Watson, October 27, 1961.
75. Interview with Frederick Flahiff. McLuhan, "Have with You to Madison Avenue."
76. Unpublished interview conducted by Kirwin Cox, NA.
77. Letter to D.C. Williams, September 8, 1971. *Letters* with Mother St. Michael (Winnifred Guinan). Interview with Walter J. Ong, S.J.
78. McLuhan, "John Dos Passos: Technique vs. Sensibility," in *Interior Landscape*.
79. Letter to Felix Giovanelli, n.d.
80. Interview with Seon Manley.
81. Letter to Elsie McLuhan, n.d. Letter to Felix Giovanelli, January 12, 1949, *Letters*, p. 209.
82. Interview with Armand Maurer, C.S.B. Interview with Seon Manley.
83. Letter to Felix Giovanelli, n.d., Indiana University Library. Letter to Felix Giovanelli, November 11, 1949, *Letters*, p. 214.
84. Letter to Ezra Pound, January 5, 1951, *Letters*, p. 217. Letter to Felix Giovanelli, November 11, 1949, *Letters*, p. 215. Letter to Felix Giovanelli, November 11, 1948, Indiana University Library.
85. "Playboy Interview." Letter to Paul Levinson, September 8, 1977.

7. The Discovery of Communications

1. Pound, *Guide to Kulchur*.
2. Letter to Ezra Pound, June 22, 1951, *Letters*, p. 227.
3. Letter to Ezra Pound, January 5, 1951, *Letters*, p. 218.
4. McLuhan, "The Later Innis."
5. Innis, *Empire and Communications*, p. 9.
6. Innis, *Bias of Communication*, p. 3.
7. Innis, *Bias of Communication*, p. 105.
8. Letter to William Wimsatt, December 20, 1973.
9. "Interview," *Maclean's*.

10. Letter to Ezra Pound, January 5, 1951, *Letters*, p. 218.
11. "Network No. 1," unpublished paper, NA.
12. Interview with Edmund Carpenter.
13. Interview with Edmund Carpenter.
14. Interview with Northrop Frye.
15. Letter to Ezra Pound, June 22, 1951, *Letters*, p. 226.
16. Interview with Omar Solandt.
17. Interview with Edmund Carpenter.
18. Interview with John Meagher.
19. Stearn, *McLuhan: Hot & Cool*, p. 265.
20. Interview with Donald Theall.
21. *Explorations: Studies in Culture and Communication* (1953–1959), University of British Columbia Library, Special Collections.
22. Letter from Felix Giovanelli to Marshall McLuhan, January 22, 1949.
23. Letter to Gershon Legman, July 16, 1969, *Letters*, p. 382.
24. Letter to Elsie McLuhan, July 4, 1951.
25. Letter to Kaj Spencer, November 8, 1977. *Playboy*, March 1969.
26. "Playboy Interview."
27. Letter to Marshall Fishwick, April 7, 1972. Letter to Rene Wellek, January 3, 1973. Letter to John Rowan, December 17, 1969.
28. McLuhan acquired this understanding of the word *satire* and its nonmoral nature from Wyndham Lewis's *Men Without Art*.
29. Letter to J.G. Keogh, July 6, 1970, *Letters*, p. 413.
30. Letter to Ezra Pound, January 5, 1951, *Letters*, p. 218. This point was emphasized by another article in the first issue by Gyorgy Kepes, professor of visual design at MIT, who argued that art tends to reify the abstractions of science through its use of imagery.
31. *Playboy*, March 1969.
32. McLuhan, Speech in *Empire Club Addresses*. Letter to Harold Rosenberg, March 1, 1965. "Probe for *Foundations*," unpublished note, NA.
33. Letter to James Carey, March 25, 1974. Letter to Tim Bost, January 28, 1974. "Playboy Interview."
34. Diary, August 4, 1978.
35. Edmund Carpenter, "Certain Media Biases," *Explorations*, no. 3, p. 74.
36. McLuhan et al., "Report of the Ford Seminar at Toronto University, 1953–55," p. 6, Ford Foundation Archives, PA53–70.
37. Interview with Edmund Carpenter.
38. Note to Bernard Berelson from H.M. McLuhan, "Inter-Disciplinary Seminar in Culture and Communications at Toronto University," June 1954, p. 4. Ford Foundation Archives, PA53–70.
39. Letter to James M. Curtis, September 27, 1972.
40. Interview with Endel Tulving.
41. Interview with Edmund Carpenter.
42. Letter to Wyndham Lewis, March 7, 1955, *Letters*, p. 247.
43. Interview with John Bassett.
44. Interview with Seon Manley.
45. Walter J. Ong, "Space and Intellect in Renaissance Symbolism," *Explorations*, February 1955.
46. Neil Compton, "Cool Revolution." *Commentary*, January 1965.
47. Culkin, "Marshall's New York Adventure," unpublished essay.
48. Letter to William Jovanovich, June 8, 1967. Interview with Daniel Cappon. Thomas Dilworth, "Remembering Marshall McLuhan for Philip Marchand," unpublished essay, n.d. Interview with Frederick Rainsberry.
49. Interview with Robert Logan.

50. Letter to Margaret Mead, January 25, 1973, *Letters*, p. 463.
51. Unpublished note, NA.
52. Letter to Frank Sheed, February 19, 1970.
53. "Playboy Interview."
54. Letter to Charles Silberman, November 26, 1964. "Marshall McLuhan, Prophet and Analyst."
55. Kostelanetz, "Marshall McLuhan," pp. 109–10.
56. McLuhan, "AAP," p. 3, unpublished essay, NA.
57. Letter to Charles Silberman, October 5, 1963. Letter to George Leonard, May 12, 1967. McLuhan, *Understanding Media*, p. 66.
58. McLuhan, "Educational Effects of the Mass Media of Communication," paper delivered at Columbia University, November 1955, unpublished, NA.
59. Interview with Louis Forsdale.
60. Interview with Neil Postman.
61. In today's real estate market in Toronto the house, which still stands, would command an astronomical price.
62. Interview with Addie Strohbach.

8. The Electronic Call Girl

1. Pound, *Jefferson and/or Mussolini*.
2. McLuhan, "Our New Electronic Culture."
3. McLuhan, "Progress Report for Project in Understanding New Media — From September to November 1959." Letter to John Wain, November 20, 1959.
4. Letter to Harry Skornia, December 1, 1958.
5. Letter to Harry Skornia, October 8, 1959.
6. Interview with Sheila Watson.
7. Letter to Harry Skornia, October 1, 1959.
8. Interview with Derrick de Kerckhove.
9. Letter to Harry Skornia, October 8, 1959.
10. Interview with Everett Munro.
11. Interview with Arthur Porter.
12. Interviews with Arthur Porter, Ernest Sirluck, and James Ham.
13. Letter to Ethel Tincher, February 7, 1960. Letter to Harry Skornia, March 24, 1960.
14. Letter to Harry Skornia, January 21, 1960. McLuhan, "New Media and the New Education."
15. In stating that the structure of the newspaper page was "auditory," for example, McLuhan explained that "any pattern in which the components co-exist without direct linear hook-up or connection, creating a field of simultaneous relations, is auditory, even though some of its aspects can be seen" ("Agenbite of Outwit"). The newspaper, that is to say, existed in auditory rather than visual space. No wonder people could not "understand" McLuhan.
16. Quoted by McLuhan in his Foreword to *Training That Makes Sense*.
17. Letter to Harry Skornia, April 4, 1960.
18. Letter to Harry Skornia, December 14, 1959.
19. Letters to Harry Skornia, March 8, April 4, 1960.
20. Letter to Claude Bissell, March 8, 1960, *Letters*, p. 265.
21. McLuhan, "Progress Report for Project in Understanding New Media — From September to November 1959." Interview with Louis Forsdale.
22. "Report on Project in Understanding New Media, June 30, 1960, from Marshall McLuhan."
23. "Report on Project in Understanding New Media, June 30, 1960, from Marshall McLuhan."
24. Letter to Elsie McLuhan, March 7, 1935, *Letters*, p. 63.
25. "Report on Project in Understanding New Media." "Playboy Interview."
26. McLuhan, "Humanities in the Electronic Age."

27. Charles Hoban, *New Teaching Aids for the American Classroom* (Stanford, Calif.: Stanford Institute for Communications Research, 1960), p. 103.
28. Interview with Scott Taylor.
29. Corelli, "Great Gulf." Deane, "Sheriff and the Lawyer."
30. McLuhan, "Prospect of America."
31. McLuhan, "Murder by Television."
32. Interview with Ralph Baldwin.
33. Interview with Ralph Baldwin.
34. Letter from John Snyder to Marshall McLuhan, August 7, 1963. Letters to John Snyder, August 4, August 14, 1963.
35. Letter to Wyndham Lewis, [Dcember 1955], *Letters*, p. 248. Letter to Sheila Watson, August 17, 1958. Letter to John Wain, November 20, 1959. Letter to Harry Skornia, December 1, 1958.
36. Letter to Sheila Watson, August 12, 1963.
37. "Misunderstanding Media?" *Toronto Telegram*, August 22, 1964.
38. Kostelanetz, "Marshall McLuhan," p. 85. Letter to Gerald Dunne, July 24, 1970.
39. Letter to Sheila Watson, August 12, 1963.
40. McLuhan, *Understanding Media*, p. 329.
41. Interview with Frank Zingrone.
42. A. Alvarez, "Evils of Literacy," *New Statesman*, December 21, 1962.
43. Frank Kermode, "Between Two Galaxies," *Encounter*, February 1963.
44. *Times Literary Supplement*, July 19, 1963, August 6, 1964.
45. Letter to Charles Silberman, October 5, 1963. Letter to S.D. Neill, n.d.
46. McLuhan, "*Life* Magazine and the Extensions of Man," NA.
47. Letter from Charles Silberman to Marshall McLuhan, October 14, 1963. Interview with Charles Silberman. Interview with Leslie Fiedler.
48. Interview with Charles Silberman.
49. McLuhan even believed that the effects of artifacts could be felt *before* the artifacts existed. Artists, for instance, perceived these effects, as Tolstoy felt the effects of movies with his cinematic narrative technique in *War and Peace* and Seurat felt the effects of television with his pointillist paintings.
50. Interview with John Kelly, C.S.B. Interview with Claude Bissell.
51. Interview with John Kelly, C.S.B.
52. Interview with Ernest Sirluck.
53. In 1969, at the height of McLuhan's fame, the total budget of the center was a little under $39,000. Four years later, with inflation beginning to accelerate, the budget was just under $47,000. Of that total, $26,513 was for McLuhan's salary, $10,752 for his assistant, $8,000 for his secretary, and the rest for office supplies and incidental expenses. Year after year this basic division of funds remained unaltered. The budget was cut only once. In the late seventies, the center received a "ceremonial" cut of five percent, when the university was under severe financial strain and the dean wanted to be able to say that he had cut everyone's budget.
54. Interview with A.E. Safarian.
55. Letter to Sheila and Wilfred Watson, July 19, 1963.
56. Letter to William S. Paley, April 13, 1964.
57. Interview with Mac Hillock.
58. Letter to Herbert Krugman, June 25, 1970.
59. Hurst perfected a design for a "stereoscopic head camera" to detect dyslexia and also devised a series of eye exercises to combat the condition. McLuhan never succeeded in obtaining the estimated $50,000 that Hurst, who became a research associate of the center, needed to get his camera built.
60. *Toronto Daily Star*, May 7, 1964.
61. Letter to Robert Gray, June 12, 1964.

62. Fulford, "On Marshall McLuhan."
63. McLuhan, *Understanding Media*, p. 15.
64. McLuhan, "Invisible Environment."
65. "Understanding Canada," *Mademoiselle*.
66. Letter to John Etenza, September 30, 1964.
67. McLuhan, "Inside Blake and Hollywood."
68. Letter to Richard Berg, November 16, 1964.
69. Rosenberg, "Philosophy."
70. McLuhan, *Understanding Media*, p. 80.
71. McLuhan, *Understanding Media*, p. 329. Letter to Herbert Mitgang, October 15, 1970.
72. Letter to Robert Manning, July 21, 1970.
73. Letter to Charles Silberman, July 2, 1964; letter to Marylois Purdy (for the editor of *Time*), August 7, 1964.

9. "Canada's Intellectual Comet"

1. McLuhan, "Television in a New Light."
2. Interview with Abraham Rogatnick.
3. *Financial Post*, April 18, 1970.
4. Wolfe, "What If He Is Right?"
5. Interview with Gerald Feigen.
6. Interview with Gordon Thompson.
7. Wolfe, "What If He Is Right?" p. 130.
8. Letter to Tom Wolfe, November 22, 1965, *Letters*, p. 330.
9. "Advertising: A Different Look at Creativity," *New York Times*, September 8, 1965.
10. Letter from Jonathan Miller to Marshall McLuhan, April 28, 1965. Letter to Harold Rosenberg, March 1, 1965. Letter to Sheila Watson, September 22, 1964.
11. McLuhan, "A Day in the Life."
12. Drucker, "Adventures of a Bystander," p. 251.
13. "All-at-once World," *Canadian*.
14. "Interview," *Miss Chatelaine*.
15. Letter to Joseph Foyle, April 4, 1972.
16. Letter from Buckminster Fuller to John Ragsdale, November 7, 1966.
17. Interview with Leslie Fiedler. Kostelanetz, "Marshall McLuhan," p. 82.
18. *New York Times*, October 5, 1970.
19. Letter to David Meynell, February 9, 1965.
20. Interview with Arthur Porter.
21. Letter to Ralph Cohen, October 1, 1964. Letter to Carroll Hollis, August 2, 1973.
22. "Table Talk," *Maclean's*.
23. Letter to Hoyt Spelman, November 14, 1966.
24. Wain, "Incidental Thoughts."
25. Letter to Sheila and Wilfred Watson, January 28, 1965.
26. "All-at-once World," *Canadian*.
27. McLuhan, "Man Who Came to Listen."
28. *New York Times*, November 22, 1966.
29. *New York Times*, June 14, 1966.
30. Interview with Gordon Dryden.
31. Letter to John Wain, June 25, 1971.
32. Interview with Gordon Dryden.
33. Wain, "Incidental Thoughts." Letter to William Jovanovich, October 21, 1965.
34. Baldwin, who had negotiated the matter with Corinne, understood that he would receive forty

percent of the fees from articles and from speaking engagements and consultations that he arranged for McLuhan. Unfortunately, McLuhan himself was not apprised of the deal, and some friction later resulted between the two men over the arrangement.

35. Interview with Frank Zingrone.
36. McLuhan, "Future of Education."
37. McLuhan, "Future of Sex."
38. Letter to Brenda Kunz, March 13, 1975.
39. Quoted in the *Edmonton Journal*, January 18, 1967.
40. Quoted in Kostelanetz, "Understanding McLuhan (in Part)."
41. Letter to Marius Bewley, November 26, 1969.
42. Letter to Alexis De Beauregard, May 11, 1972. Letter to Jacqueline Tyrwhitt, May 28, 1970.
43. Letter to Sheila Watson, October 4, 1964. Letter to John Holt, April 16, 1971.
44. Interview with David Nostbakken.
45. Letter to Max Nanny, April 8, 1971.

10. New York City

1. John Henry Newman, *Discourses Addressed to Mixed Congregations*, 1849.
2. Letter from Ralph Baldwin to Marshall and Corinne McLuhan, June 21, 1966.
3. Interview with Ralph Baldwin.
4. *New York Times*, September 19, 1967.
5. *WSN & Free Press*, December 14, 1967.
6. Interview with Philip Romano.
7. *New York Times*, September 7, 1967.
8. Interview with Harley Parker.
9. *New York Times*, October 10, 1967.
10. Hoving, untitled article.
11. Letter to Richard Kluger, February 25, 1970.
12. Interview with Tony Schwartz.
13. *Harper's Bazaar*, April 1968.
14. Letter to David Bell, July 20, 1972. Letters from Eugene Schwartz to Marshall McLuhan, July 23, November 7, 1968.
15. These centers were not such a far-fetched notion in the late sixties, especially for McLuhan, who found it hard to give up his cherished schemes for sensory quotient tests and the like. It was at least partially because of his ongoing interest in such matters that he signed on in December 1967 as a "consultant" with the Responsive Environments Corporation of Englewood Cliffs, New Jersey, marketer of the Talking Typewriter, a computer-based learning system for teaching reading and other skills.
16. Interview with Paul Klein.
17. Unpublished note by McLuhan, n.d., NA.
18. Letter to Edward T. Hall, November 3, 1969, *Letters*, p. 391.
19. Wain, "Incidental Thoughts." Interview with Marcel Kinsbourne.
20. Interview with Edmund Carpenter.
21. Interview with Gregory Baum.
22. Letter to Ted Willison, May 15, 1974. Interview with Derrick de Kerckhove.
23. Stearn, *McLuhan: Hot & Cool*, p. 261. "Playboy Interview."
24. *Playboy*, March 1969.
25. Letter to Robert Campbell, August 19, 1968. Letter to Jacques Maritain, May 6, 1969, *Letters*, p. 370.
26. Letter to M.J. Bradley, July 30, 1971. Diary, May 21, 1972.
27. Hoffman quoted in *New York*, September 2, 1968. Letter to Donna Schrader, January 16, 1969. Leary quoted in Boyle, "Marshall McLuhan."

28. Letter to John Lindsay, June 12, 1968. Letter from John Lindsay to Marshall McLuhan, June 28, 1968.
29. McLuhan, "All the Candidates."
30. "Table Talk," *Maclean's*.
31. "Table Talk," *Maclean's*.
32. Letter to J.M. Davey, May 13, 1970.
33. McLuhan, *War and Peace in the Global Village*, p. 97.
34. McLuhan, *War and Peace in the Global Village*, p. 149.
35. Quoted in Hugh Kenner, *The Pound Era* (Berkeley: University of California Press, 1971), p. 436.
36. McLuhan, *War and Peace in the Global Village*, p. 152. Letter to Philip and Molly Deane, December 15, 1967. *New York Times*, September 29, 1967.
37. Letter to Elsie McLuhan, [June 28, 1936], *Letters*, p. 86.
38. "Interview," *Miss Chatelaine*. Interview with Kirwin Cox.
39. "Understanding Canada," *Mademoiselle*.
40. "Understanding Canada," *Mademoiselle*.

11. Unsold Books

1. He offered to produce for Frank Kermode, then editing a series of books for the publisher Fontana, a small volume under that title dealing with the life of cities and neighborhoods. (See Letter to Frank Kermode, July 5, 1969, *Letters*, p. 375.)
2. McLuhan, "A Day in the Life."
3. In fact, Sheila Watson functioned that year virtually as McLuhan's portable memory, directing his attention to passages he had forgotten from books he had once known almost by heart. She had first met him in 1956, when she was a graduate student at the University of Toronto writing a doctoral thesis on Wyndham Lewis. McLuhan had directed her thesis, and she and her husband, Wilfred, had remained close friends of his.
4. Diary, November 9, 1974.
5. *New York Times*, October 5, 1970.
6. Interview with Gordon Thompson.
7. Interview with Alan Williams.
8. Letter to Frank Kermode, July 5, 1969, *Letters*, p. 375. Letter from Gerald Stearn to Matie Molinaro, October 12, 1970. Interview with Robert Silver.
9. Diary, May 17, May 19, 1974.
10. Letter to William Jovanovich, July 23, 1970. Letter to Frederic Hills, April 6, 1976.
11. Interview with David Nostbakken.
12. Interview with Derrick De Kerckhove.
13. Letter to Pierre Trudeau, November 27, 1970. Letter to James Davey, April 8, 1971. Letter to Pierre Trudeau, September 21, 1971, *Letters*, p. 439. *New York Book Review*, November 17, 1968.
14. Letter to Pierre Trudeau, February 24, 1977, *Letters*, p. 527.
15. Interview with Stanley Burke.
16. Letter from Ann Landers to Marshall McLuhan, December 29, 1969.
17. Letters to Pierre Trudeau, October 17, 1968, June 13, 1974, *Letters*, pp. 356–57, 499.
18. Interview with Patrick Watson.
19. Unpublished note, NA.
20. Interview with C. Edward McGee. Interview with Scott Taylor.
21. Letter to Joan and Stan Brasier, December 14, 1973. Diary, March 19, 1976.
22. Letter to Roger Broughton, May 15, 1968.
23. McLuhan was introduced to transcendental meditation when he gave a talk to an audience of 1,200 TM enthusiasts in July 1972. He received a standing ovation from the audience after the Maharishi himself introduced McLuhan as a "seer."

24. Letter to Betty Lewis, October 24, 1968. Letter to Ted Carpenter, April 24, 1969. Letter to Joseph Keogh, May 28, 1971. Interview with David Belyea. Interview with Bede Sullivan.
25. "Playboy Interview."
26. "Interview," *Miss Chatelaine.*
27. Letter to Ralph Nader, July 31, 1974.
28. Letter to John Rowan, December 17, 1969.
29. Letter to John Rowan, December 17, 1969. "Interview," *Miss Chatelaine.*
30. This was not the first time researchers had tried to test the validity of McLuhan's theses. Ted Carpenter had initiated such tests with his media bias experiment in 1954. In the years since then, various other experiments, contrary to Carpenter, had found no significant differences in the retention of information by users of the various media. A researcher at UCLA named Pierre Lorion took an entirely different tack in testing, focusing not on the effects of different media but on the general sensibility of pre- and post-television generations. He showed a film, edited in a random fashion from various bits of preexisting films, to 150 subjects of various ages and tested their responses. His conclusion was that the younger, presumably television-raised, subjects were indeed more attuned to "mosaic thinking" than to "lineality." Taking yet another approach, two graduate students at the Annenberg School of Communications at the University of Pennsylvania attempted to test the relative "involvement" of viewers of film and television by measuring their pupillary dilations. Involvement in film, according to this test, was greater than involvement in television, contrary to McLuhan's thesis. (See "The Avenging Eye," *Saturday Review*, May 16, 1970; Kaj Murray, "An Empirical Examination of Two Theories on the Nature and Effects of Media as Presented by Marshall McLuhan in *Understanding Media: The Extensions of Man*," unpublished thesis.)
31. Herbert Krugman, "Brain Wave Measures of Media Involvement," *Journal of Advertising Research*, February 1971.
32. Herbert Krugman, "TV vs. Print," *Newsweek*, November 2, 1970.
33. Letter to Kenneth Millar, September 26, 1969.
34. Letter to Bill Bret, March 17, 1973.
35. Letter to Claire Smith, December 16, 1969. Letter to John Bassett, March 15, 1971.
36. Letter to Frank Kermode, July 5, 1969, *Letters*, p. 375.
37. Letter to Barbara Rowes, April 15, 1976. Letter to Joseph Keogh, February 26, 1971. Letter to Sheila Watson, September 4, 1971.
38. Letter to Sheila Watson, July 27, 1972.
39. Letter to Claude Cartier-Bresson, July 19, 1972, *Letters*, pp. 452–53. Letter to Margaret Mead, January 25, 1973, *Letters*, pp. 463–64. Letter to William Jovanovich, July 27, 1972.
40. Letter to Tom Wolfe, February 4, 1971.
41. Interview with Barrington Nevitt. Unpublished Harcourt Brace Jovanovich memorandum, "Preparation of Marshall McLuhan and B. Nevitt book — *Take Today*," NA. Letter to Bernard Muller-Thym, September 2, 1970. Letter to Joseph Foyle, May 9, 1973.
42. Letter to Aubrey Williams, May 18, 1972.
43. Interview with Barrington Nevitt.
44. Letter to Thomas J. Farrell, August 3, 1973. Letter to Sam Neill, September 12, 1972.

12. The Sage of Wychwood Park

1. Ernest Gellner, *The Psychoanalytic Movement.*
2. Diary, November 7, 1972.
3. *New York Times*, September 29, 1976.
4. Letter to Sidney Halpern, January 9, 1974, *Letters*, p. 488. Diary, October 8, 1973.
5. McLuhan, "A Last Look at the Tube." Letter to L.A. Morse, December 9, 1970.
6. "Interview," *Miss Chatelaine.*

7. Letter to John Rowan, December 17, 1969. McLuhan, "Violence of the Media."

8. Diary, May 9, 1977. Letter to Robert Hittel, December 20, 1977. Letter to David Staines, May 9, 1977. Letter to Edward Wakin, July 27, 1978.

9. "Table Talk," *Maclean's*. Unpublished text of an interview with Marshall McLuhan by Professor Gary Kern, NA.

10. Letter to Donald Theall, March 17, 1977. Letter to Barbara Rowes, January 31, 1977. Letter to David Staines, January 26, 1977. Diary, January 26, 1977. Interview with Scott Taylor.

11. Letter to Gerald Piel, November 26, 1973. Unpublished note, NA.

12. Letter to Ted Carpenter, January 4, 1972. Letter to Mark Lathrop, February 19, 1974.

13. Letter to Gerald Piel, October 22, 1973. Unpublished note, "Changing Paradigms in the Arts and Sciences Since 1900," NA. Interview with Harley Parker. Interview with Betty Corson.

14. Unpublished note, NA.

15. McLuhan, *Understanding Media*, p. 12.

16. *Extra!* April 1978.

17. Letters to Paul Levinson, September 8, 1977, February 10, 1978. McLuhan, "Laws of the Media." In the book published after McLuhan's death, *Laws of Media* (University of Toronto Press, 1988), by Marshall and Eric McLuhan, the authors flatly state, "The laws of media in tetrad form belong properly to rhetoric and grammar, not philosophy." (So much for the scientific method.) It is interesting to note that Aristotle, in those sections of his book *On Rhetoric* that amount to the first systematic treatment of human psychology, organizes his discussion in a four-part, chiastic structure curiously reminiscent of the structure of McLuhan's tetrad. That is, Aristotle discusses various emotions in a positive/negative, negative/positive sequence. McLuhan's tetrad could also be read this way: positive (enhance)/negative (reverse), negative (make obsolete)/positive (retrieve).

18. Letter to Marshall Fishwick, November 12, 1973.

19. Letter to Louis Forsdale, May 9, 1978.

20. Interview with Betty Corson.

21. Letter to Richard Kostelanetz, April 9, 1975. Letter to Thomas Stanley, May 9, 1974. Diary, November 11, 1974, December 31, 1974.

22. Diary, March 25, 1972.

23. Unpublished interview (1976) with McLuhan, conducted by John Ayre, author of "McLuhan Revisited."

24. Letter to Christine Breech, November 9, 1976. Letter to Richard Berg, September 14, 1976. Letter to Murray Koffler, March 30, 1977.

25. Letter to David Nostbakken, August 10, 1976. Letter to David Staines, August 4, 1976. "Interview," *Maclean's*.

26. Interview with Marcel Kinsbourne.

27. Letter to Nathan Cervo, May 4, 1976. Letter to John Cambus, May 10, 1974. Letter to Gertrude LeMoyne, August 3, 1976. McLuhan, "English Literature as Control Tower in Communication Study."

28. Letter to Edward T. Hall, September 5, 1973, *Letters*, pp. 480–81. Letter to Sheila Watson, September 4, 1971.

29. Diary, January 4, 1975.

30. Diary, May 25, June 7, 1974.

31. Diary, November 11, 1974.

32. Letter to Malcolm Ross, July 6, 1970.

33. Letter to Bill and Iris Key, September 17, 1975.

34. Diary, March 14, March 26, March 28, August 7, September 8, September 15, October 29, 1974, December 27, 1975, March 29, 1976, June 11, 1977. Letter to Philip Chamberlin, September 30, 1977.

35. Grady, "They Loved Her."

36. Interview with Maurice Farge. Mr. Farge unfortunately cannot remember the name of the daughter who voiced this complaint. It might have been any one of them.
37. Diary, July 1, 1975.
38. "Report of the Review Committee for the Centre for Culture and Technology, University of Toronto School of Graduate Studies," May 1980. Powe, "Marshall McLuhan, the Put-on."
39. Ayre, "McLuhan Revisited."
40. In some respects, the innovations of the microchip computer — its invasion of other media such as the telephone, its dramatically enlarged frames of reference, and its enhanced abilities to recognize the context of data — validated much of McLuhan's thinking, such as his emphasis on the nonsequential nature of information in the electric age. They passed him by, in any case.
41. Letter to Frederick Wilhelmsen, December 20, 1971, February 29, 1972.
42. Letter to Frederick Wilhelmsen, July 31, 1975.
43. Letter to J.M. Davey, March 8, 1971, *Letters*, p. 427. Letter to Archie Malloch, January 5, 1976.
44. Letter to John Atkin, March 16, 1971.
45. "McLuhan McLuhan McLuhan."
46. McLuhan, "Politics as Theatre."
47. Letter to Vincent Lackner, June 4, 1973. Diary, April 30, 1973. Unpublished note, NA.
48. Diary, February 22, 1973. Letter to Sheila Watson, February 23, March 22, 1973.
49. Letter to R. Drake Will, February 6, 1976.
50. Letter to Joseph Foyle, July 18, 1979. Letter to Clare Boothe Luce, January 7, 1976. Letter to William Ewald, June 6, 1968.
51. Letter to Eric McLuhan, September 24, 1975. This letter is evidence of McLuhan's deep concern over Eric's progress at the University of Dallas, and particularly over the fate of his M.A. thesis. At one point, McLuhan consulted one of his old nonadmirers in the English department, F.E.L. Priestley, concerning some point in Eric's thesis on which Priestley was an authority. McLuhan was never shy about enlisting the aid even of people he knew had no reason to love him if the matter were sufficiently important.
52. Letter to A.E. Safarian, November 10, 1975. Letter to Amleto Lorenzini, September 30, 1977. Diary, October 18, 1977.
53. Diary, May 2, May 15, 1976.
54. Letter to the *Toronto Globe and Mail*, June 22, 1976. Diary, July 18, 1976.
55. Interview with Robert Logan.
56. Letter to Pierre Trudeau, February 24, 1977, *Letters*, p. 528.
57. Diary, August 11, 1976.
58. Interview with Scott Taylor.
59. Diary, May 10, 1976.
60. Diary, September 23, 1976.
61. Letter to Sister Geraldine, November 16, 1976. Letter to Tony Schwartz, November 25, 1976. Letter to Frederick Wilhelmsen, November 5, 1976. Letter to David Staines, November 8, 1976. McLuhan, "The Debates."
62. Canadian Press Obituary.
63. Letter to Anna Childers, November 21, 1974. *The Australian*, June 20, 1977.
64. Letter to Marshall Fishwick, August 1, 1974, *Letters*, p. 506.
65. Interview with Derrick De Kerckhove.
66. Interview with Peter Newman.
67. Interview with Neil Postman.
68. Letter from Corinne McLuhan to Mrs. Wyndham Lewis, October 18, 1976, Cornell University Library.
69. Diary, July 12, 1977. Letter from Leslie Brinegar to Marshall McLuhan, September 7, 1977. Letter to Leslie Brinegar, September 16, 1977.

70. Letter to Ethel Tincher, January 10, 1977. Diary, March 21, 1978.
71. Letter to Tom and Dorothy Easterbrook, March 3, 1972.
72. Interview with Gerald Goldhaber.
73. Interview with Gerald Goldhaber.
74. "A Day in the Life," *Weekend*.
75. Letter to Pierre Trudeau, September 7, 1979, *Letters*, p. 545.
76. Letter to Mel James, June 27, 1979. Letter to Gerald Goldhaber, February 5, 1979. Letter to Marshall McLuhan, August 24, 1979.
77. Diary, October 17, 1975.
78. McLuhan, "One Wheel, One Square." Letter to Sister St. John O'Malley, October 17, 1975. Interview with Scott Taylor.
79. Letter to John Greer Nicholson, February 7, 1979.
80. Assessor "A," Canada Council Document, File: 410–78–0395, NA.
81. Letter to Pierre Trudeau, July 26, 1978. Letter to Jim Coutts, January 29, 1979.
82. Interview with Dennis Duffy.
83. Letter to Pierre Trudeau, September 7, 1979, *Letters*, p. 546.

13. Silence

1. Interview with Armand Maurer, C.S.B.
2. Interview with Marcel Kinsbourne.
3. Interview with Patrick Watson.
4. Interview with John Leyerle.
5. "Report of the Review Committee for the Centre for Culture and Technology, University of Toronto School of Graduate Studies," May 1980.
6. Letter from Pierre Trudeau to Marshall McLuhan, April 28, 1980.
7. Interview with John Leyerle.
8. Interview with Margaret Stewart.
9. Interview with Barrington Nevitt.
10. Interview with Frank Stroud.
11. Howard, "Oracle of the Electric Age."
12. McLuhan, *Understanding Media*, p. 291.
13. Letter to Vlada Petric, January 14, 1975.

Bibliography

Books by Marshall McLuhan

The Mechanical Bride: Folklore of Industrial Man. New York: Vanguard Press, 1951.

Selected Poetry of Tennyson. Edited by Marshall McLuhan. New York: Rinehart, 1954.

Report on Project in Understanding New Media. National Association of Educational Broadcasters, 1960.

Explorations in Communication. Edited by Edmund Carpenter and Marshall McLuhan. Boston: Beacon Press, 1960.

The Gutenberg Galaxy: The Making of Typographic Man. Toronto: University of Toronto Press, 1962.

Understanding Media: The Extensions of Man. New York: McGraw-Hill, 1964.

Voices of Literature. Edited by Marshall McLuhan and Richard J. Schoeck. New York: Holt, Rinehart and Winston, 1964, 1965.

The Medium Is the Massage: An Inventory of Effects. With Quentin Fiore and Jerome Agel. New York: Bantam, 1967.

Verbi-Voco-Visual Explorations. New York: Something Else Press, 1967. (Reprint of *Explorations,* no. 8.)

War and Peace in the Global Village. With Quentin Fiore and Jerome Agel. New York: Bantam, 1968.

Through the Vanishing Point: Space in Poetry and Painting. With Harley Parker. New York: Harper & Row, 1968.

Counterblast. With Harley Parker. New York: Harcourt, Brace and World, 1969.

The Interior Landscape: The Literary Criticism of Marshall McLuhan, 1943–1962. Edited by Eugene McNamara. New York: McGraw-Hill, 1969.

From Cliché to Archetype. With Wilfred Watson. New York: Viking, 1970.

Culture Is Our Business. New York: McGraw Hill, 1970.

Take Today. With Barrington Nevitt. New York: Harcourt Brace Jovanovich, 1972.

City as Classroom: Understanding Language and Media. With Eric McLuhan and Kathryn Hutchon. Toronto: Book Society of Canada Limited, 1977.

Laws of Media: The New Science. With Eric McLuhan. Toronto: University of Toronto Press, 1988.

The Global Village: Transformations in World Life and Media in the 21st Century. With Bruce R. Powers. New York: Oxford University Press, 1989.

Essential McLuhan. Edited by Eric McLuhan and Frank Zingrone. Toronto: Anansi, 1995.

Articles by Marshall McLuhan

To list all the articles written by McLuhan would take many pages. The following are some of the more significant, including most of those cited in this book.

"Macaulay: What a Man!" *The Manitoban* (University of Manitoba student newspaper), October 28, 1930.

"George Meredith as a Poet and Dramatic Novelist." M.A. thesis, University of Manitoba, 1934.

"G.K. Chesterton: A Practical Mystic." *Dalhousie Review*, January 1936, pp. 455–64.
"The Cambridge English School." *Fleur de Lis* (St. Louis University student literary magazine), 1937, pp. 21–25.
"Peter or Peter Pan." *Fleur de Lis*, May 1938, pp. 4–7.
Review of *The Culture of Cities* by Lewis Mumford. *Fleur de Lis*, December 1938, pp. 38–39.
Review of *Art and Prudence* by Mortimer J. Adler. *Fleur de Lis*, October 1940.
"Apes and Angles." *Fleur de Lis*, December 1940, pp. 7–9.
Review of *Poetry and the Modern World* by David Daiches. *Fleur de Lis*, March 1941.
Review of *American Renaissance* by F.O. Matthiessen. *Fleur de Lis*, October 1941.
"Aesthetic Patterns in Keats' Odes." *University of Toronto Quarterly*, January 1943.
"Education of Free Men in Democracy: The Liberal Arts." *St. Louis Studies in Honor of St. Thomas Aquinas*, 1943, pp. 47–50.
"The Place of Thomas Nashe in the Learning of His Time." Ph.D. thesis, Cambridge University, April 1943.
"Dagwood's America." *Columbia*, January 1944, pp. 3, 22.
"Edgar Poe's Tradition." *Sewanee Review*, January 1944, pp. 24–33.
"Poetic vs. Rhetorical Exegesis." *Sewanee Review*, April 1944, pp. 266–76.
"Wyndham Lewis: Lemuel in Lilliput." *St. Louis University Studies in Honor of St. Thomas Aquinas*, 1944, pp. 58–72.
"Another Aesthetic Peep-Show." *Sewanee Review*, Autumn, 1945, pp. 674–77.
"The Analogical Mirrors." In *Gerald Manley Hopkins*. Kenyon Critics Edition. Norfolk, CT: New Directions Books, 1945, pp. 15–27.
"An Ancient Quarrel in Modern America." *Classical Journal*, January 1946, pp. 156–62.
Review of *William Ernest Henley* by Jerome Hamilton Buckley. *Modern Language Quarterly*, September 1946.
"Out of the Castle into the Counting-House." *Politics*, September 1946, pp. 277–79.
"Footprints in the Sands of Crime." *Sewanee Review*, October 1946, pp. 617–34.
"Mr. Connolly and Mr. Hook." *Sewanee Review*, July 1947, pp. 167–72.
"The Southern Quality." *Sewanee Review*, July 1947, pp. 357–83.
"Inside Blake and Hollywood." *Sewanee Review*, October 1947, pp. 710–15.
"American Advertising." *Horizon*, October 1947, pp. 132–41.
Introduction to *Paradox in Chesterton* by Hugh Kenner. London: Sheed and Ward, 1948.
"The 'Colour-Bar' of BBC English." *Canadian Forum*, April 1949, pp. 9–10.
"Mr. Eliot's Historical Decorum." *Renascence*, Autumn 1948, pp. 9–15.
"Pound's Critical Prose." In *Examination of Ezra Pound: A Collection of Essays*, edited by Peter Russell. New York: New Directions, 1950.
Review of eleven books on T.S. Eliot. *Renascence*, Autumn 1950, pp. 43–48.
"Joyce, Aquinas, and the Poetic Process." *Renascence*, Winter 1951, pp. 3–11.
"John Dos Passos: Technique vs. Sensibility." In *Fifty Years of the American Novel*, edited by Charles Gardiner. New York: Scribner's, 1951.
"Advertising as a Magical Institution." *Commerce Journal*, January 1952, pp. 25–29.
Review of *The Poetry of Ezra Pound* by Hugh Kenner. *Renascence*, Spring 1952, pp. 215–17.
"Defrosting Canadian Culture." *American Mercury*, March 1952, pp. 91–97.
"Technology and Political Change." *International Journal*, Summer 1952, pp. 189–95.
"The Aesthetic Moment in Landscape Poetry." In *English Institute Essays*, edited by Alan Downe. New York: Columbia University Press, 1952.
"From Eliot to Seneca." *University of Toronto Quarterly*, January 1953, pp. 199–202.
"Comics and Culture." *Saturday Night*, February 28, 1953, pp. 1, 19–20.
"James Joyce: Trivial and Quadrivial." *Thought*, Spring 1953, pp. 75–98. Reprinted in *The Interior Landscape: The Literary Criticism of Marshall McLuhan, 1943–1962*, edited by Eugene McNamara. New York: McGraw-Hill, 1969.

"The Age of Advertising." *Commonweal,* September 11, 1953, pp. 555–57.

"Maritain on Art." *Renascence,* Autumn 1953, pp. 40–44.

"The Later Innis." *Queen's Quarterly,* Autumn 1953, pp. 385–84[check pg.nos].

"Wyndham Lewis: His Theory of Art and Communication." *Shenandoah,* Autumn 1953, pp. 77–88.

"Culture without Literacy." *Explorations: Studies in Culture and Communication,* no. 1, December 1953, pp. 117–27.

"Joyce, Mallarmé, and the Press." *Sewanee Review,* Winter 1954, pp. 38–55.

"Media as Art Forms." *Explorations,* no. 2, April 1954, pp. 6–13.

"New Media as Political Forms." *Explorations,* no. 3, August 1954, pp. 120–26.

"Counterblast." Privately published booklet, 1954.

"Space, Time, and Poetry." *Explorations,* no. 3, August 1954, pp. 120–26.

"Five Sovereign Fingers Taxed the Breath." *Explorations,* no. 4, February 1955. (Reprinted in *Shenandoah,* Autumn 1955, pp. 50–52.)

"Nihilism Exposed." *Renascence,* Spring 1955, pp. 97–99.

"Radio and Television vs. The ABCED-Minded." *Explorations,* no. 5, June 1955, pp. 12–18.

"Educational Effects of Mass Media of Communication." *Teachers College Record,* March 1956, pp. 400–403.

"The Media Fit the Battle of Jerico." *Explorations,* no. 6, July 1956, pp. 15–19.

"The Effect of the Printed Book on Language in the 16th Century." *Explorations,* no. 7, March 1957, pp. 99–108.

"Classrooms Without Walls." *Explorations,* no. 7, March 1957, pp. 22–26.

"David Riesman and the Avant-Garde." *Explorations,* no. 7, March 1957, pp. 112–16.

"Jazz and Modern Letters." *Explorations,* no. 7, March 1957, pp. 74–76.

"Brain Storming" (and other essays). *Explorations,* no. 8, October 1957.

"Have with You to Madison Avenue, or The Flush Profile of Literature." Unpublished review of *The Anatomy of Criticism* by Northrop Frye, 1957, NA.

"One Wheel, One Square." *Renascence,* Summer 1958.

"Our New Electronic Culture: The Role of Mass Communications in Meeting Today's Problems." *NAEB Journal,* October 1958, pp. 19–20, 24–26.

"Myth and Mass Media." *Daedalus,* Spring 1959, pp. 339–48.

"Myth, Oral and Written." *Commentary,* July 1960, pp. 90–91.

"Inside the Five Sense Sensorium." *Canadian Architect,* June 1961, pp. 49–51.

"The Humanities in the Electronic Age." *Humanities,* Fall 1961, pp. 3–11.

"The New Media and the New Education." *Basilian Teacher,* December 1961, pp. 93–100.

"Prospect of America." *University of Toronto Quarterly,* October 1962, pp. 107–8.

"The Agenbite of Outwit." *Location,* Spring 1963.

"Printing and the Mind." *Times Literary Supplement,* July 19, 1963, pp. 517–18.

"Murder by Television." *Canadian Forum,* January 1964, pp. 222–23.

"Notes on Burroughs." *Nation,* December 28, 1964, pp. 517–19.

Introduction to *The Bias of Communication* by Harold A. Innis. Reprint edition. Toronto: University of Toronto Press, 1964.

"Wordfowling in Blunderland." *Saturday Night,* August 1965, pp. 23–27.

"Address at Vision 65." *American Scholar,* Spring 1966, pp. 196–205.

"Electronics and the Psychic Drop-Out." *This,* April 1966.

"The Invisible Environment." *Canadian Architect,* May 1966, pp. 71–74.

"Great Change-overs for You." *Vogue,* July 1966, pp. 62–63, 114–15, 117.

"The Relation of Environment to Anti-Environment." *University of Windsor Review,* Fall 1966.

"Questions and Answers with Marshall McLuhan." *Take One,* November/December 1966, pp. 7–10.

"Television in a New Light." In *The Meaning of Commercial Television*, edited by Stanley T. Donner. Austin: University of Texas Press, 1966, pp. 87–107.

"Cybernation and Culture." In *The Social Impact of Cybernetics*, edited by Charles Dechert. South Bend, IN: University of Notre Dame Press, 1966.

"Art as Anti-Environment." *Art News Annual*, 1966.

"The Future of Education" (with George B. Leonard). *Look*, February 21, 1967, pp. 23–25.

"Love." *Saturday Night*, February 1967, pp. 25–28.

"Marshall McLuhan Massages the Medium." *Nation's Schools*, June 1967, pp. 36–37.

"The Future of Sex" (with George B. Leonard). *Look*, July 25, 1967, pp. 56–63.

"All the Candidates Are Asleep." *Saturday Evening Post*, August 1968, pp. 34–36.

Review of *Federalism and the French Canadians* by Pierre Trudeau. *New York Times Book Review*, November 17, 1968.

"The Reversal of the Overheated Image." *Playboy*, December 1968.

"Wyndham Lewis." *Atlantic Monthly*, December 1969, pp. 93–98.

"The Man Who Came to Listen" (with Barrington Nevitt). In *Peter Drucker: Contributions to Business Enterprise*, edited by Tony Bonaparte and John Flaherty. New York: New York University Press, 1970, pp. 35–55.

Foreword to *Training That Makes Sense* by A.J. Kirshner. San Rafael, CA: Academic Therapy Publications, 1972, pp. 5–7.

"Politics as Theatre." *Performing Arts in Canada*, Winter 1973, pp. 14–15.

Speech included in *The Empire Club Addresses: 1972–1973*. Toronto: Empire Club Foundation, 1973.

"Do Americans Go to Church to Be Alone?" *Critic*, January/February 1973, pp. 14–23.

"Mr. Nixon and the Dropout Strategy." *New York Times*, July 29, 1973.

"McLuhan McLuhan McLuhan." *New York Times*, May 10, 1974.

"English Literature as Control Tower in Communication Study." *English Quarterly* (University of Waterloo), Spring 1974, pp. 3–7.

"A Media Approach to Inflation." *New York Times*, September 21, 1974.

"McLuhan's Laws of the Media." *Technology and Culture*, January 1975, pp. 74–78.

"The Debates." *New York Times*, September 23, 1976.

"The Violence of the Media." *Canadian Forum*, September 1976, pp. 9–12.

"Laws of the Media." *Et Cetera*, June 1977, pp. 173–78.

"A Last Look at the Tube." *New York*, April 3, 1978, p. 45.

"A Day in the Life: Marshall McLuhan," *Weekend*, June 10, 1978, p. 10.

Published Interviews with Marshall McLuhan

"Understanding Canada and Sundry Other Matters: Marshall McLuhan." *Mademoiselle*, January 1967, pp. 114–15, 126–30.

"Playboy Interview: Marshall McLuhan." *Playboy*, March 1969, pp. 26–27, 45, 55–56, 61, 63.

"The Table Talk of Marshall McLuhan" by Peter C. Newman. *Maclean's*, June 1971, pp. 42, 45.

"An Interview with Marshall McLuhan: His Outrageous Views about Women" by Linda Sandler. *Miss Chatelaine*, September 3, 1974, pp. 58–59, 82–87, 90–91.

"'It Will Probably End the Motor Car': An Interview with Marshall McLuhan" by Kirwan Cox and S.M. Crean. *Cinema Canada*, August 1976, pp. 26–29.

"Interview with Professor Marshall McLuhan." *Maclean's*, March 7, 1977.

Books about Marshall McLuhan

Benedetti, Paul and Nancy DeHart, editors. *Forward Through the Rearview Mirror: Reflections On and By Marshall McLuhan*. Toronto: Prentice Hall Canada, 1996.

Crosby, Harry H., and George R. Bond, editors. *The McLuhan Explosion: A Casebook on Marshall McLuhan and Understanding Media.* New York: American Book Company, 1968.

Duffy, Dennis. *Marshall McLuhan.* Toronto: McClelland and Stewart, 1969.

Finkelstein, Sidney. *Sense and Nonsense of McLuhan.* New York: International Publishers, 1968.

Gordon, W. Terrence. *Marshall McLuhan: Escape into Understanding.* Toronto: Stoddart, 1997.

———. *McLuhan for Beginners.* New York and London: Writers & Readers Publishing, 1997.

Grosswiler, Paul. *The Method Is the Message.* Montreal: Black Rose Books, 1998.

Kroker, Arthur. *Technology and the Canadian Mind: Innis/McLuhan/Grant.* Montreal: New World Perspectives, 1984.

Miller, Jonathan. *McLuhan.* London: William Collins, 1971.

Molinaro, Matie, Corinne McLuhan, and William Toye, editors. *Letters of Marshall McLuhan.* Toronto: Oxford University Press, 1987.

Neill, S. D. *Clarifying McLuhan: An Assessment of Process and Product.* Westport, CT.: Greenwood Press, 1993.

Nevitt, Barrington, and Maurice McLuhan, editors. *Who Was Marshall McLuhan?* Toronto: Stoddart, 1995.

Patterson, Graeme. *History and Communications: Harold Innis, Marshall McLuhan, and the Interpretation of History.* Toronto: University of Toronto Press, 1990.

Rosenthal, Raymond, editor. *McLuhan: Pro and Con.* Baltimore: Penguin, 1968.

Sanderson, George, and Frank Macdonald, editors. *Marshall McLuhan: The Man and His Message.* Golden, Colorado: Fulcrum, 1989.

Stearn, Gerald Emanuel, editor. *McLuhan: Hot & Cool.* New York: Dial, 1967.

Theall, Donald F. *The Medium Is the Rear View Mirror: Understanding McLuhan.* Montreal: McGill-Queen's University Press, 1971.

Willmott, Glen. *McLuhan, or Modernism in Reverse.* Toronto: University of Toronto Press, 1996.

Other Writings about Marshall McLuhan

"The All-at-once World of the Management Hootenanny Starring Marshall McLuhan." *Canadian,* August 6, 1966.

Ayre, John. "McLuhan Revisited." *Weekend,* April 10, 1976.

Boyle, Harry J. "Marshall McLuhan." *Weekend,* March 19, 1967.

Callwood, June. "What Are You Doin' Mr. McLuhan." *Globe and Mail,* January 17, 1975.

Cooper, Thomas William. "Pioneers in Communication: The Lives and Thought of Harold Innis and Marshall McLuhan." Unpublished Ph.D. thesis, University of Toronto, 1980.

Corelli, Rae. "The Great Gulf Between U.S., Russ Spies." *Daily Star,* June 2, 1960, p. 25.

Deane, Philip. "The Sheriff and the Lawyer." *Globe and Mail,* October 15, 1960.

Dobbs, Kildare. "What Did You Say, Professor?" *Star Weekly,* March 10, 1962.

Finkleman, Danny. "Marshall McLuhan and Tom Easterbrook." In *Speaking of Winnipeg,* edited by John Parr. Winnipeg: Queenston House, 1974.

Fulford, Robert. "On Marshall McLuhan: What One Communications Expert Discerns — but Has Trouble Getting Across." *Maclean's,* June 20, 1964.

Grady, Wayne. "They Loved Her in Cape Breton." *Today,* March 7, 1981.

Harding, D.W. "Trompe l'oeil." *New York Review of Books,* January 2, 1969.

Hoving, Thomas P.F. Untitled article, *Park East,* October 19, 1967.

Howard, Jane. "Oracle of the Electric Age." *Life,* February 25, 1966.

Kettle, John. "Marshall McLuhan, Prophet and Analyst of the Age of Instant Knowledge: Easing the Technological Burden of Western Man." *Canada Month,* October 1965.

Kostelanetz, Richard. "Marshall McLuhan: High Priest of the Electronic Village." In *Master Minds.* New York: Macmillan, 1967.

———. "Understanding McLuhan (in Part)." *New York Times Magazine,* January 29, 1967.

Kritzwiser, Kay. "The McLuhan Galaxy." *Globe and Mail Magazine*, January 4, 1960.

Krugman, Herbert. "Brain Wave Measures of Media Involvement." *Journal of Advertising Research*, February 1971.

"McLuhan Asks for Shakeup in Church." *Companion*, April 1977.

"On the Scene." *Playboy*, February 1967.

Powe, B.W. "Marshall McLuhan, the Put-on." In *A Climate Charged*. Oakville, Ont.: Mosaic Press, 1984.

Rosenberg, Harold. "Philosophy in a Pop Key." *New Yorker*, February 27, 1965.

Schickel, Richard. "Marshall McLuhan: Canada's Intellectual Comet." *Harper's*, November 1965.

Wain, John. "The Incidental Thoughts of Marshall McLuhan." *Encounter*, June 1985.

Wolfe, Tom. "What If He Is Right?" In *The Pump House Gang*. New York: Farrar, Straus and Giroux, 1968.

Literary Sources

Chesterton, G.K. *What's Wrong with the World*. London: Cassell and Company, 1910.

Eliot, T.S. *The Sacred Wood*. 1920. London: Methuen, 1960.

———. *The Use of Poetry and the Use of Criticism*. London: Faber and Faber, 1933.

Empson, William. *Seven Types of Ambiguity*. 1930. New York: Meridian, 1955.

Giedion, Sigfried. *Mechanization Takes Command*. 1948. New York: Norton, 1969.

———. *Space, Time, and Architecture: The Growth of a New Tradition*. Cambridge: Harvard University Press, 1954.

Innis, Harold. *Empire and Communications*. Oxford: Oxford University Press, 1950.

———. *The Bias of Communication*. Toronto: University of Toronto Press, 1951.

Joyce, James. *Finnegans Wake*. London: Faber and Faber, 1939.

Leavis, F.R. *New Bearings in English Poetry*. London: Chatto and Windus, 1932.

———. With Denys Thompson. *Culture and Environment*. London: Chatto and Windus, 1933.

———. *Revaluation*. London: Chatto and Windus, 1936.

Lewis, Wyndham. *Time and Western Man*. London: Chatto and Windus, 1927.

———. *The Diabolical Principle and the Dithyrambic Spectator*. London: Chatto and Windus, 1931.

———. *Men Without Art*. 1934. Santa Rosa, CA: Black Sparrow Press, 1987.

———. *America and Cosmic Man*. Garden City, NY: Doubleday, 1949.

———. *Self Condemned*. 1954. Santa Barbara, CA: Black Sparrow Press, 1983.

Maritain, Jacques. *Art and Scholasticism and the Frontiers of Poetry*. 1920. Translated by Joseph W. Evans. New York: Scribner's, 1962.

Mumford, Lewis. *Technics and Civilization*. 1934. New York: Harcourt, Brace and World, 1963.

Ong, Walter J. *Ramus, Method, and the Decay of Dialogue*. Cambridge: Harvard University Press, 1958.

Pound, Ezra. *Guide to Kulchur*. 1938. New York: New Directions, 1970.

Richards, I.A. *Practical Criticism*. 1929. New York: Harcourt, Brace and World.

———. *The Philosophy of Rhetoric*. New York: Oxford University Press, 1936.

Index